Essential Trends in Inorganic Chemistry

D. M. P. Mingos

Sir Edward Frankland BP Professor of Inorganic Chemistry, Imperial College, London; and Dean of the Royal College of Science

Oxford   New York   Tokyo

OXFORD UNIVERSITY PRESS

1998

Oxford University Press, Great Clarendon Street, Oxford OX2 6DP

Oxford New York
Athens Auckland Bangkok Bogota Bombay
Buenos Aires Calcutta Cape Town Dar es Salaam
Delhi Florence Hong Kong Istanbul Karachi
Kuala Lumpur Madras Madrid Melbourne
Mexico City Nairobi Paris Singapore
Taipei Tokyo Toronto Warsaw
and associated companies in
Berlin Ibadan

Oxford is a trade mark of Oxford University Press

Published in the United States
by Oxford University Press, Inc., New York

A catalogue record for this book is available from the British Library

Library of Congress Cataloging in Publication Data
Mingos, D. M. P., 1944–
Essential trends in inorganic chemistry / D.M.P. Mingos.
1. Periodic law.   2. Chemical elements.   I. Title.
QD467.M64   1997   546—dc21   97–35359
ISBN 0 19 850109 9 (Hbk)
ISBN 0 19 850108 0 (Pbk)

Typeset by the author
Printed in Great Britain by
The Bath Press, Bath

Preface

The explosion in chemical knowledge during the last fifty years poses great problems for the chemical educator. How is one to provide a textbook which accurately reflects modern inorganic chemistry when there is at least one major book on all the naturally occurring elements and sometimes a monograph on the derivatives of one particular compound? For example, whole books have been published recently on the chemistries of C_{60} and the sandwich compound ferrocene, $Fe(\eta-C_5H_5)_2$. Collective works such as *Comprehensive Inorganic Chemistry* and the *Encyclopaedia of Inorganic Chemistry* provide invaluable up-to-date information on recent developments in inorganic chemistry, but also serve as constant reminders of how quickly the subject is developing. The textbook writer has an almost impossible task of attempting to distil all this information into a form which is suitable for an undergraduate course. The ideal book has to contain an accurate representative selection of fundamental facts. These must include a detailed knowledge of the essential properties of traditional compounds, so that the reader does not join that apocryphal group of students who think that lithium is a green gas. The book must transmit the important concepts and bonding models in inorganic chemistry so that the student can interpret the facts in a modern fashion. This creates certain difficulties since chemistry is a subtle science and it is not always possible to interpret a specific chemical property in terms of a single atomic parameter.

The aim of this book is to present the subject in an alternative fashion which will make it more amenable to the student and make the factual information form part of a more general pattern. Therefore, its central feature is a detailed discussion of the important trends in the properties of the atoms of the elements and their compounds.

All inorganic textbooks pay lip-service to the Mendeleev periodic classification, but most do little to illustrate how a chemist uses the Periodic Table to organize and assimilate the mass of information which currently exists. It has been frequently said that the Mendeleev Periodic Table is the Rosetta Stone of modern inorganic chemistry. This analogy does provoke a wonderfully seductive image for the student since it suggests that once he/she has the Periodic Table in his hand he/she can decipher the subject. However, the reality proves to be a mirage, because even when a student is provided with a Periodic Table he/she is rarely able to provide a coherent and reasoned account of the chemistries of a group of related elements. Part of the reason for this is the downgrading of essential chemical facts in modern pre-university courses, but, in my opinion, the major reason is that undergraduate textbooks do not inform the student how to use the Periodic Table to inter-relate chemical facts. In general, textbooks tell the student that the Mendeleev Periodic Table is a wonderful thing, but then leave the student to discover in isolation how it works in practice.

The Rosetta Stone analogy is misleading. The knowledge of the structure of the Periodic Table does not lead to a direct interpretation of chemical phenomena. The Mendeleev Periodic Table is more closely related to a family tree. A family tree defines very precisely the dates of birth of the family members and the parent–children vertical connections and the brother–sister horizontal connections, but it says nothing about the actual relationships between the family members. Do the children and parents resemble each other—do the brothers and sisters have friendly relationships or did they argue continuously?

The quantum mechanical interpretation of the Mendeleev Periodic Table is a triumph of modern physics and it provides a very precise description of the electronic configurations of the individual atoms in terms of a generalized shell model. It also demonstrates that the most important parameter in determining the formulae of compounds is the number of valance electrons of the central atom. However, a chemist is interested in other features of a compound besides its formula—is it an oxidizing agent or reducing agent, is it an acid or base, does it have a high melting point or is it a gas, etc. These chemical properties, of course, depend on other variables besides the formula and the Mendeleev Table does not provide the necessary parameters to interpret the variations of properties within a family of compounds with a similar formula. These other parameters include the sizes of the atoms, the availability of the electrons for bond formation, the total number of electrons in the molecule, the structure of the molecule, and the strengths of the chemical bonds within the molecule.

This book explores and interprets in some detail the properties of the elements in the family tree, which has been defined by the Periodic Table. It achieves this exploring the relationships between the atomic and molecular properties of the individual members in the family tree. Therefore, two large chapters are used to define vertical and horizontal and diagonal trends in the properties of the elements and their compounds and a further chapter discusses isoelectronic and iso-stoichiometric relationships. A systematic approach is followed and the trends in the atomic properties of the atoms are discussed first and these properties are used to interpret the physical properties of the elements in their standard states. Then the chemical properties of the elements are described, with particular emphasis on their reduction and oxidation properties. Finally, the chemical and physical properties of the compounds themselves are discussed. Each of these chapters contains a great deal of factual information, but it is introduced in a way which emphasizes the inter-relationships. The approach is essentially qualitative, but hopefully at the end of the book the basic methodology, which has been developed, is sufficiently well established that the student can apply the basic principles to other problems and read detailed accounts of modern inorganic chemistry in a more intelligent and meaningful fashion.

The book, which is intended for first and second year undergraduates, does assume some prior knowledge of quantum theory and bonding and it may be helpful for the student to have at hand *Essentials of Inorganic Chemistry 1*, Oxford University Press, 1995, and *Essentials of Inorganic Chemistry 2* which is due to be published in 1998, for definitions of some of the basic concepts which are used in the book.

I should like to take this opportunity of thanking all those who have helped me with this book. Bill Griffith, Paul Lickiss, Julian Gale, Paul Dyson, Phil Dyer, Nick Long, and Paul Sharpe read individual chapters and gave me many timely criticisms and comments, and Simon Cotton was particularly helpful in critically reading the section on lanthanides and actinides. Tony Downs and Ged Parkin kindly provided me with data on specific aspects of the book. Courtney Phillips and Nick Norman read the whole book in draft form for Oxford University Press and the former provided some fundamental criticisms of the methodology which made me rethink much of what I had written. Most of all, however, I should like to thank Jack Barrett for the dedication he has put into preparing the book in a camera ready form. When we agreed with OUP that we would produce a camera ready version of the manuscript in order to prevent some of the production difficulties we had experienced with *Essentials of Inorganic Chemistry 1* little did we know how much work was going to be involved. Thank you Jack so much for your patience and attention to detail—two admirable characteristics which are not usually associated with the author.

London   D.M.P.M.
September, 1997

Contents

Dedicated to the essential elements of my family—
Stacey, Zoë, and Adam

1 The quantum mechanical basis of the Periodic Table

1.1 Introduction

Mendeleev's original classification of the elements was based upon the formulae of the compounds which the elements formed and the physical properties of the elements and their compounds. The explosion of knowledge associated with each element since his time has made it even more necessary to have a conceptual framework in order to manage the amount of factual information which has to be remembered. At the same time the development of the quantum mechanical model for the atom has led to a much deeper understanding of the underlying factors responsible for Mendeleev's classification and its limitations. The modern use of the Periodic Table also utilizes horizontal and diagonal trends as well as the vertical relationships which were originally highlighted by Mendeleev. Therefore, the aim of this book is to illustrate how a modern chemist utilizes the Periodic Table to classify and interpret chemical phenomena.

1.2 Atomic orbitals

An atomic orbital is the wave function (Ψ) of an electron in an atom. Its square (Ψ^2) gives the probability of finding the electron at that point.[*]

The Schrödinger solution defines the wave equation for the electron in the hydrogen atom as a product of a radial and angular part as follows:

$$\Psi_{n,l,m}(r,\theta,\phi) \quad = \quad \underset{\text{Radial part}}{R_{n,l}(r)} . \underset{\text{Angular part}}{Y_{l,m}(\theta,\phi)}$$

where the radial part is governed by the quantum numbers n and l and the angular part by the quantum numbers l and m. In addition, if the hydrogen atom is solved with relativistic corrections for the motion of the electron around the nucleus, a fourth quantum number s emerges which can take values of $+\frac{1}{2}$ or $-\frac{1}{2}$. This quantum number is described as the spin quantum number and is pictorially associated with the notion that the electron can spin on its own axis.

---

[*] If this starting point is found to be too advanced the book *Introduction to Quantum Theory and Atomic Structure* by P. A. Cox (OUP) will provide an excellent introduction to the basis of quantum theory and its applications to atoms.

The angular part of the wave function defines the shape of the orbital by specifying the number of nodal planes or cones which pass through it. The radial part of the wave function defines the variation in the electron density probability around the nucleus.

A node in a wave function is a point, plane or surface where the electron probability is zero, because it represents the intersection of regions of the wave function with + and – signs. The difference between radial and angular nodes may be emphasized by the following analogies: if a piece of string is placed around a balloon and pulled tightly then eventually two smaller and approximately spherical shapes are generated. These two regions of higher pressure are separated by a point where there is no air if the string is pulled very tightly. Therefore the string defines an angular nodal plane between the two lobes of the balloon and the balloon now resembles a p orbital rather than the spherical s orbital shape it started with initially. The use of two pieces of string could similarly produce a d orbital shape if they are tightened in two mutually perpendicular directions.

Imagine a pool of mercury in a saucer. The process of bringing down onto the pool a round pastry cutter which breaks the surface and then makes contact with the bottom of the surface makes two distinct regions of liquid mercury. These are separated by the nodal circle of the cutter, where there is no mercury. This line represent a radial node. Clearly it represents a higher energy state because energy was required to push the cutter through the mercury pool. These analogies emphasize the important general point that the introduction of radial and angular nodes results in higher energy states.

Since a particle can rotate either clockwise or anticlockwise about a rotation axis there are two quantized states with quantum numbers of $+\frac{1}{2}$ or $-\frac{1}{2}$. The electron has an intrinsic magnetic moment and in the presence of an external magnetic field the energies of the $+\frac{1}{2}$ and $-\frac{1}{2}$ quantized states no longer remain equal, i.e. it behaves like a compass does in the Earth's magnetic field.

The quantum number n is the principal quantum number and governs the quantized energy states which are the solutions of the Schrödinger equation for the hydrogen atom.

$$E_n = -\frac{hcR}{n^2}$$

where R is the Rydberg constant, c is the speed of light and h is Planck's constant.

n can take any positive integer values: 1, 2, 3,... and is independent of l, m, and s.

l is the orbital angular momentum quantum number (sometimes called the azimuthal quantum number in older textbooks) and can take up integer values from 0 to a maximum of $(n-1)$. The total angular momentum associated with the wave function $\Psi_{n,l,m}(r,\theta,\phi)$ is $\{l(l+1)\}^{1/2}h/2\pi$ and is related to the number of nodes in the angular part of the wave function. The shape of the orbital is defined by the number of angular nodes.

The s orbital ($l = 0$) has no angular nodes and is therefore spherical, the three p orbitals ($l = 1$) each have a single angular node. The five d orbitals ($l = 2$) have two angular nodes each and the seven f orbitals ($l = 3$) have three angular nodes each.

m is the magnetic quantum number and defines the components of the angular momentum, $mh/2\pi$, along the z axis and is restricted to having $2l+1$ integer values, 0, ±1, ...±l. This quantum number defines the direction of the orbital in space by specifying the number of nodal planes which coincide in their direction with the z axis. For example, p_x, p_y, and p_z all have the 'dumbbell' shape illustrated in Fig. 1.1 and a single nodal plane, but only p_x and p_y (for which $m = \pm1$) have a single nodal plane coincident with the z axis, p_z (for which $m = 0$) has its nodal plane in the xy plane. The choice of direction is arbitrary because all the orbitals are equivalent, but if a magnetic field is applied in a specific direction then this equality is lost and the interaction of the orbital angular momentum with the magnetic field depends on the quantum number m.

Since there are not five independent and doubly noded functions the $m = 0$ solution has two nodal cones intersecting at the origin and spreading out along the $+z$ and $-z$ directions. The remaining four d functions have two nodal planes intersecting at right angles at the nucleus. The total number of angular nodes is therefore equal to l. These shapes are illustrated in Fig. 1.1.

Each energy state of the electron is therefore labelled by four quantum numbers. Three quantum numbers, n, l, and m, define an orbital which can accommodate an electron with spin s of $+\frac{1}{2}$ or $-\frac{1}{2}$. All orbitals with the same quantum number n are said to belong to the same *shell* and all orbitals sharing the same quantum number l are said to belong to the same *subshell*.

Therefore, each shell has a total of n^2 orbitals all with the same energy and each subshell has $2l + 1$ orbitals. The subshells are designated by the letters s, p, d, f, etc., to describe their common quantum number, l, as shown in Table 1.1.

Table 1.1 The designations of values of the l quantum number by letters

l	0	1	2	3	4
Designation	s	p	d	f	g
Number of orbitals	1	3	5	7	9

In any one shell there is one s orbital, three p orbitals (if $n > 1$), five d orbitals (if $n > 2$) and seven f orbitals (if $n > 3$). The components of the subshells vary with n as shown in Table 1.2 for $n = 1$–4.

Table 1.2 The numbers of orbitals in the first four quantum shells

n	l	Subshell	m	No. of orbitals in subshell
1	0	1s	0	1
2	0	2s	0	1
	1	2p	1, 0, –1	3
3	0	3s	0	1
	1	3p	1, 0, –1	3
	2	3d	2, 1, 0, –1, –2	5
4	0	4s	0	1
	1	4p	1, 0, –1	3
	2	4d	2, 1, 0, –1, –2	5
	3	4f	3, 2, 1, 0, –1, –2, –3	7

The shapes of the subshell orbitals with different numbers of nodes are illustrated in Fig. 1.1. The schematic illustrations shown are pictorial representations of the spherical harmonic functions $Y_{l,m}(\theta,\phi)$ which are obtained from the solution of the Schrödinger equations.

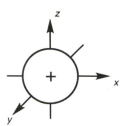

s orbital, $l = 0$; no angular nodes

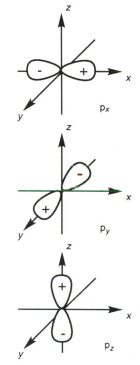

p orbitals, $l = 1$; one angular node

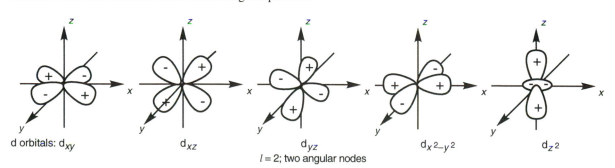

d orbitals: d_{xy} d_{xz} d_{yz} $d_{x^2-y^2}$ d_{z^2}

$l = 2$; two angular nodes

Fig. 1.1 Schematic illustrations of the angular parts of the wave functions for the hydrogen atom for $l = 0$, 1, and 2. d_{z^2} represents an abbreviation for $d_{2z^2-x^2-y^2}$, i.e. a combination of $d_{z^2-x^2}$ and $d_{z^2-y^2}$

These are shown in their real forms and chosen because they are helpful visualizations for many chemically interesting problems. There are, of course, alternative ways of representing the orbitals which place the nodal planes or cones in different locations.

The radial part of the wave function $R_{n,l}(r)$ defines the probability of finding the electron in the region of the atom. More specifically the radial distribution function $4\pi r^2 \Psi^2 dr$ is the probability of finding the electron in the shell of thickness dr located at r from the nucleus (see Fig. 1.2). The surface area of the spherical shell increases with r, hence the probability density is zero at the nucleus, passes through one or more maxima and decays exponentially at longer distances. Each time the wave function passes through zero along the r axis this corresponds to a radial node in the wave function (except when $r = 0$). The total number of radial nodes of an orbital with the quantum numbers n and l is $n - l - 1$. A radial node is therefore a spherical surface separating parts of the radial wave function having opposite signs.

All the s wave functions have $(n - 1)$ radial nodes, the p wave functions have $(n - 2)$ radial nodes, and the d wave functions $(n - 3)$ radial nodes.

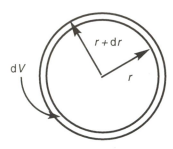

Fig. 1.2 A diagram representing a spherical shell around the nucleus; the volume of the shell of thickness dr is given by $4\pi r^2 dr$

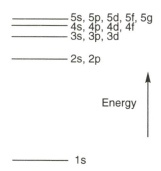

Energy

1s

Fig. 1.3 A schematic illustration (not to scale) of the energy states of the hydrogen atom, $E_n \propto -\frac{1}{n^2}$

> For the hydrogen atom the energies of the orbitals are determined solely by the total number of nodes. The total number of radial and angular nodes for a shell is $n - 1$ and therefore the subshell orbitals ns, np, nd,... have the same energy. The shells, subshells, and energies of the solutions of the Schrödinger equation are illustrated in Fig. 1.3.

Orbitals which have the same energies are described as **degenerate**. Therefore orbitals belonging to the same subshell are always degenerate in the isolated atom. The unique feature of the hydrogen atom is that all the orbitals sharing a common principal quantum number, n, are also degenerate, i.e. 2s and 2p; and 3s, 3p, and 3d, etc. have equal energies. NB, the energy levels become more closely spaced as n increases.

The single electron possessed by the hydrogen atom occupies the lowest energy (most stable) 1s orbital in its ground state. The higher energy states are only accessible when electromagnetic energy of the appropriate energy is absorbed by the electron in the 1s orbital and thereby it is promoted to the $n = 2, 3, 4,...$ etc. shells. In general the orbital energy is given by:

$$E_n = -\frac{hcR}{n^2}$$

so, the energy differences between the excited states and the ground state are given by the equation:

$$\Delta E = E_2 - E_1 = hcR\left[\frac{1}{n_1^2} - \frac{1}{n_2^2}\right]$$

Using the Planck relationship:

$$\Delta E = h\nu$$

calculations such as that above account for the frequencies, ν, of the line spectra of the hydrogen atom. The model may also be used to calculate the ionization energy of the hydrogen atom, since the loss of an electron from the ground state corresponds to the energy difference, $\Delta E = E_1 - E_\infty$, i.e. from $n_1 = 1$ to $n_2 = \infty$, which agrees with the observed value, 1312 kJ mol^{-1}.

The Schrödinger solution for the hydrogen atom generates a spectrum of atomic orbitals with well-defined energy states. Each orbital can in principle accommodate two electrons with opposite spins, although for the hydrogen atom only a single electron is present (see discussion of the Pauli principle on page 8).

It is interesting to explore whether the energy level scheme in Fig. 1.3 may be extended to atoms with more than one electron. Specifically, it is important to establish whether the degeneracies of the (2s, 2p), (3s, 3p, 3d) orbitals are maintained in polyelectronic atoms.

For polyelectronic atoms the energy levels are filled according to the *aufbau* procedure. The *aufbau* procedure means that the energy levels in Fig. 1.3 are filled by locating electrons in the most stable orbitals first. The second electron will therefore pair with the first electron in the 1s orbital by occupying the same orbital but with an opposite spin. The third electron would occupy the higher energy 2s orbital. The fourth to eighth electrons would occupy the 2s and 2p orbitals to eventually complete the shell. In this imaginary process the addition of each electron is accompanied by the addition of a proton to the nucleus in order to maintain an uncharged atom. In summary, the hypothetical ordering which results is based on a simple increment in n:

$$1s \rightarrow (2s, 2p) \rightarrow (3s, 3p, 3d) \rightarrow (4s, 4p, 4d, 4f) \rightarrow \text{etc.}$$

Since each row is associated with $2n^2$ elements the resulting Periodic Table has the unfamiliar form, illustrated schematically in Fig. 1.4. The first and second n shells are completed at the noble gases He and Ne. For the third shell the filling of 3s and 3p orbitals would lead to the elements Na–Ar and then filling of the 3d orbitals would require the elements K, Ca, Sc, Ti, V, Cr, Mn, Fe, Co, and Ni to be considered together. The fourth shell would start with Cu, corresponding to the filling of the 4s orbital. The hypothetical Periodic Table separates Li and Na from K and Rb. Since the Periodic Table does not have this structure this means that the energy states for polyelectronic atoms must differ from those of hydrogen. In order to understand this more fully it is necessary to describe in more detail the radial distribution functions, $4\pi r^2 \Psi^2 dr$, for the hydrogen atom.

The radial distribution functions for the hydrogen atom for $n = 1$ (1s), 2 (2s and 2p), and 3 (3s, 3p, and 3d) are shown in Fig. 1.5.

			Nd	U
			Pr	Pa
			Ce	Th
			La	Ac
			Ba	Ra
			Cs	Fr
			Xe	Rn
			I	At
			Te	Po
			Sb	Bi
			Sn	Pb
			In	Tl
			Cd	Hg
			Ag	Au
		Ni	Pd	Pt
		Co	Rh	Ir
		Fe	Ru	Os
		Mn	Tc	Re
		Cr	Mo	W
		V	Nb	Ta
		Ti	Zr	Hf
		Sc	Y	Lu
		Ca	Sr	Yb
		K	Rb	Tm
	Ne	Ar	Kr	Er
	F	Cl	Br	Ho
	O	S	Se	Dy
	N	P	As	Tb
	C	Si	Ge	Gd
	B	Al	Ga	Eu
He	Be	Mg	Zn	Sm
H	Li	Na	Cu	Pm
2	8	18	32	50

Fig. 1.4 A hypothetical Periodic Table based upon the Schrödinger solution to the hydrogen atom. The number of elements which can fit into each column is indicated.

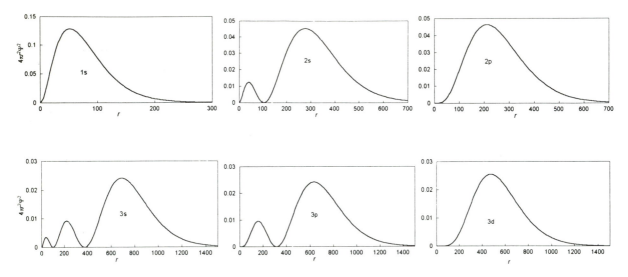

Fig. 1.5   The radial distribution functions for the 1s, 2s, 2p, 3s, 3p, and 3d orbitals of the hydrogen atom

The following points are noteworthy.

1.   For a given l quantum number value the maximum in the radial distribution function moves progressively to longer distances from the nucleus. This may be seen most clearly for the 1s, 2s, and 3s orbitals in Fig. 1.5. The number of nodes increases with n reflecting the larger kinetic energies of the electron in the higher energy states.

2.   For a given n quantum number value the maximum in the radial distribution function is farthest from the nucleus for s orbitals and comes progressively closer to the nucleus for p and d orbitals. However, the differences are smaller than those noted in 1.

3.   For the 3s, 3p, and 3d orbitals the total electron density within 400 pm of the nucleus is s > p > d. These differences in electron distribution in the proximity of the nucleus are very important for understanding the different energies of s, p, and d orbitals in polyelectronic atoms.

1.3 Polyelectronic atoms

In polyelectronic atoms it is assumed that the angular parts of the wave functions are identical to those of the hydrogen atom described above, but the radial part is altered to take into account the differences in nuclear charge and electron repulsion experienced by the electrons. In such atoms the nuclear charge is $+Z$, which balances the total charges associated with the electrons. The orbitals with identical n values, but with different l values, have different energies in a polyelectronic atom because of the unequal screening and penetration effects which their electrons experience (point 3 above).

In polyelectronic atoms the wave functions of the orbitals radiate away from the nucleus, but unlike those for hydrogen they have to pass through the regions of electron density associated with the core shells of electrons. Since the extent to which they penetrate these shells differs for ns, np, and

nd orbitals, their relative energies are affected. This results in the ns, np, and nd subshells losing their common **degeneracy**.

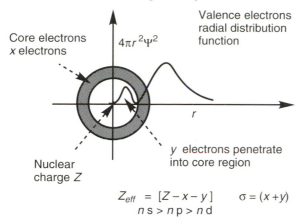

$$Z_{eff} = [Z - x - y] \qquad \sigma = (x + y)$$
$$n\,\mathrm{s} > n\,\mathrm{p} > n\,\mathrm{d}$$

Fig. 1.6 A diagrammatic representation of the penetration of the core electron density by a valence electron function

The discussion given here represents a gross simplification of the complex situation in a polyelectronic atom, where each electron is represented by a wave function whose characteristics are dependent not only on the nuclear charge it experiences, but the effects of the wave functions associated with the other electrons. However, the crude concepts which emerge— *screening* and *penetration* —provide a convenient means of rationalizing the relative energies of orbitals in polyelectronic atoms.

The radial distribution function of a valence electron, shown schematically in Fig. 1.6, has a part which lies on the outside of the spherical shell representing the core electrons and a part which penetrates the spherical shell. That part which remains outside the core experiences an effective nuclear charge: $Z_{eff} = Z - x - y$, where Z is the nuclear charge, x is the number of electrons which make up the core, and y is the proportion of electron density associated with the radial distribution function of the outer valence electrons which penetrate the core region. For a row of s and p block atoms Z_{eff} increases by approximately +0.7e from atom to atom. For the d block elements the corresponding increase in Z_{eff} is approximately 0.8e.

The outer electrons in an atom do not therefore experience the full effect of the nuclear charge associated with the nucleus because of the 'screening' by the shells and subshells of electrons lying between them and the nucleus. Thus, the outer valence electrons experience an effective nuclear charge, $Z_{eff} = Z - \sigma$, where σ is the screening constant. This screening constant depends on the degree of penetration of the orbital inside the radial distribution functions of the core electrons. The differences in the radial distribution functions for ns, np, nd, and nf orbitals arising from the $(n - l - 1)$ radial nodes (see Fig. 1.5) means that their relative penetrating power is s > p > d > f and therefore the effective nuclear charge experienced by the orbitals is ns > np > nd > nf (see Table 1.3 for specific examples). The higher effective nuclear charge experienced by ns relative to np is clearly visible. Consequently in polyelectronic atoms the subshells no longer are of equal energy and their relative stabilities are ns > np > nd > nf. This change in the relative ordering of the atomic orbitals has a very profound influence on the structure of the Periodic Table. The differences in screening and penetration leads to a shifting of the p, d, and f orbital energies relative to that of s, as shown schematically in Fig. 1.7.

Table 1.3 Effective nuclear charges (Z_{eff}) experienced by electrons in valence orbitals (Clementi–Raimondi values)

	ns	np
Li	1.28	
Na	2.51	
K	3.50	
B	2.58	2.42
Al	4.12	4.07
Ga	7.07	6.22
C	3.22	3.14
Si	4.90	4.29
Ge	8.04	6.78
N	3.85	3.83
O	4.49	4.45
F	5.13	5.10

The effective nuclear charge for ns is greater than that for np and the difference increases as n becomes larger. Also, the effective nuclear charge increases across the series Li–F because as the additional electrons are introduced they do not screen the valence electrons from the nucleus with perfect efficiency.

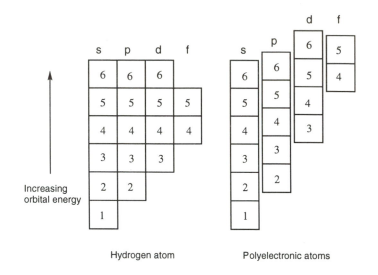

Increasing
orbital energy

Hydrogen atom   Polyelectronic atoms

Fig. 1.7 A diagram comparing the order of orbital energies of the hydrogen atom with those of polyelectronic atoms

Table 1.4 The maximum occupancies of subshells

Subshell	Total number of electrons
s	2
p	6
d	10
f	14

The universal energy level diagram which results from the differences in screening and penetration noted above and shown in Fig. 1.7 accounts for most of the features of the Periodic Table as we currently know it. The energy levels shown on the right-hand side of Fig. 1.7 are filled according to the *aufbau principle*, i.e. orbitals with the lowest energy are filled first. The *Pauli exclusion principle* which concludes that no two electrons may have the same set of values of the four quantum numbers limits the total number of electrons which may enter a subshell to $2(2l + 1)$, i.e. 2 in the s subshell, 6 in the p subshell, 10 in the d subshell and 14 in the f subshell (summarized in Table 1.4).

Within the framework of this simplified universal energy diagram the relative shifting of orbitals with the same quantum number leads to the following general order of filling: 1s → 2s → 2p → 3s → 3p → 4s → 3d → 4p → 5s → 4d → 5p → 6s → 4f → 5d → 6p etc. This means that 3d is only filled after 4s, 4d after 5s, 4f after 6s, 5d after 4f, and 5f after 7s. Therefore, in polyelectronic atoms the relative orbital energies are no longer dominated by n, the principal quantum number.

Those elements which result from the filling of the nd orbitals between the $(n +1)$s and $(n +1)$p subshells are described as the transition elements. The elements which result from the filling of the 4f and 5f subshells are called the lanthanides and actinides and occur after the 6s and 7s subshells respectively and before the 5d and 6d subshells.

The orbital filling sequence which accounts for the major features of the Periodic Table is illustrated schematically in Fig. 1.8. The orbitals in polyelectronic atoms are filled in a sequence where $(n + l)$ rather than n is incremented. For example, the sequence 3d → 4p → 5s has $n + l = 5$; and 4f → 5d → 6p → 7s has $n + l = 7$. Fig. 1.8 retains the easy-to-remember energy level scheme for the hydrogen atom, but introduces an *aufbau* sequence based on filling the energy levels according to $n + l$, rather than n.

Hence, the orbitals are filled along diagonal lines where $n+l$ is successively 1, 2, ... 8. For a given value of $n+l$ the energy levels are filled in order of increasing value of n.

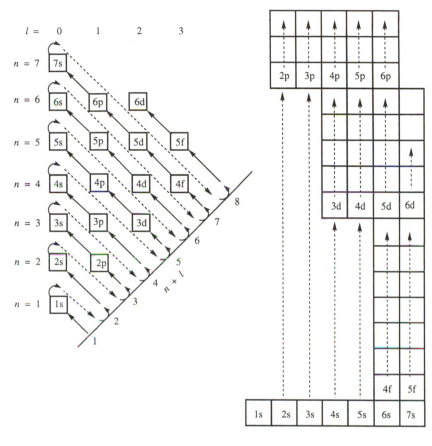

A modern multi-storey car park provides a more familiar analogy for the *aufbau* principle. The cars entering at the bottom of the building when it is busy tend to fill the bays at the lowest level first and only begin to drive to a higher level when all the bays in the lower level are filled. Also the cars tend to line up in a parallel fashion, because the drivers prefer not to expend the additional energy required to reverse the cars into the bays. Electrons also prefer to enter their orbitals with parallel spins, because more energy is required to make them align in an anti-parallel fashion. In addition the filled car park has two lines of cars pointing in opposite directions because they utilize bays adjacent to the parallel walls. A complete subshell of electrons has pairs of anti-parallel electrons in each orbital.

The above analogy is misleading in the sense that it suggests that the energies of the orbitals remain constant for a specific subshell and do not change as additional electrons are introduced. The height of the floor of the car park is not affected by the introduction of cars, but the orbital energies of atoms are affected by the introduction of additional electrons. Going across a row of the Periodic Table the nuclear charge is increased and the effective nuclear charge also increases, but in a somewhat dampened manner because of screening effects. Therefore, the universal energy level scheme shown in Fig. 1.7 provides a convenient basis for accounting for the gross shape of the Periodic Table, but does not accurately represent the orbital energies of any specific atom. The orbital energies change across a row and the relative orderings may even change.

Fig. 1.8 The order of orbital filling and the Periodic Table based on non-hydrogen orbital energies

The relationship between this universal energy level ordering scheme and the Periodic Table which results from placing electrons into the energy levels according to the *aufbau* principle is also shown in Fig 1.8. The energy levels of the atoms are indicated on a vertical scale and filled in the sequence indicated by the arrows. The box diagram on the right-hand side of Fig. 1.8 illustrates the number of orbitals associated with each subshell and each box represents an orbital which can accommodate two electrons with opposite spins in accordance with the Pauli principle.

The box diagram suggests that the first complete shell is associated with two elements which contribute one and two electrons to the 1s shell, i.e. H and He. The next shell corresponds to the filling of the 2s and 2p orbitals and begins with Li ($1s^2 2s^1$) and ends with Ne ($1s^2 2s^2 2p^6$), and the third shell starts with Na ($1s^2 2s^2 2p^6 3s^1$) and ends with Ar ($1s^2 2s^2 2p^6 3s^2 3p^6$). These shells can therefore each accommodate a total of 8 electrons.

For example, at the beginning of the first transition series the 4s orbital lies at a lower energy than the 3d, but the effective nuclear charge experienced by the latter increases more as the transition series progresses and by the end the 3d orbitals are significantly more stable than the 4s. The relative ordering of orbitals may also be influenced by the charge on the atom. For example, for the first row transition elements the 4s orbital is more stable than the 3d in the neutral atom, but in the positively charged ions, M^{n+}, the 3d orbitals are more stable than the 4s.

The fourth shell begins with the occupation of the 4s orbitals (K and Ca), but then the 3d orbitals are occupied before the 4p orbitals. Thus Sc has the electronic configuration $(1s^2 2s^2 2p^6 3s^2 3p^6 4s^2 3d^1)$ and begins the transition series and Zn with the electronic configuration $(1s^2 2s^2 2p^6 3s^2 3p^6 4s^2 3d^{10})$ completes it. The next element gallium results from the addition of an electron to the 4p subshell and the shell is completed at krypton $(1s^2 2s^2 2p^6 3s^2 3p^6 4s^2 3d^{10} 4p^6)$. The next shell follows a similar sequence, i.e. the filling of 5s, then 4d and finally 5p, and starts at Rb and finishes at Xe. These shells are both associated with a total of 18 elements. The next two shells have a total of 32 elements because the atoms contribute the necessary electrons to fill the subshells 6s, 4f, 5d, and 6p and 7s, 5f, 6d, and 7p. The former starts with Cs and finishes with Rn and the latter starts with Fr and remains incomplete.

Each row of the Periodic Table starts with an alkali metal with a $ns^1 np^0$ electronic configuration and ends with a noble gas with a configuration $ns^2 np^6$. The number of elements in successive rows is:

Row:	1	2	3	4	5	6	7
No. of elements	2	8	8	18	18	32	32*

(*currently 26 are known)

A general formula which generates the numbers of elements in successive rows is:

$$2\left\{ \mathrm{INT}\left(\frac{N+2}{2} \right) \right\}^2$$

where N is the number of the row ($N = 1$–7) and the function $\mathrm{INT}(x)$ represents the rounded-down integer of x.

It is more conventional to represent the Periodic Table in a horizontal rather than vertical manner and such a Table is illustrated in Fig. 1.9. This differs from the orbital box diagram shown in Fig. 1.8 in one important respect. The elements associated with filling the 4f and 5f orbitals have been detached from the main part of the Periodic Table and placed below the transition elements. This is done partly for aesthetic reasons, because the long form of the Table which includes the lanthanides and actinides looks unbalanced when represented on the written page. However, this form of the Periodic Table also emphasizes that at this point the orbital filling model developed above has its limitations.

The universal energy level scheme illustrated in Fig. 1.8 is not quite as perfect for predicting the precise electronic configurations of all atoms as the discussion above has suggested. When the sixth shell is being completed the 5d orbital lies below 4f initially and therefore lanthanum has the electronic configuration $[Xe]6s^2 5d^1$, but on moving to the next element, cerium, the 4f orbital energy falls below 5d and it has the configuration $[Xe]6s^2 4f^2$. Therefore, the exact location of the start of the lanthanides and the third transition series remains ambiguous. Similar complexities occur for the 5f elements and therefore it is helpful to indicate these anomalies by the Periodic Table representation shown in Fig. 1.9.

H																	He
Li	Be											B	C	N	O	F	Ne
Na	Mg											Al	Si	P	S	Cl	Ar
K	Ca	Sc	Ti	V	Cr	Mn	Fe	Co	Ni	Cu	Zn	Ga	Ge	As	Se	Br	Kr
Rb	Sr	Y	Zr	Nb	Mo	Tc	Ru	Rh	Pd	Ag	Cd	In	Sn	Sb	Te	I	Xe
Cs	Ba	* Lu	Hf	Ta	W	Re	Os	Ir	Pt	Au	Hg	Tl	Pb	Bi	Po	At	Rn
Fr	Ra	** Lr	Rf	Db	Sg	Bh	Hs	Mt	110	111	112						

	*	La	Ce	Pr	Nd	Pm	Sm	Eu	Gd	Tb	Dy	Ho	Er	Tm	Yb
	**	Ac	Th	Pa	U	Np	Pu	Am	Cm	Bk	Cf	Es	Fm	Md	No

Fig. 1.9 The modern Periodic Table. Elements that are shaded only have radioactive isotopes

1.4 The modern Periodic Table

A detailed version of the modern Periodic Table of the elements is shown in Fig. 1.9 and on the inside front cover. The names of the elements, their atomic numbers, and their ground state electronic configurations are given on the inside back cover. The synthesis of elements during the evolution of the Universe has resulted in 81 naturally occurring elements existing on Earth, others having decayed completely. These elements have one or more isotopes which do not undergo spontaneous radioactive decay. For the elements above bismuth ($Z = 83$), no stable isotopes exist and two earlier elements technetium and promethium have only short-lived isotopes which are radioactive. Uranium and thorium are found in nature as naturally occurring radioactive isotopes because they have long decay lifetimes. The question of nuclear stability is discussed in more detail in Section 1.11. The Periodic Table on the inside front cover also includes the IUPAC recommended numbering of the groups. The s block elements (alkali metals and alkaline earths) make up groups 1 and 2. The d block elements (the transition metals) make up groups 3–12 and are associated with the filling of the d shells. The p block elements (the post-transition elements of which the non-metals are a subgroup) represent columns 13–18. The f block elements are not numbered in columns within the recommended IUPAC classification primarily because the lanthanides and actinides have relatively similar properties. The older 'A' and 'B' classification of the periodic groups is also illustrated in this version of the Table.

The chemical implications of the Periodic Table arise primarily from the fact that elements with the same numbers of valence electrons occupying orbitals with the same *l* quantum number values can form compounds with identical stoichiometries.

Regions of the Periodic Table

Different regions of the Periodic Table are summarized in Table 1.5. They are defined either by the orbitals occupied by the valence electrons or by more traditional names, as indicated in Table 1.6 for the main group elements and Table 1.7 for the transition elements, the lanthanides, and the actinides.

Table 1.5 Regions of the Periodic Table and their specific descriptions

Element groups	Specific name
Groups 1 and 2 with Groups 13–18	The representative elements or the s and p block elements or main group elements.
Groups 1 and 2	The s block elements.
Groups 13–18	The p block elements. The heavier elements in these groups are also sometimes referred to as the post-transition elements, and the metals of this subgroup as the B-Group metals.
Groups 3–12	The transition elements or the d block elements. 3d elements are also referred to as the first transition series; 4d elements as the second transition series; 5d elements as the third transition series.
Lanthanides	The elements with electrons occupying 4f valence orbitals.
Actinides	The elements with electrons occupying 5f valence orbitals.

Table 1.6 Names of specific groups and their common electronic configurations

Group or series	Elements	Traditional name	Electronic configurations
Group 1	Li–Fr	Alkali metals	$ns^1 (n = 2 - 7)$
Group 2	Be–Ra	Alkaline earths	$ns^2 (n = 2 - 7)$
Post-transition elements	B–Ne		$2s^2 2p^1 - 2s^2 2p^6$
	Al–Ar		$3s^2 3p^1 - 3s^2 3p^6$
	Ga–Kr		$4s^2 4p^1 - 4s^2 4p^6$
	In – Xe		$5s^2 5p^1 - 5s^2 5p^6$
	Tl–Rn		$6s^2 6p^1 - 6s^2 6p^6$
Group 15	N–Bi	Pnictogens	$ns^2 np^3$
Group 16	O–Po	Chalcogens	$ns^2 np^4$
Group 17	F–At	Halogens	$ns^2 np^5$
Group 18	He–Rn	Inert (noble) gases	$ns^2 np^6$

Table 1.7 The elements of the four transition series, the lanthanides and the actinides and their electronic configurations*

Element series	Electronic configurations
First transition series Sc–Zn	$4s^2 3d^1 – 4s^2 3d^{10}$
Second transition series Y–Cd	$5s^2 4d^1 – 5s^2 4d^{10}$
Third transition series La–Hg	$6s^2 5d^1 – 6s^2 5d^{10}$
Fourth transition series Lr to element 112	$7s^2 6d^1 – 7s^2 6d^{10}$
Lanthanides Ce–Lu	$4f^1 – 4f^{14}$
Actinides Th–Lr	$5f^1 – 5f^{14}$

* See Table 1.9 for examples of atoms which do not have these generalized electronic configurations based on the *aufbau* principle

The important areas of the Periodic Table are also illustrated schematically in Fig. 1.10. The definition of the rows of the Periodic Table used in this book are defined in Fig. 1.11. Hydrogen and helium which constitute the first row have been separated to place hydrogen above the alkali metals and helium above the other noble gases.

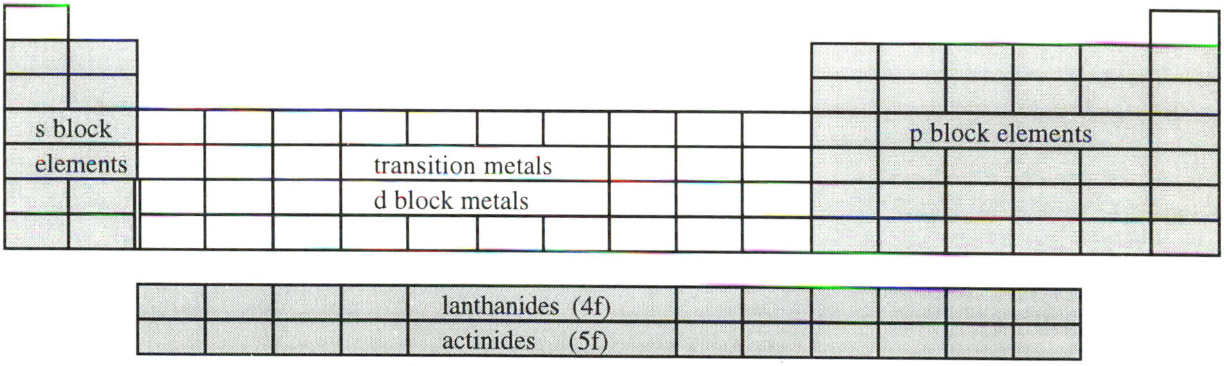

Fig. 1.10 Important areas of the Periodic Table

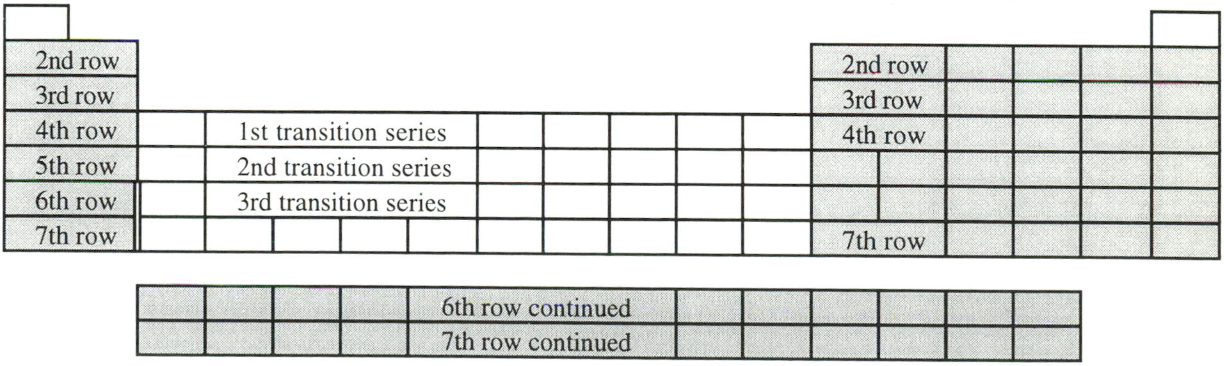

Fig. 1.11 Definitions of the rows of the Periodic Table as used in this book (the elements hydrogen and helium form the first row)

1.5 Exchange energies

The *aufbau* principle used above is based on the premise that successive orbitals have quite large energy separations and electron–electron repulsion effects are unimportant. When these conditions are not met, electron configurations which diverge from those predicted strictly according to the *aufbau* principle are observed.

A pair of electrons with parallel spins experience less repulsion than a pair in separate orbitals with anti-parallel spins. The difference is described as the exchange energy, K. This difference results from spin correlation, i.e. the quantum mechanical effect which states that two electrons in an orbital do not move in a completely independent fashion. Two electrons with opposite spins occupy a smaller volume than two electrons with parallel spins and therefore experience more repulsion. The relative exchange energies for alternative electron configurations is estimated by calculating the number of pairs of electrons with parallel spins and multiplying it by K.

For three electrons the alternative configurations shown in the boxes in the margin have exchange energies of $3K$ and K respectively. Therefore the former, with the maximum spin multiplicity is more stable by $2K$ plus the additional repulsion energy resulting from putting two electrons in the same orbital. For n parallel spins there are $n(n-1)/2$ pairs of electrons with parallel spins. This is the basis of *Hund's first rule* which states that the ground state of an atom is that having the greatest spin multiplicity, i.e. that having most unpaired electrons. The exchange energies for some common electronic configurations are summarized in Table 1.8.

The value of K varies from atom to atom and also as the charge on the atom is changed. Since the radial distribution functions for orbitals with different n and l quantum numbers vary, K is also different for different orbitals.

Examples of exceptions to the *aufbau* principle are given in Table 1.9. The atoms listed in Table 1.9 in italics represent exceptions which can be attributed to exchange energy effects and are associated with either full or half full d and f shells.

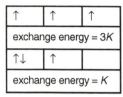

↑	↑	↑
exchange energy = 3K		
↑↓	↑	
exchange energy = K		

The multiplicity of a configuration is given by the value of $2S + 1$, where S is the total spin

Table 1.8 Exchange energies for atoms of the p and d blocks

p shell			
p^1	0	p^2	K
p^3	$3K$	p^4	$3K$
p^5	$4K$	p^6	$6K$

d shell			
d^1	0	d^2	K
d^3	$3K$	d^4	$6K$
d^5	$10K$	d^6	$10K$
d^7	$11K$	d^8	$13K$
d^9	$16K$	d^{10}	$20K$

Table 1.9 Some examples of exceptions to the *aufbau* principle

Cr $3d^54s^1$	La $5d^16s^2$
Cu $3d^{10}4s^1$	Ce $4f^15d^16s^2$
Gd $4f^75d^16s^2$	Pr $4f^36s^2$
Mo $4d^55s^1$	Nb $4d^45s^1$
Pd $4d^{10}5s^0$	Ru $4d^75s^1$
Ag $4d^{10}5s^1$	Rh $4d^85s^1$
Au $5d^{10}6s^1$	Th $6d^27s^2$
Cm $5f^76d^17s^2$	Pa $5f^26d^17s^2$
Lr $5f^{14}6d7s^2$	Np $5f^{14}6d^17s^2$
Au $5d^{10}6s^1$	Pt $5d^96s^1$

The atoms with their electronic configurations listed in *italics* may be attributed to exchange energy effects

The remainder are associated with the similar orbital energies of 4d and 5s, 5d and 6s, 5d and 4f, and 6d and 5f orbitals. For these atoms the energy separations between the orbitals is approximately the same as the interelectronic repulsion energies.

Repulsion between two electrons in different orbitals is smaller than that between two electrons in the same orbital. Therefore, if the energy difference between two orbitals is smaller than the difference in electron repulsion for the two electron configurations then there is energy stabilization by the two electrons occupying different orbitals. They still maintain parallel spins, however, in order not to lose the favourable exchange energy contribution.

1.6 Closed shells and half-filled shells

The quantum mechanical description of the atom is based on a spherical shell model and this suggests that atoms with complete shells may have greater stabilities than those with incomplete shells. Complete shells have no inherent stability and an atom with a closed shell configuration is only more stable than the adjacent atoms if its electrons are less available for bonding. The energy of the electron in an orbital may be associated with its ionization energy and therefore a plot of ionization energy versus atomic number such as that shown in Fig. 1.12 provides a basis for defining the applications and limitations of the shell model.

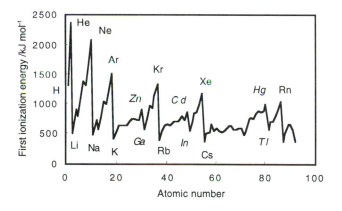

Fig. 1.12 A plot of the first ionization energies of the elements against their atomic numbers

Fig. 1.12 clearly demonstrates that the 'inert' or 'noble' gases which have completed shells have significantly higher ionization energies than those of the adjacent atoms and therefore the orbital separations are sufficiently large to provide a well-defined shell structure corresponding to a filling of the rows in the Periodic Table.

The filling of the shell coincides with the effective nuclear charge reaching a maximum and consequently the orbitals are contracted and more energy is required to remove an electron from the influence of the nucleus.

Fig. 1.12 also suggests that the orbital energies become less well separated as the atomic number increases and that the closed shell concept becomes less valid down a column of the Periodic Table. This accords with the observation that the 'inert' gases no longer remain inert down the column and indeed the chemistry of xenon is now very well developed and many Xe compounds with fluorine and oxygen are known.

Atoms with a single electron outside the closed shells have the lowest ionization energies and correspond of course to the strongly reducing alkali metals Li, Na, K, Rb, and Cs. For these atoms the valence electron resides in an orbital with a new principal quantum number, n, which has a radial distribution function with a maximum well separated from the core electrons. The large average distance of the electron from the nucleus makes it relatively easy to remove the electron from the atom.

The electronic stabilities of atoms with closed shell electronic configurations carries over to their chemical compounds. The formation of chemical bonds is commonly associated with the attainment of inert gas configurations by the atoms within the molecule. In idealized ionic compounds the transfer of an electron from one atom to another leads to a pair of oppositely charged ions, both of which attain inert gas configurations as a result of the electron transfer process, e.g.

$$Na\ [Ne]3s^1\ +\ Cl\ [Ne]3s^23p^5\ \rightarrow\ Na^+\ [Ne]\ +\ Cl^-\ [Ar]$$

The electrostatic attraction between cation and anion compensates for the energy required for the electron transfer, particularly in the solid state where each ion is surrounded by 6 or 8 oppositely charged nearest neighbours.

In covalent compounds the inert gas configuration is achieved by the sharing of electron pairs. This sharing leads simultaneously to both atoms achieving an inert gas configuration.

In the majority of real molecules the bonds are neither fully ionic or covalent and a continuum of bond types between these extremes occurs, but since both bond types are associated with the attainment of the inert gas configuration the concept of achieving a closed shell is a key concept in inorganic chemistry. The important bonding generalization is described as the effective atomic number (EAN) rule which is discussed in detail in Chapters 4 and 5. However, in common with atoms the attainment of the inert gas configuration by an atom in a molecule is not the only consideration in the formation of stable compounds. The inert gas configuration only represents a thermodynamically stable state if the energy gaps between the highest occupied and lowest unoccupied orbitals within the molecule are large, i.e. the molecule has a well-defined shell structure. There are therefore many exceptions to the inert gas rule.

Although the largest atomic energy level separations in atoms occur when a row of the Periodic Table is complete, there are other energy separations associated with the filling of subshells. For example, a filled s shell, a filled d shell, and a filled f shell can lead to atoms with ionization energies larger than those of neighbouring atoms. The data shown in Fig. 1.12 show that the elements with a completed ns^2 subshell have relatively high ionization energies, particularly if they occur after a filled d shell, i.e. Zn, Cd, and Hg. In molecules also, there are relatively stable examples which can be associated with atoms with electron configurations which result from completed subshells. For example, although the alkali metals form a wide range of ionic compounds which result from the ionization of an electron from the outer ns shell, more recently, it has been established that they are also able to form anions which result from the filling of the s shell, e.g. Na^-, K^-, Rb^-. Similarly, gold, which has the electronic configuration $[Xe]5d^{10}6s^1$, is capable of forming a wide range of compounds where the formal oxidation state of the metal is +1, but can also form compounds with auride ions Au^- when reduced in liquid ammonia.

The elements in Groups 13–15 not only form compounds where the formal oxidation state of the central atom corresponds to the ionization of all the electrons from the valence orbitals, e.g. Al^{III}, Si^{IV}, P^V, but also compounds where the electrons are removed from the p shell, but with the s electrons retained. Specifically, the oxidation states Tl^I, Sn^{II}, Pb^{II}, As^{III}, and Sb^{III} are commonly observed in compounds of these elements. In all such ions the central ion retains an $n s^2$ configuration.

The post-transition elements of rows 4, 5, and 6 of Groups 11–18 do not form compounds where all the valence electrons are ionized from the s shell, because the resultant high oxidation states are chemically unsustainable and therefore the highest oxidation state observed corresponds to a partially filled shell which retains a fully completed d shell. For example, Cd^{II}, In^{III}, Sn^{IV}, Sb^V, Te^{VI}, I^{VII}, and Xe^{VIII} are all isoelectronic ions with $5d^{10}$ configurations and are the limiting highest oxidation states found in compounds of these elements.

The transition metals, the lanthanides, and the actinides may form compounds where the central metal ion has an incomplete d or f shell. However, the redox properties of the compounds do not vary in a linear fashion across a series; compounds where the central metal ion has an electron configuration which corresponds to a half-filled shell, i.e. d^5 (transition metal) and f^7 (lanthanides and actinides), are less readily oxidized than those of the adjacent metal ions. However, it must be remembered that the half-filled shell is not stable in its own right. The ionization energies of atoms with these electronic configurations tend to be higher than those of the adjacent atoms. The exchange energy effects (see page 14) reach a maximum for d^5, d^{10}, f^7, and f^{14} and cause atoms with these configurations to have slightly higher ionization energies and this carries over to the properties associated with the compounds containing these ions. For example, aqueous solutions of Mn^{2+} (d^5) are more difficult to oxidize than Fe^{2+} (d^6).

1.7 Orbital types and the Periodic Table

The quantum mechanical description of the polyelectronic atom leads to regions of the Periodic Table which arise from the filling of s, p, d, or f orbitals by the valence electrons. This raises the question of whether these blocks of elements have similar chemical and physical properties.

s block elements: alkali and alkaline earth metals

The alkali metals are soft, low melting point metals which are very strongly reducing and reactive. They react very rapidly and at times explosively with

Oxidation state: this is a formal device for partitioning electrons in a molecule in a chemically intelligent way. The oxidation state of an atom in a compound is the charge which would result if the electrons in each bond to that atom were assigned to the more electronegative atom. The more electronegative atom is thereby made to complete its octet of electrons. e.g. The ion MnO_4^- can be regarded as $Mn^{7+} + 4O^{2-}$, the oxidation states are therefore Mn^{VII}, O^{-II}

Valency: this is the number of homopolar covalent bonds which an atom is capable of forming. The majority of atoms exhibit multiple valencies corresponding to alternative ways of utilizing their valence electrons, e.g. S ($3s^23p^4$) exhibits valencies of two in SH_2, four in SF_4, and six in SF_6.

H_2O to liberate H_2 and form the basic hydroxides (MOH). They are so strongly reducing that they can only be isolated from their salts by electrolysis. The alkaline earth metals are more brittle, but remain malleable and extrudable and are harder and have higher melting points than the alkali metals. They are good electrical conductors. They are strong reducing agents, but react less rapidly with H_2O than the alkali metals.

d block elements: transition metals

The transition metals have high melting points, high densities, high thermal and electrical conductivities. They are malleable, ductile, and have a metallic lustre. The chemical properties and densities of the metals are sufficiently different for them to be readily separated, but nonetheless they exhibit a range of similar chemical properties. They form compounds in a wide range of oxidation states, and form an extensive series of complexes many of which are highly coloured and paramagnetic. They are less reducing than the s block elements. They, and their compounds, are effective catalysts in both industrial and biological processes. This property arises because they are able to coordinate and enhance the reactivities of small molecules such as H_2, CO, C_2H_2, and O_2 and break C–H, Si–H, H–H, and C–X (X = Cl, Br, or I) bonds.

f block elements: lanthanides and actinides

The lanthanides are silvery white, dense, metals and are good conductors and have high melting points. Chemically, they are quite electropositive and form compounds usually in the +3 oxidation state. The compounds in this oxidation state are so similar that separation of the individual metals is difficult. It requires either multiple recrystallization of soluble salts or more conveniently ion exchange processes. The compounds are lightly coloured and commonly paramagnetic. The actinide elements are less similar and in some respects resemble the transition metals more closely. They form stronger and more stable complexes than the lanthanides, they are slightly less electropositive, but remain pyrophoric (spontaneously inflammable) as fine powders. Although the +3 oxidation state is common, they also form compounds in +4, +6, and +7 oxidation states. Both the lanthanides and actinides have large ionic radii and consequently form compounds with high coordination numbers (generally 6–14). The organometallic complexes of the actinides are more stable and less reactive than those of the lanthanides.

p block elements: post-transition or main group elements

The three groups discussed above are sufficiently similar for some credence to be given to the view that there is a simple correlation between the orbital type and chemical properties. For the p block elements there is such a divergence of chemical properties that this view has to be abandoned. This group of elements has a diagonal line passing through it such that, under normal conditions of temperature and pressure, elements above the line, i.e. B, C, N, O, F, Ne, Si, P, S, Cl, Ar, As, Se, Br, Kr, I, and Xe are typical non-metals and those below the line are typical metals. However, the metals below the line are not strongly reducing. They are 'base metals' which require oxidizing acids to dissolve them. Even amongst the non-metals the range of reactivities is enormous. The noble gases are very unreactive and only Xe forms a wide range of compounds. For example, it reacts with fluorine to form XeF_2, XeF_4, and XeF_6. In contrast, fluorine is the most reactive of elements, which oxidizes every other element except the lighter noble gases.

From the brief summaries above, it is evident that elements with valence electrons occupying orbitals with the same angular momentum quantum numbers have similarities which diminish in the order: f > d > s >> p and a consistent account of the comparative properties of the elements requires a more detailed analysis of the electronic properties of specific groups of atoms.

This book attempts to provide such an account.

1.8 Angular part of the wave function

In the first section of this chapter the l and m quantum numbers were shown to define the shapes of the atomic orbitals illustrated in Fig. 1.1. These shapes have an important influence on the number and types of covalent bonds which an atom is capable of forming. The shape of an orbital is decided by the total number of angular nodes (l quantum number) and its bonding properties by the m quantum number which specify the number of nodes along the z axis. These shapes are very important because the overlap of atomic orbitals to form chemical bonds requires that the orbitals on the atoms have matching nodal properties. The wave nature of the electron requires the constructive overlap of the wave functions on adjacent atoms leading to enhanced regions of bonding electron density between the nuclei which can only occur if the individual wave functions have matching signs. The s ($l = 0$), p ($l = 1$), and d ($l = 2$) orbitals in Fig. 1.13 all have $m = 0$ and are capable of overlapping with orbitals also lying along the z axis and with m values of zero. The constructive overlap of the orbitals results in covalent bond formation.

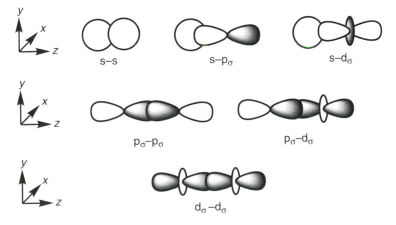

Fig. 1.13 The range of possible overlaps leading to the formation of σ bonds (all directed along the z axis)

Chemical bond formation also requires that the electrons in the two overlapping orbitals pair up their spins. This might seem strange at first, since for atoms the lower energy state is one with parallel spins. The relative spin directions of the electrons affects their distribution in space. Electrons with parallel spins tend to move apart and those with paired spins tend to move towards each other. In atoms it is energetically favourable for electrons to be as far apart as possible since this reduces their mutual repulsion. In a covalent bond the pairing of spins allows the electrons to occupy a smaller volume of space between the nuclei and thereby reduce their potential energy.

The overlaps which are associated with orbitals with no nodes along the internuclear axis are described as **sigma** (σ) bonds. The range of permutations for s, p, and d orbitals are illustrated in Fig. 1.13. These bonds are found in covalent compounds of all the elements. The out-of-phase overlap of orbitals leads to a corresponding set of less stable antibonding orbitals, which have a nodal line perpendicular to the bond direction.

For orbitals with the quantum number $m = \pm 1$ overlap occurs in a sideways fashion which retains a single node along the internuclear axis. The resultant permutations are illustrated in Fig. 1.14 and are described as pi (π) bonds. Since m can take up the values ± 1, two equal bonds of this type may be found.

In the examples provided, the two π bonds which are illustrated when taken together with the σ bond lead to a triple bond, such as those found in N_2 and CO.

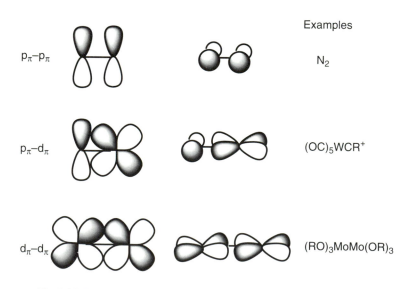

Examples

p_π–p_π N_2

p_π–d_π $(OC)_5WCR^+$

d_π–d_π $(RO)_3MoMo(OR)_3$

Fig. 1.14 Overlaps which give π bonds and examples of π-bonded compounds

In a simple diatomic molecule both π bonds can occur simultaneously but in lower symmetry molecules one of the orbitals may become involved in σ bonding with adjacent atoms and leaving only a single π bond. For example, as shown in Fig. 1.15 in CH_2O one of the $m = \pm 1$, p orbitals is used for bonding to hydrogen and only the remaining component is utilized for π bonding.

For orbitals with the quantum number $m = \pm 2$, the sideways overlap leads to covalent bonds with two nodes along the internuclear axis as shown in Fig. 1.16. Such bonds are described as delta (δ) bonds.

The two d_δ–d_δ bonds illustrated in Fig. 1.16 when taken together with two d_π–d_π and one d_σ–d_σ can lead to a total bond order of 5. However, as with π bonds, if one of the d_σ orbitals is used for σ bonding a bond order of 4 results. The ion $[Re_2Cl_8]^{2-}$, whose structure is shown in the margin, has such a quadruple bond with one σ, two π and one δ contributions.

The angular part of the wave function also decides the maximum number of two-centre two-electron (2c–2e) covalent bonds which an atom is capable of forming. If an atom has s and p valence orbitals which enable it to overlap effectively with the orbitals of neighbouring atoms the maximum number of covalent (2c–2e) bonds it can form is four, i.e. the total number of valence orbitals that are available.

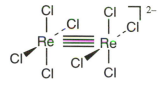

σ bonding   π bonding

Fig. 1.15 Bonding in the formaldehyde molecule

A two-centre two-electron (2c-2e) bond is equivalent to the conventional covalent bond between two atoms

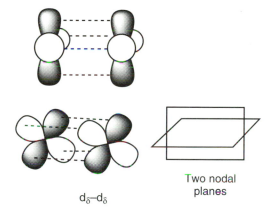

d_δ–d_δ

Two nodal planes

Fig. 1.16 The formation of a δ bond by the sideways overlap of two d_{xy} orbitals

For example, Si which has a $3s^2 3p^2$ configuration is capable of forming a total of four 2c–2e bonds which utilize the s and three p valence orbitals, e.g. in the molecules $SiCl_4$, SiH_4, and $SiMe_4$, the orbitals **hybridize** to form four equal orbitals which point towards the vertices of a tetrahedron.

A hybrid orbital is a linear combination of atomic orbitals centred on a single atom. This mixing of atomic orbitals has the effect of concentrating the electron density of the resultant hybrid into more specific regions of space and shifting the nodal surfaces. The mixing has the net effect of improving the overlap with orbitals of an adjacent atom. These effects are illustrated for sp hybrids in Fig. 1.17 and for sp^2 and sp^3 hybrids in Fig. 1.18.

The sp hybridization leads to a pair of combined orbitals which make an angle of 180°. Each hybrid is able to overlap more effectively with an atom which lies along the z axis than the isolated s or p_z orbitals. Each hybrid orbital may form either a two-centre two-electron bond with the orbital of another atom or, when occupied by an electron pair, form a non-bonding orbital. If n atomic orbitals contribute to the hybridization n orthogonal

The % orbital character is obtained by squaring the coefficients of the contributions to the hybrid orbitals:

$0.5\,s + 0.866\,p_z \equiv 1/4\,s + 3/4\,p$
(sp^3 hybrid)

$0.574\,s + 0.819\,p_z \equiv 1/3\,s + 2/3\,p$
(sp^2 hybrid)

$0.707\,s + 0.707\,p_z \equiv 1/2\,s + 1/2\,p$
(sp hybrid)

$0.408\,s + 0.707\,p_z + 0.577\,d_{z^2}$
$\equiv 1/6\,s + 3/6\,p_z + 2/6\,d_{z^2}$
(sp^3d^2 hybrid)

(non-overlapping) orbitals are formed. Therefore sp^2 hybridization leads to three orthogonal hybrids which point towards the vertices of a triangle and sp^3 hybrids point towards the vertices of a tetrahedron (see Fig. 1.18). Atomic d orbitals may also be introduced into the hybridization process and for example sp^3d^2 hybrids have six orthogonal wave functions pointing towards the vertices of an octahedron. Some of the more common hybridization schemes are summarized in Table 1.10.

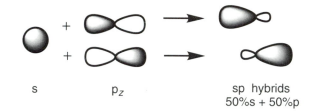

Fig. 1.17 Illustration of sp hybridization

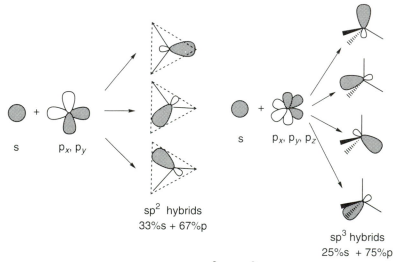

Fig. 1.18 Illustrations of sp^2 and sp^3 hybridization

The hybrid orbitals described above may be used to generate σ-bonding frameworks through the sharing of electrons in covalent bond formation, to act as acceptor orbitals in a coordinate bond, or to accommodate lone pairs of electrons. For example, ammonia may be described initially in terms of four sp^3 hybridized orbitals pointing towards the vertices of a tetrahedron with three hybrids used for N–H covalent bonds and the fourth accommodating the lone pair.

A transition metal which has nd, $(n+1)$s, and $(n+1)$p valence orbitals is able to form a maximum of nine (2c–2e) bonds to other atoms, e.g. Re[VII] in $[ReH_9]^{2-}$ which utilizes the s, three p, and five d orbitals. They hybridize to form nine orbitals which point towards the vertices of a tricapped trigonal prism. This complete utilization of valence orbitals has important consequences for the effective atomic number (EAN) rule which is discussed in Chapters 4 and 5.

Table 1.10 Hybridization schemes for common coordination polyhedra

Coordination number	Arrangement	Composition
2	*Linear*	sp, pd, sd
	Angular	sd
3	*Trigonal planar*	sp^2, p^2d
4	*Tetrahedral*	sp^3, sd^3
	Square planar	sp^2d
5	*Trigonal bipyramidal*	sp^3d, spd^3
	Square pyramidal	sp^2d^2
	Pentagonal planar	p^2d^3
6	*Octahedral*	sp^3d^2
	Trigonal prismatic	spd^4
7	*Pentagonal bipyramidal*	sp^3d^3
8	Square antiprismatic	sp^3d^4
	Dodecahedral	sp^3d^4
9	Tricapped trigonal prismatic	sp^3d^5

Those in *italics* are the more commonly utilized hybridization schemes

In order to obtain effective hybridization of atomic orbitals for chemical bond formation the following criteria must be met:

1. The atomic orbitals which are being hybridized must have similar energies and maxima in their radial distribution functions.

2. The s, p, d, and f atomic orbitals which are being combined to generate the hybrids must overlap equally well with the orbitals of the atoms which are bonding to the central atom.

The strength of a homonuclear chemical bond depends primarily on the efficiency of overlap between the two contributing orbitals and the bond order. The bond is little affected by the behaviour of electrons not involved directly in the bond, unless the atoms are very close together and electron–electron repulsions destabilize the bond. The weak bond observed in F_2 has been attributed to these indirect effects.

For a post-transition element such as tin the form of the Periodic Table may suggest that similar d, s, p hybridization effects may be important but the 4d orbitals are so much smaller than those for the 5s and 5p orbitals that they may be considered as core electrons and the bonding description is focused completely on the 5s and 5p valence orbitals.

The core-like nature of the 4f and 5f orbitals of the lanthanide and actinides means that these elements do not form a wide range of compounds which conform to an effective atomic number rule. Another factor which works against such compounds, which would involve the maximum formation of sixteen (2c–2e) bonds, is the unrealistically large coordination number which would result.

1.9 Radial part of the wave function

The radial part of the wave function influences the overall form of the Periodic Table and is very important in deciding the extent to which orbitals overlap in molecules. Therefore, differences in the radial part of wave functions for a column of elements have very important chemical consequences because they influence the relative strengths of the bonds formed between that element and others.

The wave nature of the electron means that it is not possible to define the precise size of an orbital. The radial distribution function $4\pi r\,^2\Psi^2$ defines the probability of finding the electron in the shell of thickness dr located r from the nucleus and the most probable radius r_{max}, of an orbital is given by:

$$r_{max} = \int \frac{\Psi r \Psi}{\Psi\Psi}\mathrm{d}r$$

The value of r_{max} gives a reliable indicator of the relative sizes of orbitals. For the 1s orbital of the hydrogen atom $r_{max} = 52.918$ pm, which corresponds to the Bohr radius of the hydrogen atom.

The Bohr radius is that which was derived by Bohr for the planetary model of the hydrogen atom prior to the development of quantum mechanics. Although it gave the correct radius for the electron, the model violates the Heisenberg uncertainty principle which implies that an atomic particle, such as an electron, cannot have a fixed path round the nucleus.

Table 1.11 illustrates how the 1s orbital contracts under the experience of the increased nuclear charge for hydrogen and the alkali metals. For lithium r_{max} for the 1s orbital contracts to approximately one third of the size and the valence orbital (2s) has r_{max} almost 10 times farther from the nucleus and is therefore much more able to overlap with orbitals of adjacent atoms. In the caesium atom the radius of the 1s orbital is one sixtieth of that of the hydrogen atom and very close to the nucleus. The r_{max} clearly illustrates the transition of orbitals from the valence region into the core region and some representative relative data are given in Table 1.11.

Table 1.11 Values of r_{max} [relative to that of H ($r_{max} = 52.918$ pm) =1] for the ns orbitals of hydrogen and the alkali metals

Atom	1s	2s	3s	4s	5s	6s
H	1					
Li	0.364	**3.101**				
Na	0.093	0.607	**3.387**			
K	0.053	0.317	1.078	**4.330**		
Rb	0.026	0.149	0.450	1.270	**4.650**	
Cs	0.017	0.095	0.272	0.643	1.562	**5.138**

The dominant role of the electron distribution of the valence orbitals in deciding the strengths and lengths of bonds is illustrated by Fig. 1.19 which illustrates the close correlations between r_{max} for valence orbitals and the metallic radius r_{met} for the alkali metals.

The metallic radius is half the internuclear distance between neighbouring metal atoms in the solid state structure. Therefore, it is not unreasonable that maximum overlap between the orbitals of adjacent metal atoms occurs when the metals are separated by a distance of $2r_{max}$. The correlation between r_{max} and r_{met} is not quite linear, but the average line has a slope close to 1.0.

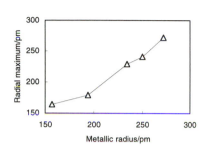

Fig. 1.19 A graph of r_{max} against r_{met} for the Group 1 elements. r_{max} is the most probable radius of the ns valence orbitals and r_{met} is the metallic radius of the metal derived from the solid state structure.

For the Group 13–18 elements the atoms have s and p valence orbitals available for bonding and therefore it is useful to contrast r_{max} for the ns and np valence orbitals and the relevant data are presented in Table 1.12.

Table 1.12 Comparison of r_{max} and r_{cov} for the Group 14 elements (the radii are given in pm)

Element	r_{cov}	r_{max} (ns)	r_{max} (np)
C	77	65	64
Si	117	95	115
Ge	122	95	119
Sn	140	110	137
Pb	144	107	140

The 2s and 2p orbitals of carbon have very similar r_{max} values and therefore they overlap almost equally well with the valence orbitals of atoms bonded to carbon. For the heavier Group 14 elements the ns orbital becomes progressively more contracted relative to the np orbital (see Table 1.12). In Table 1.12, the covalent radii of these elements are also presented. They mirror r_{max} for the p valence orbitals much more closely. It follows that in covalent compounds of these elements the bonds have a higher proportion of p orbital character and the sp hybridization is less effective than for carbon. These differences have important consequences for the geometries and reactivities of the main group elements. Table 1.13 illustrates the way in which the 2s and 2p orbitals contract under the influence of the increasing effective nuclear charge for the second row elements.

Schematic illustration of the effect of increasing the principle quantum number on the radial properties of the atomic orbitals. The introduction of radial nodes leads to the wavefunctions changing sign as they radiate from the nucleus. Positive regions of the wavefunctions are indicated by shading and negative regions by the unshaded areas.

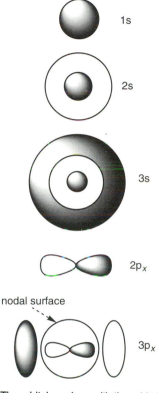

The orbital overlaps with the orbitals of neighbouring atoms are dominated by r_{max}, but there is also a contribution from the destructive overlap with the inner components of the wavefunction lying inside the nodal surface.

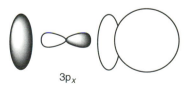

Table 1.13 Values of r_{max} (pm) for the valence orbitals of the second row elements

	Li	Be	B	C	N	O	F	Ne
2s	164	109	81	65	54	43	41	36
2p			84	64	52	44	38	34

The r_{max} drops to approximately one fifth of its initial value from Li to Ne and the orbitals become increasingly core-like. The pattern is repeated for subsequent rows of elements. These orbital contractions reinforce the shell structures associated with rows of the Periodic Table because the inert gas atom has contracted orbitals which are generally unavailable for chemical bonding. These horizontal size-contractions have very important implications and these are discussed in Chapter 3.

Table 1.14 summarizes the r_{max} values for the Group 6 transition elements chromium, molybdenum, and tungsten.

Table 1.14 Values of r_{max} (pm) and metallic radii, r_{met} (pm), for the nd and (n+1)s valence orbitals of the Group 6 transition metals

	r_{met}	r_{max} (nd)	r_{max} (($n + 1$)s)
Cr	129	46	161
Mo	140	74	168
W	141	79	147

Clearly the sizes of the nd and ($n+ 1$)s orbitals differ significantly and the nd orbitals are much more core-like. The 3d orbitals of the first row transition element, Cr, are particularly contracted relative to 4s and for distances such as those found in the metal the overlap between the 4s orbitals is much more significant that those originating from the 3d orbitals. Indeed, the participation of the 3d orbitals in covalent bonding is small for the 3d elements but increases for the 4d and 5d elements.

The 5s orbital of molybdenum is more diffuse than that of chromium and, more surprisingly, also larger than that of the 6s orbital of tungsten. This contraction has been attributed to relativistic effects. The contraction of core orbitals has been discussed above, but its consequences on the motion of the electrons have not been analysed. As an electron is constrained to move in a smaller and smaller volume its velocity, v, is increased until eventually it begins to approach the speed of light, c. In these circumstances it is necessary to correct the mass of the electron for relativistic effects according to the equation:

$$m = \frac{m_0}{\sqrt{1 - \dfrac{v^2}{c^2}}}$$

where m is the mass at velocity, v, and m_0 is the rest mass.

When these mass effects are taken into account and the wave functions of the atomic orbitals are calculated, their primary influence is on the core orbitals, but they also exert a secondary effect on the valence orbitals.

Specifically, the valence s orbitals are contracted and the valence d and p orbitals are expanded. Therefore, the contraction of the tungsten 6s orbital is attributed to these relativistic effects. The relativistic effects are particularly significant for the sixth and seventh row elements and account for a range discontinuities associated with the properties of these elements.

The 4f and 5f orbitals of the lanthanide and actinide elements are even more core-like than the 3d orbitals and their participation in covalent bonding is negligible. An expansion occurs for the 5f orbitals of the actinides, but the extent of overlap between these orbitals and the orbitals of adjacent atoms remain small. Some relevant data are summarized in Table 1.15.

Table 1.15 r_{max} (pm) for typical lanthanide and actinide elements

Ce[Xe]$6s^2 4f^2$		U[Rn]$7s^2 5d^1 4f^3$	
4f	38	5f	56
5s	73	6s	72
5p	82	6p	84
5d	116	6d	129
6s	207	7s	194

1.10 Commonality of electronic configurations

The quantum mechanical interpretation of the Periodic Table has highlighted the important fact that the adoption of common valencies and oxidation states by elements is associated with the elements having closely similar electronic configurations. Specifically, all the elements of Groups 1 and 2 have electronic configurations $ns^x np^0$ ($x = 1$ or 2 respectively and $n = 2$–7), and the main group elements (Groups 13–18) have configurations $ns^2 np^y$ ($y = 1$–6; $n = 2$–6). Therefore, columns of elements show common oxidation states and valencies because they share generic electronic configurations. For the transition elements, the lanthanides, and the actinides, the situation is less clear cut because the energies of the valence orbitals are sufficiently similar that elements within the same column exhibit different ground state electronic configurations. For example, the Group 10 elements Ni, Pd, and Pt, have the ground state electronic configurations:

Ni	$3d^8\ 4s^2\ 4p^0$
Pd	$4d^{10}\ 4s^0\ 5p^0$
Pt	$5d^9\ 5s^1\ 6p^0$

Despite these differences, the compounds of these elements show many similarities because the elements have the same number of valence electrons occupying valence orbitals with the same l quantum number values (l; $l = 0$, s; $l = 1$, p; $l = 2$, d). In compounds of these elements the properties and structures of the molecules are determined primarily by the combination of valence orbitals used to form the bonds and the number of electrons occupying the orbitals. For example, Ni, Pd, and Pt all form square planar complexes when the metals are in their +2 oxidation states. This is achieved by utilizing dsp^2 hybrid orbitals which point towards the ligands which define the square plane; Ni: $3d4s4p^2$, Pd: $4d5s5p^2$, Pt: $5d6s6p^2$. The fact that the values of the n quantum number are different does not change the orientation of these orbitals. The hybrids are empty and can accommodate lone pairs of electrons from the ligands. Since the metals are in the +2 oxidation state the M^{2+} ions have the common electronic configuration $nd^8(n+1)s^0(n+1)p^0$ ($n = 3, 4,$ or 5). When the atoms are ionized the nd orbitals are stabilized more than the $(n+1)$s and $(n+1)$p orbitals and consequently the ground state is more stable with the d orbitals occupied. Indeed, the energy gaps between the valence orbitals, nd, $(n+1)$s, and $(n+1)$p, in the positively charged ions are sufficiently large that the *aufbau* principle may be used reliably to predict the ground state configurations.

The bonding in the square planar complexes can therefore be represented schematically as shown in Fig. 1.20.

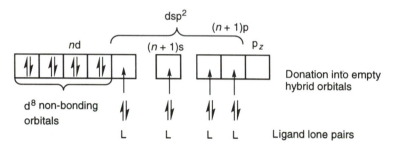

Fig. 1.20 A representation of bonding in square planar complexes

The square planar complexes of Ni, Pd, and Pt, therefore, are related because the metals are capable of forming similar empty hybrid orbitals from valence orbitals with common values of the l quantum number and have common d^8 electronic configurations associated with M^{2+} ions. Furthermore, the $[ML_4]^{2+}$ complexes are each associated with a total of 16 valence electrons, d^8 from the metals and eight donated by the four ligands.

In summary, the chemical and structural properties of the elements are determined mainly by the total number of valence electrons. The precise electron occupations of the $(n+1)$s, $(n+1)$p and nd orbitals for a column of elements is relatively unimportant. Therefore, the Mendeleev periodic relationships are not undermined by atoms in a column which do not have the same number of electrons occupying orbitals with the same l quantum number.

1.11 Nuclear stabilities

The discussion above has focused on the electronic structure of the atom and little has been said about the nucleus and the relative abundance of the elements in nature. Just as the polyelectron atomic model accounts for the gross features of the Periodic Table there is a corresponding shell model which accounts for the relative stabilities of nuclei. The atomic radius of an atom and its associated electron clouds is approximately 2×10^{-10} m, but the nucleus of the atom has a radius of only $\sim 10^{-14}$ m and therefore the protons and neutrons which make up the nucleus are concentrated in a very small part of the total volume. The situation is analogous to the Sun in the planetary system. This raises the question 'How can protons bearing the same positive charge could possibly be packed in such a small volume without repelling each other to such an extent that the nucleus flies apart?'. The protons do indeed repel each other with a force which is inversely proportional to the square of their separation ($\propto 1/r^2$), but this force is overwhelmed by an attractive force which occurs between all nucleons, independent of whether they are neutrons or protons. This attractive force is a very short range force and therefore is most effective when the nucleus is small and the nucleons are relatively close together. As the nucleus becomes larger then the nucleons are on average farther apart and the repulsive forces between the protons become more significant and the nuclear stability diminishes.

Two important results emerge from this simple model. Firstly, the stability of the nucleus as measured by the binding energy per nucleon reaches a maximum at approximately $Z = 56$, i.e. the second half of the transition series and, secondly, the ratio of neutrons to protons in the nucleus increases progressively with Z in an attempt to counteract the effect of the increasing repulsive energy terms associated with the protons. For the lightest elements ($Z < 20$) the number of protons and neutrons is almost equal, at the end of the first transition series it is approximately 1.2, at the end of the second transition series it is approximately 1.3, and it is 1.5 by the end of the third transition series. The additional neutrons contribute to the short range nucleon–nucleon attractive interactions without contributing the repulsive interactions associated with the protons and thereby help stabilize the nucleus. However, even with the additional neutrons the nuclei become progressively less stable after $Z = 56$ and more susceptible to nuclear decay via α-particle loss leading to nuclei with fewer protons and neutrons, or β-particle loss. Since the β-particle is equivalent to an electron, β-decay corresponds to the conversion of protons into neutrons. This decay process does not change the mass number, however.

The radius of the nucleus is given approximately by $R = R_0 A^{1/3}$, where R_0 is a constant and A is the *mass number* of the nucleus, which suggests that each nucleon is occupying almost equal space and the protons and neutrons are approximately close packed. The mass of the nucleus is always smaller than the sum of the masses of the nucleons which contribute to the nucleus and the mass difference Δm corresponds to the binding energy, E, of the nucleus as calculated from Einstein's famous equation: $E = \Delta mc^2$.

Radioactive decay processes

The emissions of α and β (positive or negative) particles by radioactive nuclei are commonly followed by emissions of γ-ray quanta as the product nuclei achieve their ground states.

α particles are helium nuclei, β particles are positve or negative electrons. α emission produces a product nucleus with values of Z and N which are both two units smaller than those of the parent nucleus. The mass number A is reduced by four units.

β^+ or β^- emissions do not change the mass number, but they cause the Z value of the product nucleus to be one unit smaller or greater, respectively, than the Z value of the parent nucleus. γ-ray quanta are high energy photons with frequencies higher than those of X-rays.

The nucleus of an atom is identified by the number of protons, Z, and the number of neutrons, N, contained within the nucleus. The element is characterized by the number of protons in the nucleus—the *atomic number*. *Isotopes* of an element have the same atomic number, but different numbers of neutrons, N. The *mass number* of a nucleus, A, is the sum of Z and N. For example, ^{238}U has $Z = 92$, $N = 146$ and a mass number of 238.

All atomic mass numbers from 1 to 138 are found in the natural state on earth except for masses 5 and 8. Approximately 280 stable and 67 naturally radioactive elements occur naturally on earth.

Since the speed of light c is such a large number the energies involved in binding the nucleons are orders of magnitude larger than those involving the atoms in chemical bonds.

The close-packing of nucleons in the nucleus presents a similar quantum mechanical problem to that solved above for the polyelectron atom, and when solved leads to a shell structure that has many similarities. It differs in detail, however, because the nucleons are not being attracted by a central field, but experiencing mutual attractions and repulsions. Nevertheless, the quantum mechanical solution leads to a shell structure which is represented by the same spherical harmonic functions as those used in the hydrogen atom solution (see Section 1.2). Therefore, the quantized energy states of the nucleus have the following symmetry labels and degeneracies (given in brackets): 1s(1), 1p(3), 1d(5), 2s(1), 1f(7),... which leads to the following closed shell configurations: 2, 8, 20, 28, 50, and 82 for both neutrons and protons and 114 or 126 depending on whether the nucleons are protons or neutrons. Nuclei which possess completed shells have the same status as the inert gases in the polyelectron atomic model. Nuclei which satisfy these requirements include He ($Z = 2$, $N = 2$), O ($Z = 8$, $N = 8$), Ca ($Z = 20$, $N = 20$), Pb ($Z = 82$, $N = 114$), where N = number of neutrons. The current research activity associated with the synthesis of new elements is driven in part by the belief that there might be a group of more stable elements based on a closed shell structure with $Z = 114$ and $N = 126$ ($A = 240$).

Just as the ionization energies of electrons from atoms progressively decrease as the atomic number increases, a similar phenomenon occurs with nuclei. The corresponding process is the loss of an α-particle (a helium nucleus) from the nucleus, which becomes progressively easier as the atomic number increases. The nucleons progressively occupy higher energy states in the shell model and therefore it becomes easier to lose an α-particle. For example, ^{235}U ($Z = 92$) has a half-life for α-particle decay of 5×10^9 years, for ^{247}Cm ($Z = 96$) it is 4×10^7 years and for ^{252}Fm ($Z = 100$) it is 21 hours.

As with the polyelectron atom, the spins of the nucleons play an important role in influencing the properties of nuclei. The protons and neutrons have a spin of $\frac{1}{2}$ and the protons or neutrons in an unfilled shell can form a stable paired state with their spins pointing in opposite directions. Since it is easier to remove a proton or neutron from the nucleus if it is unpaired, an alternation in the stabilities of nuclei is observed. If the difference in the binding energies between successive nuclei are plotted against the value of Z, this alternation effect is clearly visible. There is a larger discontinuity for those nuclides with a closed shell, e.g. when either $Z = 50$ or $N = 50$. Furthermore, the separation energies associated with the removal of either a proton or a neutron from the nucleus also show regular alternations with the separation energies being larger for nuclides with an even number of the corresponding nucleon than those with an odd number.

It follows from the shell model that even–even nuclei are more stable than odd–odd nuclei, with even–odd and odd–even nuclei having intermediate stabilities, and the commonest situation with even mass number is that there are two stable nuclides with the same value of A, each with even values of Z and N. Only in a few cases of light nuclides, e.g. ^{14}N, do the stable nuclides have odd values of Z and N. This generalization goes some way to explaining the pattern of observed stable isotopes found in nature for the elements. Ninety-nine percent of the universe is made up of hydrogen and helium which formed in the first few instants of the Universe after the 'Big Bang', and the remaining elements which make up the remaining 1% have been synthesized by nuclear reactions which have occurred in the centre of stars. For example, studies of the solar spectrum have identified 67 elements in the Sun's atmosphere. These nuclear reactions have led to the distribution of elements in the Periodic Table. In general the heavier nuclei are less abundant than the lighter nuclei and generally nuclear stability diminishes with atomic number.

The elements Fe, O, Si, Mg, Ca, Al, S, and Ni all have average abundances of more than 1% in nature. The large binding energies associated with the nuclides and also the considerable 'activation energies' required to bring positively charged nuclei together in a fusion process to build up larger nuclides requires very large kinetic energies and therefore such processes occur only in the centres of stars where the temperatures are sufficiently high. Common metals such as Fe, Cr, Ni, Cu, Ti, and Zn were probably formed early in the history of our galaxy. As a result of these nuclear reactions, elements with odd Z usually only have one or two stable isotopes, usually with odd masses and with N even. Elements with even Z often have several isotopes and those with an even mass, which implies an even N, are more common. Elements where Z is a magic number associated with completing a shell frequently have a large number of stable isotopes. Also there may be a large number of stable nuclides with the same neutron number N, when this is a magic number. The range of possible stable isotopes does not guarantee that each element must have a stable isotope and in fact two elements with $Z \leq 83$, i.e. technetium and promethium, have no stable isotopes.

In the minerals which make up the earth's crust oxygen amounts to about 47% by mass, silicon 28%, aluminium about 8%, and Fe, Ca, Na, K, and Mg contribute collectively about 16%. The remaining elements represent, therefore, only about 1% of the total mass of the crust. It is fortunate for mankind that these elements are not distributed evenly throughout the crust. Apart from technetium, element number 43, and promethium, element number 61, elements up to and including uranium, element number 92, are found in nature, albeit in very different amounts. The elements with atomic numbers greater than 92 have been synthesized by scientists since 1940. Since that time approximately 20 elements have been made and characterized. Therefore in the last half of the twentieth century the Periodic Table has been expanded by approximately 20%. Each of these elements has been synthesized by a range of transmutation reactions involving the bombardment of heavy nuclei by either neutrons or charged particles. These elements have about 30 isotopes which are sufficiently stable, i.e. have half-lives long enough that they can be studied by conventional chemical techniques. This means that they can be weighed and stay around for sufficiently long periods of time for their reactions to be studied.

Technetium was first discovered by Perrier and Segre in 1937 from the bombardment of a molybdenum target with deuterons (D^+). The most useful isotope is ^{99}Tc ($t_{1/2} = 2.1 \times 10^5$ yr). Technetium is widely used in nuclear medicine as an imaging agent for looking at human hearts and bones in patients suspected of having medical problems.

Although technetium does not occur naturally, 1800 kg of the element are produced annually in nuclear reactors, making it more abundant than naturally occurring rhenium and at ~$60 (£36) per gram significantly cheaper.

Promethium occurs only in nature in trace amounts in uranium ores, where it results from the spontaneous fission of ^{238}U. Milligram quantities are obtained by ion exchange of the fission products of nuclear reactors. The isotope ^{147}Pm decays by β-emission and its half-life is 2.64 years.

The elements following uranium (the transuranium elements) were initially made by the neutron irradiation of target materials such as uranium itself:

$$^{238}U \quad + \quad ^1n \quad \rightarrow \quad ^{239}U \quad \rightarrow \quad ^{239}Np + \beta^- \quad \rightarrow \quad ^{239}Pu + \beta^-$$

Isotopes of the elements einsteinium ($Z = 99$) and fermium ($Z = 100$) were first isolated from the debris following the 'Mike' (Mike for Megaton) H-bomb test in 1952

The ^{239}U isotope decays by β-decay initially to neptunium and then to plutonium. The high neutron fluxes in nuclear reactors enables these transformations to be undertaken on reasonably large samples and resulted in the formation of new elements in chemically meaningful quantities. However, the method suffers from the disadvantage that the mass number is increased only by one at a time and the resulting isotopes are relatively neutron deficient. The isotope ^{257}Fm (fermium) represented an upper limit of this method of synthesis.

Many of the transuranium elements have been isolated in large quantities by neutron bombardment reactions in nuclear fission reactors. Plutonium in ton quantities, neptunium, curium, and americium in kilogram quantities, berkelium in 100 gram quantities, californium in gram quantities and einsteinium in milligram quantities. The remainder have only been obtained in trace amounts and at times almost down to the atomic level. The large-scale plutonium production of course originated from its applications in nuclear weapons, but it has subsequently been used in electricity generation.

Elements with $Z > 100$ may be made in particle accelerators by colliding accelerated positively charged nuclei. The high velocity of impact ensures that there is sufficient energy to overcome the electrostatic repulsions between the positively charged nuclei which are being brought together, e.g.

$$^{246}U \quad + \quad ^{12}C \quad \rightarrow \quad ^{254}No \quad + \quad 4\,^1n$$

The new elements are effectively produced as single atoms and identified by mass spectrometric techniques and radio-physical measurements which depend on the decay pathways of the nuclides. These fusion reactions produce nuclides in such highly excited states that they undergo fission or emit neutrons and gamma rays. For elements with $Z > 106$ the kinetic energy of the lighter bombarding particle is adjusted to produce just the correct energy for fusion. The newly synthesized nuclide then has minimal kinetic energy and therefore decay processes involving fission and neutron loss are minimized. Pb and Bi are now used as targets since they have magic or near magic numbers and they are bombarded by positively charged nuclei which also have stable nuclides (e.g. ^{58}Fe and ^{64}Ni). In this way a large part of the excess kinetic energy associated with the colliding particles is used to break up the stable closed shells and the new nuclide does not have a great excess of kinetic energy. Examples of successful syntheses based on this strategy are:

$$^{209}Bi \quad + ^{58}Fe \quad \rightarrow \quad ^{265}109 \quad + \quad 2\,^1n$$

$$^{208}Pb \quad + \quad ^{64}Ni \quad \rightarrow \quad ^{271}110 \quad + \quad ^1n$$

These techniques have been used to synthesize new elements with atomic numbers up to 112.

These transmutation experiments have led to the completion of the actinide series (the filling of the 5f shell) and the discovery of the majority of the elements of the 4th transition series. The most recently discovered elements are those with $Z = 110$, 111, and 112. The naming of the trans-lawrencium elements has generated some controversy and the Periodic Table on the inside front cover gives the current recommendations. This research still continues in the expectation that there might be an island of stability associated with elements with atomic numbers in the region of 120.

1.12 Summary

The quantum mechanical interpretation of the Periodic Table represents one of the major triumphs of modern quantum theory. The exact solution of the wave equation of the hydrogen atom leads to a shell structure based on orbitals with characteristic quantum numbers which define the number of radial and angular nodes associated with each wave equation solution. For polyelectronic atoms the angular solutions of the wave equation for the hydrogen atom are retained, but the radial parts are modified to take into account the different effective nuclear charges experienced by the orbitals. Therefore, the basic shell structure based on s, p, d, and f orbitals remains, and the orbitals are filled according to the *aufbau* principle, i.e. the lowest energy orbitals are filled first. The orbitals are filled in a sequence based on the magnitude of $(n + l)$, i.e. 1s, 2s, 2p, 3s, 3p, 4s, 3d, 4p, etc. The filling sequence leads to the current shape of the Periodic Table. The numbers of elements in successive rows is given by the formula:

$$2\left\{\mathrm{INT}\left(\frac{N+2}{2}\right)\right\}^2$$

where N is the number of the row ($N = 1$–7) and the function $\mathrm{INT}(x)$ represents the rounded-down integer of x.

The Periodic Table is organized so that elements with similar electronic configurations occur in columns and therefore it readily accounts for the common oxidation states and valencies observed in compounds of these elements.

The shell and subshell structure derived from the quantum mechanical model leads naturally to an interpretation of ionic and covalent bonds which result from the completion of closed shell and closed subshell electronic structures. The shapes of the orbitals are defined by the angular part of the wave function and determine the type of bonds which the atom is capable of forming. The radial part of the wave function influences the extent of overlap between the orbitals on adjacent atoms and therefore determines the strengths of covalent bonds.

Further reading

P. W. Atkins, *The Periodic Kingdom*, Weidenfeld and Nicolson, 1995, London.

S. A. Cotton, After the Actinides, Then What?, *Chem. Soc. Rev.*, 1996, **25**, 219.

P. A. Cox, *The Elements on Earth*, Oxford University Press, 1995.

R. L. DeKock and H. B. Gray, *Chemical Structure and Bonding,* Benjamin-Cummings, Menlo Park, 1980.

J. P. Desclaux, Relativistic Dirac-Fock Expectation Values for Atoms with $Z = 1$ to $Z = 120$, *Atomic Data and Nuclear Data Tables*, **12**, 311–406, 1973.

N. Kaltsoyannis, *Relativistic Effects in Inorganic and Organometallic Chemistry*, *J. Chem. Soc., Dalton Trans.*, 1997, 1.

J. Emsley, *The Elements*, 3rd Edn, Clarendon University Press, 1998.

W. G. Richards and P. R. Scott, *Energy Levels in Molecules*, Oxford University Press, Oxford, 1995.

G. T. Seaborg, Transuranium Elements; Past, Present and Future, *Acc. Chem. Res.*, 1995, **28**, 257.

G. T. Seaborg and W. D. Loveland, *The Elements Beyond Uranium*, John Wiley, New York, 1990.

T. P. Softley, *Atomic Spectra*, Oxford University Press, Oxford, 1994.

A. Streitwieser, Jr. and P. H. Owens, *Orbital and Electron Density Diagrams: An Application of Computer Graphics*, Macmillan, New York, 1973.

M. J. Winter, *Chemical Bonding*, Oxford University Press, 1994.

2 Vertical trends for the s and p block elements

2.1 Trends associated with properties of isolated atoms

Introduction

Chapter 1 contains a description of the essential features of the Periodic Table and the quantum mechanical basis of its shell structure. Mendeleev's original proposals were based on the classification of elements, which showed common valencies in their compounds, into columns and ordered them according to their RAM values (atomic weights). This chapter provides a modern extension of this methodology and relates the chemical and physical properties of the s and p block elements to their atomic properties. The relationship is not a simple one, because the trends in physical and chemical properties of compounds frequently depend on a very fine balance of several factors which have comparable energies. Chemistry is thus a subtle science and its interpretation rests on a range of concepts which are used only in a semi-quantitative manner.

Atomic sizes

For atoms with a common electronic configuration in a periodic column, the increment in the principal quantum number (n) generally results in an increase in the size of the atom. It is difficult to define the radius of an isolated atom, but for metallic elements the metallic radius is defined as half the distance between the nuclei of nearest neighbour atoms in the metal lattice of the solid. The metallic radius may then be used to define trends in the relative sizes of atoms.

The variation in metallic radii for elements in Groups 1, 2, and 13 are shown in Fig. 2.1 and for Groups 3 to 12 in Fig. 2.2.

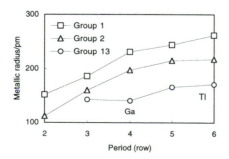

Fig. 2.1 Metallic radii for Groups 1, 2, and 13

For Groups 1 and 2 the metallic radii increase with principal quantum number, although the effect is not linear, but approximately parabolic, with the incremental increase becoming smaller as *n* increases. For the Group 13 elements a discontinuity is observed at Ga, resulting from the filling for the first time of a d shell (3d) (**transition metal contraction**) and the radius expansion at Tl is smaller than predicted and results from the filling for the first time of an f shell (4f) (**lanthanide contraction**).

The elements comprising the rows or periods of the modern Periodic Table are outlined below:

First row elements
H, He

Second row elements
Li, Be, B, C, N, O, F, Ne

Third row elements
Na, Mg, Al, Si, P, S, Cl, Ar

Fourth row elements
K, Ca
Sc, Ti, V, Cr, Mn, Fe, Co, Ni, Cu, Zn
Ga, Ge, As, Se, Br, Kr

Fifth row elements
Rb, Sr
Y, Zr, Nb, Mo, Tc, Ru, Rh, Pd, Ag, Cd
In, Sn, Sb, Te, I, Xe

Sixth row elements
Cs, Ba
La, Ce, Pr, Nd, Pm, Sm, Eu
Gd, Tb, Dy, Ho, Er, Tm, Yb
Lu, Hf, Ta, W, Re, Os, Ir, Pt, Au, Hg
Tl, Pb, Bi, Po, At, Rn

Seventh row elements (unfinished)
Fr, Ra
Ac, Th, Pa, U, Np, Pu, Am
Cm, Bk, Cf, Es, Fm, Md, No
Lr, Db, Jl, Rf, Bh, Hn, Mt,...

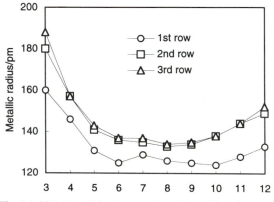

Fig. 2.2 Metallic radii for Groups 3 to 12 (transition elements)

Across any series the effective nuclear charge of the atoms increases because the valence electrons are not effectively screened from the positively charged nucleus by the electrons in the same valence shell. The effective nuclear charge experienced by the valence electrons of Ga is larger than might have been anticipated from an extrapolation from B and Al because, prior to Ga, the 3d shell has been filled for the first time. The effective nuclear charge experienced by the 4s electrons of gallium is therefore considerably larger than that for aluminium because of the presence of the 3d shell. Therefore, the valence electrons of the Ga atom experience a relative contraction in size. Similarly the effective nuclear charge experienced by the valence electrons of Tl is larger because of the inefficient screening by the 4f shell which is filled for the first time in the sixth row of the Periodic Table.

For the transition elements the change in radius in going from the 3d to the 4d metals is much larger than that from 4d to 5d. This effect has also been attributed to the 'lanthanide contraction,' because it results from the filling of the 4f shell prior to the occupation of the 5d orbitals. The similar radii of the 4d and 5d metals are illustrated in Fig 2.2 which compares the metallic radii of the first, second, and third row transition metals.

For non-metallic elements their covalent radii may be used as an indicator for the variation of atomic sizes. The covalent radius of an atom is its contribution to a covalent bond length where the bond order is 1. Specifically, if A and B have covalent radii r_A and r_B the A–B bond length is $r_A + r_B$. Fig. 2.3 illustrates the variation in covalent radius for the elements in Groups 14–17. The radii in general increase with *n*, the principal quantum number, but the **transition metal contraction** is again evident for the 3rd period elements. The transition metal contraction is more pronounced for the Group 14 and 15 atoms than the Group 16 and 17 atoms.

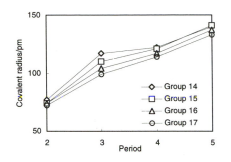

Fig. 2.3 Covalent radii for elements of Groups 14–17

The relative sizes of atoms in a particular group has important implications for their structural and chemical properties. For example, coordination numbers generally become larger for compounds of the heavier elements because the radius of the central atom increases. The sizes of the atoms also have an indirect effect on the lattice energies of ionic compounds, the hydration enthalpies of the ions and the rates of reaction.

Ionization energies of atoms

The first ionization energy, I_1, is the minimum energy required to remove an electron from an atom in the gas phase and in its ground state (i.e. its lowest possible electronic energy state).

$$A(g) \rightarrow A^+(g) + e(g) \quad \text{Energy required} = I_1$$

The second, third, etc., ionization energies can be similarly defined as:

$$A^+(g) \rightarrow A^{2+}(g) + e(g) \qquad\qquad I_2$$

$$A^{2+}(g) \rightarrow A^{3+}(g) + e(g) \qquad\qquad I_3, \text{etc.}$$

Down a given column of elements the first ionization energies generally decrease. As the principal quantum number increases the maximum in the radial distribution function for the valence orbital moves progressively farther from the nucleus. For a given effective nuclear charge it is easier to remove an electron from an orbital which on average is farther from the nucleus. For a column of elements both the effective nuclear charge and the average radius of the valence orbital increase. The radius of the valence orbital proves to be the dominant factor and a valence electron is more readily ionized from a larger atom where on average it lies farther from the nucleus. Fig. 2.4 illustrates the variation in first ionization energies (I_1) for the alkali and alkaline earth metals (Groups 1 and 2). The decrease in ionization energies is particularly noticeable for the alkaline earth metals of Group 2.

Although the variations in ionization energies discussed above refer to the gas phase metal atoms, they have significant implications on the ability of elements form the ions M^{n+} in solution and the solid state. They also influence the electropositive character of the metal—the lower the ionization

The elements comprising the Main Groups (s and p blocks) of the Periodic Table are shown below:

Group 1 (The alkali metals)
Li, Na, K, Rb, Cs, Fr
H is formally a member of Group 1 but its chemistry is usually dealt with separately from the other elements.

Group 2 (The alkaline earths)
Be, Mg, Ca, Sr, Ba, Ra

Group 13
B, Al, Ga, In, Tl

Group 14
C, Si, Ge, Sn, Pb

Group 15 (The pnictogens)
N, P, As, Sb, Bi

Group 16 (The chalogens)
O, S, Se, Te, Po

Group 17 (The halogens)
F, Cl, Br, I, At

Group 18 (The noble or inert gases)
He, Ne, Ar, Kr, Xe, Rn

energy in general the more electropositive the metal. The electropositive character of a metal is reflected in the ionic character of the compounds formed and the reducing ability of the metal. For example, the electropositive character increases in the order: Be < Mg < Ca < Sr < Ba.

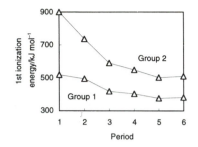

Fig. 2.4 Variations of I_1 for Groups 1 and 2

The second row elements are exceptional because their core electrons are restricted only to the filled 1s shell. Their valence electrons experience more fully the nuclear charge. Consequently they have higher ionization energies and their orbitals are more contracted and have the potential to overlap more effectively with the orbitals of like atoms at short internuclear distances.

For the Group 13 elements (see Fig. 2.5) the variation of ionization energies shows anomalies at Ga and Tl resulting from **transition metal** and **lanthanide contraction** effects noted in the previous section. Gallium and thallium have smaller radii than expected from a linear extrapolation and therefore it is more difficult to remove an electron from their valence orbitals; as a result, the ionization energies are higher than a simple extrapolation may have suggested. Similar anomalies are observed in the ionization energies of the Group 14 and 15 elements. Fig. 2.5 also illustrates the variations in I_1 for the Group 17 elements. Clearly, by the end of the series the effects of the transition metal and lanthanide contractions have been reduced and the ionization energies decrease in a more regular fashion.

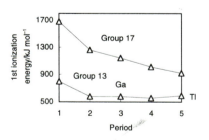

Fig. 2.5 Variation of I_1 for Groups 13 and 17

Fig. 2.6 illustrates how the cumulative sum of the ionization energies changes as progressively more highly ionized atoms are formed from the elements of Group 14, i.e. C, Si, Ge, Sn, and Pb. The manner in which the sum of the ionization energies for carbon are significantly larger than those for the remaining elements of the group is particularly noteworthy. This pattern is reproduced for the other groups of p block elements. The chemistries of the second row elements B, C, N, O, and F are frequently found to be significantly different from those of the later elements and this anomaly has its origins in the ionization energy variations.

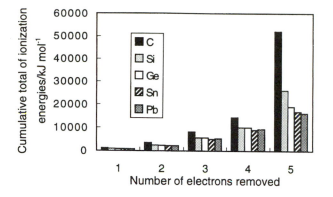

Fig. 2.6 The cumulative totals of ionization energies of the Group 14 elements plotted against the total number of electrons removed

Electronegativities

The electronegativity coefficient, χ, of an element is a measure of the power of an atom of that element to attract electrons towards itself in a chemical bond which it forms. An element which has a low electronegativity is described as **electropositive**. The major scales of electronegativity are those proposed by Pauling, Mulliken, and Allred and Rochow. They show the same broad trends. The Allred–Rochow scale is used in this book and relates electronegativity to the force exerted by the nuclear charge of the atom in the region where the electron density of the bond is largest, i.e. $Z_{eff}e^2/r^2$ where Z_{eff} is the effective atomic number, e is the unit of electronic charge, and r is the covalent radius of the atom. The Allred–Rochow electronegativity coefficients for the s and p block elements are given in Table 2.1 and illustrated in Fig. 2.7.

The electronegativity difference between atoms influences the polarities of covalent bonds. Two atoms which have similar electronegativities form covalent bonds which have very little ionic character. In contrast, a large electronegativity difference leads to bonds which have little covalent character and are essentially ionic.

The following trends in electronegativities down columns of the Periodic Table are noteworthy.

(a)   The electronegativities of the atoms generally decrease down a column of the Periodic Table. However, the transition metal and lanthanide contractions cause variations in sizes and effective nuclear charges which lead to clear discontinuities particularly for the Groups 13–16 elements. Specifically, Ga, Ge, As, Se, and Br have higher electronegativities than expected because of the transition metal contraction and Tl, Pb, and Bi because of the lanthanide contraction.

(b)   Generally there is a large decrease in electronegativity between the second and third row elements, e.g. C to Si, N to P, O to S, and F to Cl. The electronegativities change less dramatically for subsequent rows. This variation has major implications for the anomalous behaviour of the second row elements B, C, N, O, and F.

(c)   The electronegativities increase across a period because Z_{eff} increases and r decreases.

Chemists have realized for more than a century that the chemical and physical properties of compounds are influenced by the relative abilities of the constituent atoms to attract electrons towards themselves. However, turning this into a concept which may be rigorously defined and accurately quantified has proved to be problematical. The alternative electronegativity scales which have been developed have used different simplifications in order to define a scale which relates the ability of an atom in a molecule to attract electrons towards it in terms of physical properties of the isolated atom, e.g. in Mulliken's method the ionization and electron attachment enthalpies, or a property of the molecule, e.g. the mean bond enthalpy in Pauling's method. The resulting scales agree well for atoms which either have high or low electronegativities but differ in detail for some atoms in the middle of the range.

R. S. Drago and N. M. Wong, *J. Chem. Educ.*, 196, **73**, 123

For a detailed discussion of a more quantitative approach see D. Bergman and J. Hinze, *Angew. Chem. Int. Ed.*, 1996, **35**, 150.

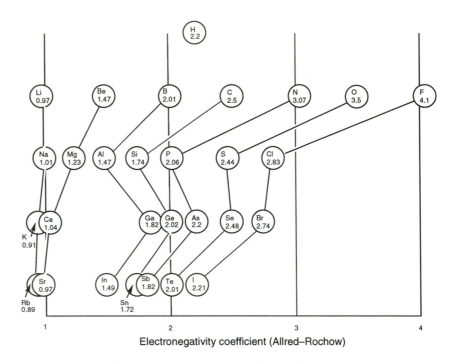

Electronegativity coefficient (Allred–Rochow)

Fig. 2.7 A diagram showing the variations in the electronegativity coefficients of the s and p block elements

The dipole moments (see page 93) for the following molecules in the gas phase illustrate the way in which the increased electronegativity difference leads to more polar molecules.

Molecule	$\mu(D)$
Cl_2	0
ClBr	0.52
ClLi	7.09
ClK	10.70

Table 2.1 Allred–Rochow electronegativity coefficients for the s and p block elements

H						
2.20						
Li	Be	B	C	N	O	F
0.97	1.47	2.01	2.50	3.07	3.50	4.10
Na	Mg	Al	Si	P	S	Cl
1.01	1.23	1.47	1.74	2.06	2.44	2.83
K	Ca	Ga	Ge	As	Se	Br
0.91	1.04	1.82	2.02	2.20	2.48	2.74
Rb	Sr	In	Sn	Sb	Te	I
0.89	0.97	1.49	1.72	1.82	2.01	2.21
Cs	Ba	Tl	Pb	Bi	Po	At
0.86	0.97	1.44	1.55	1.67	1.76	1.96

The electronegativity differences between atoms influence the polarities of bonds and therefore determine not only their ionic character but also their chemical reactivities. The following examples illustrate the way in which the electronegativity concept may be applied and its limitations.

The electronegativity difference between atoms A and B in the compound AB is a useful starting point for defining the bonding model which is to be used to interpret its properties. If A and B have identical electronegativities the bonding is exclusively covalent and the even sharing of electrons between the atoms leads to a symmetrical build up of electron density between the atoms. If the electronegativity difference lies approximately between 0 and 2 the atom with the larger electronegativity provides a greater pull on the electron density in the internuclear region and a charge separation develops along the bond. The bonds in such molecules are described as *polar covalent* and the magnitude of the charge separation depends on the electronegativity difference ($\chi_A - \chi_B$). The charge separation may be related to the progressive increase in the ionic character of the bond and may be indicated by signs above the atoms as follows:

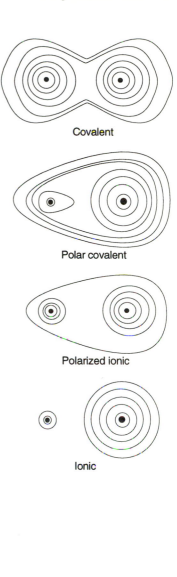

Covalent

Polar covalent

Polarized ionic

Ionic

		$\delta\delta+ \ \delta\delta-$	$\delta+ \ \delta-$	$\delta+ \ \delta-$	$+ \ -$
	H–H	H–C	H–O	H–F	Cs F
$\chi_A - \chi_B$	0	0.3	1.3	1.9	3.24
% covalent character	100				0

The maximum electronegativity difference occurs for the combination CsF ($\chi_A - \chi_B = 3.24$) and it is assumed that in such a bond the charge separation is complete and the bonding is 100% ionic. In an idealized ionic bond the electron density around each nucleus is high and is spherical in shape, but falls to a very low value between ions. Molecules such as CsF are known at high temperatures in the gas phase, but in the solid state they condense to form infinite structures where each ion is surrounded by ions with opposite charges. As the electronegativity decreases from 3.24 more covalency is introduced into the bond, the spherical cloud of electron density around the anion is distorted, and the bonding may be described as *polarized ionic*. The distinction between *polarized ionic* and *polar covalent* is more a question of one's starting point than something which could be distinguished experimentally. Diagrams illustrating these points are shown in the margin.

An inorganic chemist would relate these differences in physical properties to a progressive transition from ionic to covalent bonding which is expressed in the solid state by the adoption of molecular rather than infinite structures. This simplified interpretation has its dangers because other atomic or molecular variables may also contribute to the observed differences, but nonetheless it is a useful starting point for discussing the factors responsible for the trend.

1. AlF_3, $\Delta\chi = 2.63$ (m.p. 1200°C), has an infinite structure with all the aluminium ions octahedrally coordinated whereas $AlCl_3$, $\Delta\chi = 1.36$ (m.p. 192°C), has a layered polymeric structure. Both compounds when melted are electrical conductors confirming the presence of mobile ions in this phase. $AlBr_3$, $\Delta\chi = 1.27$ (m.p. 98°C), and AlI_3, $\Delta\chi = 0.74$ (m.p. 191°C), have molecular structures based on the dimeric $[Al_2X_6]$ units shown in the margin and they are non-conducting when molten.

2. The hydrides of the p block elements become progressively acidic towards the right-hand side of the Periodic Table:

$$CH_4 < NH_3 < OH_2 < HF$$

This may be related to the increasing electronegativity of the non-hydrogen atom which causes the hydrogen to become increasingly positively charged in the molecule and therefore more readily solvated as H^+ in solution. The limitations of this naive interpretation may be appreciated by the additional knowledge that the acidities of the haloacids (HF < HCl < HBr < HI) does not follow that anticipated on the basis of electronegativities. The strength of an acid in solution therefore cannot be related simply to a molecular property because the strengh of the bond to hydrogen and the solvation energies of the ions formed in solution are also important factors.

In summary, electronegativity differences represent a useful starting point for discussing bond types but they must be used cautiously in the interpretation of physical and chemical properties of series of compounds. These properties rarely depend exclusively on the charge separations in the bonds.

The polarizability also depends on the availabliity of low lying electronic excited states of the atom or ion. Mixing in of excited states with *ungerade* symmetry into a ground state with *gerade* symmetry has the effect of distorting the spherical cloud of electron density.

Polarizabilities and polarizing powers

Atoms and ions are not hard undeformable spheres. The electron clouds which surround the nucleus may be distorted by the electrostatic fields generated by electric charges on neighbouring atoms. The ease with which the electron density may be deformed is related to the volume of the atom or ion and the number of electrons associated with the atom and electron density probability function. Large atoms or ions with low electronegativities are generally highly polarizable, whereas small atoms with high electronegativities are not highly polarizable.

One of the most important polarizing influences is that arising from positively charged ions which are adjacent to negatively charged anions, because the resulting distortion of the electron cloud corresponds to the formation of partially covalent bonds. This is shown by the diagram in Fig. 2.8. Since electron density is transferred from the anion to the region between and cation and anion this relocation of electron density corresponds to partial covalent bond formation and is represented diagramatically in the margin.

Intuitively the polarizability of an anion A^{m-} may be related approximately to the product of its charge and volume, $\frac{4}{3}\pi r_A^3$:

$$\text{polarizability} \propto me \times \frac{4}{3}\pi r_A^3$$

The calculated polarizabilities for some common anions, given in Table 2.2, vary with anion charge and radius.

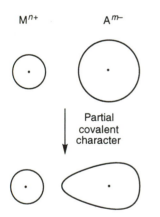

Fig. 2.8 A diagram illustrating the production of partial covalent character as the result of the polarization of an anion by a cation

Table 2.2 Polarizabilities of some common ions ($C\,m^3 \times 10^{-13}$)

Ion	Polarizability	Ion	Polarizability	Ion	Polarizability
F^-	13	O^{2-}	30	N^{3-}	54
Cl^-	37	S^{2-}	68	P^{3-}	103
Br^-	47				
I^-	69				

The polarizing power of a cation M^{n+} may be intuitively considered to be proportional to its charge $(n+)$ but is inversely proportional to its volume, i.e. the smaller the ion, the greater is the electrostatic field which it exerts on a neighbouring anions:

$$\text{polarizing power (pp)} \propto ne\, /(^4/_3\pi r_M{}^3)$$

Table 2.3 summarizes the calculated polarizing powers of some common cations, and gives an indication of the magnitude of the differences induced by the changes in volume and charge.

Table 2.3 Polarizing powers (pp) of some inorganic cations ($C\,m^{-3} \times 10^{-25}$)

Ion	pp	Ion	pp	Ion	pp	Ion	pp
Li^+	52	Be^{2+}	1109	B^{3+}	1664		
Na^+	24	Mg^{2+}	120	Al^{3+}	364	Si^{4+}	970
K^+	11	Ca^{2+}	52	Ga^{3+}	261	Ge^{4+}	508
Rb^+	8	Sr^{2+}	33	In^{3+}	138	Sn^{4+}	268
Cs^+	6	Ba^{2+}	23	Tl^{3+}	105	Pb^{4+}	196

From the semi-quantitative calculations provided above it is evident that the covalent contribution to a bond formulated initially as an ionic bond increases in the following manner.

1. As the size of the cation decreases its polarizing power increases and therefore the covalent contribution becomes more significant:

$$LiF > NaF > KF > RbF > CsF$$

$$\longleftarrow$$
more covalent

2. As the size of the anion becomes smaller its polarizability decreases and therefore the bonding becomes more ionic, i.e. the covalent contribution decreases:

$$NaI > NaBr > NaCl > NaF$$

$$\longleftarrow$$
more covalent

3. As the charges on the cation and anion are varied the following trend in covalent character is observed:

$$AlN > MgO > NaF$$

$$\longrightarrow$$
more ionic

The chemical implications of these variations in ionic/covalent character are discussed in more detail in Section 2.4 and in Chapter 4.

The polarizabilities of ions also have significant implications on the stabilities of complexes which are formed between anions and cations. More generally, the hard and soft character of Lewis acids and bases are related to

In this simplified model the polarization of the cation by the anion has been ignored because for the majority of cations, and especially those with a completely empty valence shell, the polarization effects do not make a significant contribution. These ions, e.g. Na^+, Ca^{2+}, Al^{3+}, do not have electrons occupying valence orbitals and consequently the polarization of their completed shells of core electrons is not significant. In contrast, cations with partially filled shells, e.g. Sn^{2+}, Pb^{2+} ($nd^{10}(n+1)s^2$) and Zn^{2+}, Cd^{2+} ($nd^{10}(n+1)s^0$) have significant polarizabilities and it is important to consider the polarizabilities and polarizing powers of both the cations and the anions. The enhanced polarizabilities of these cations also may be related to the presence of low lying excited states with the appropriate symmetry which can mix into the ground state and thereby distort the spherical electron cloud. The excited states of $nd^{10}(n+1)s^2$ [$nd^{10}(n+1)s^1(n+1)p^1$] and nd^{10} [$nd^9(n+1)p^1$] mix p orbital character (*ungerade* symmetry) into the *gerade* ground state and thereby distort the spherical electron cloud. These additional polarizability terms significantly influence the lattice enthalpies and structures of compounds of the heavier Group 11, 12, 13, and 14 metals. (P. A. Madden and M. Wilson, *Chem. Soc. Rev.*, 1996, **25**, 339.)

Ionic bond formation is favoured by large cations and small anions which bear unit charges. Partial covalent character results when the cation and anion sizes are mismatched and a polarizing cation is placed with a highly polarizable anion.

the polarizabilities of ions and the topic is discussed in more detail on page 65 and in Chapter 5.

> The electronegativity and polarizability concepts presented above represent alternative, but complementary, ways of defining qualitatively the polarities of bonds. The electronegativity concept takes as its starting point the covalent bond, whereas the polarizability concept takes the ionic extreme as its starting point. Of course, both approaches, if used properly, should give the same qualitative guide to the relative contributions of ionic and covalent bonding in a specific bond.

Relative orbital energies and overlaps

The increase in principal quantum number down a column of the Periodic Table leads to an increase in the sizes of the valence orbitals and this aspect has been discussed in Chapter 1 on the basis of the calculated maxima in the radial distribution functions (r_{max}) of the orbitals (see the data given in the margin). It was also noted that on descending the column the s orbital contracts relative to the p and participates less in covalent bonding. The strengths of covalent chemical bonds depend on the extent of overlap between the orbitals on adjacent atoms in the molecule. In general the overlaps between atoms separated by the sum of their covalent radii decrease down a column.

Comparison of r_{max} of the ns and np orbitals for the Group 14 elements (in pm, see Table 1.12)

Element	r_{max} (ns)	r_{max} (np)
C	65	64
Si	95	115
Ge	95	119
Sn	110	137
Pb	107	140

Table 2.4 Average orbital ionization energies/kJ mol^{-1} for some p block elements

	2nd row	3rd row	4th row
Group 13	B	Al	Ga
ns	1351	1088	1220
np	801	574	574
Group 14	C	Si	Ge
ns	1888	1447	1507
np	1029	753	730
Group 15	N	P	As
ns	2464	1806	1698
np	1268	981	873
Group 16	O	S	Se
ns	3122	1997	2009
np	1531	1124	1041
Group 17	F	Cl	Br
ns	4473	2440	2320
np	1806	1328	1208

The relative energies of the ns and np valence orbitals of the p block elements also vary across the periods and down the groups of the Periodic

Table. Table 2.4 gives the average orbital ionization energies for some 2nd, 3rd, and 4th row elements. For all these elements the ns orbital is significantly more stable than np because of the greater effective nuclear charges it experiences (see Chapter 1).

The s (and p) orbital ionization energies are very similar for the third and fourth row elements, but are significantly smaller than those for the second row elements. The fact that the chemical properties of the second row elements appear to be 'anomalous' and quite different from the subsequent elements in a column has been noted above. The high ionization energies of the 2s and 2p orbitals of the second row elements results from the contracted nature of these orbitals. These orbitals also overlap very strongly with the orbitals of other atoms at short internuclear distances. So, these elements form particularly strong multiple bonds. Their small size also makes the strengths of bonds sensitive to the presence of electron–electron repulsion effects between lone pairs of electrons on adjacent atoms. For the heavier elements the relatively expanded valence orbitals lead to poorer overlaps and generally weaker bonds. The strength of the π bonding interactions deteriorate particularly rapidly down the group.

The separation between s and p orbitals also becomes larger across a row of elements and the ionization energies of the s orbitals are particularly high for the Group 17 elements. This makes the s orbital less available for bonding and the chemical bonding depends increasingly on the overlapping abilities of the p orbitals. The s orbitals are essentially core-like for the Group 17 and 18 elements. The high orbital ionization energy of the 2s orbital in fluorine is particularly noteworthy.

The **transition metal contraction** also influences the 4s orbital ionization energies for Groups 13 and 14, where Ga and Ge have higher ns orbital energies than a simple extrapolation may have suggested and the 4s orbitals of these elements are more core-like and are less involved in bonding.

Some preliminary comments on the Mendeleev classification

Central to the Mendeleev periodic classification was the idea that the properties of a particular element and its compounds could be interpolated by reference to the properties of the elements above and below it in the Periodic Table. This is an essentially linear view of the origins and evolution of chemical properties, which would only be valid if the atomic properties varied in a linear fashion. The brief description of the periodic trends in atomic sizes, ionization energies, and electronegativities presented above suggests that the trends in these parameters are reasonably regular, but are rarely simple linear functions. The atomic properties vary in a reasonably linear manner for the s block elements and for Groups 16–18, but there are many anomalies associated with the atomic properties of the remaining p block groups. Many of these anomalies arise from the **transition-metal** and **lanthanide contractions** which result from the large increase in effective nuclear charges when the

3d and 4f orbitals are filled for the first time. Therefore, the modern applications of the Periodic Table are not based on simple interpolations, but rather on an attempt to interpret the trends in particular properties in terms of the atomic properties introduced in this chapter.

This methodology, if used intelligently, not only accounts for the gross trends, but also the anomalies. It should be emphasized that the approach is neither rigorous, nor quantitative, but provides a flexible framework which may be used to assimilate and interpret the multitude of facts currently available on the elements.

In the following sections it should be emphasized that the basic facts responsible for the trend being discussed are likely to remain unchanged, but the interpretation which has been given to account for the trend may be an oversimplification and may not last the test of time. Chemistry is a subtle science and the interpretation should be seen as an *aide memoire* rather than the statement of an absolute truth. Although chemical properties represent the balance of many energetic terms, the accumulation of a body of related data results in the emergence of a pattern which provides the semblance of order within that subsection of the subject. In these circumstances it is useful to associate that pattern with a specific interpretation or parameter so long as one recognizes that it may eventually prove to be an over-simplification of the underlying electronic factors responsible for the trend.

2.2 Trends associated with the physical properties of elements

Melting points and enthalpies of atomization

When the elements of a particular column of the Periodic Table have similar structures the melting and boiling points reflect the strength of interaction between the atoms, or if the element is molecular, e.g. S_8 and I_2, the strength of the intermolecular interactions.

The Group 1 elements all have body-centred cubic metallic structures and the melting points in Table 2.5 suggest that the metal–metal bonding interactions become progressively weaker as the Group is descended.

Standard state
The standard state of a substance is that of the pure substance at a pressure of 101.325 kPa. The standard states for thermodynamic quantities are denoted by the symbol$^{\oplus}$. Standard states may be defined for any temperature, but in this book they refer to 298K, 25°C. For standard states at any other temperatures the temperatures are specifically included with the symbol.

Table 2.5 Melting points and enthalpies of atomization, $\Delta_{at}H^{\oplus}$, of Group 1 elements

	Li	Na	K	Rb	Cs
m.p. /°C	181	98	64	39	29
$\Delta_{at}H^{\oplus}$ / kJ mol^{-1}	159	107	89	81	76

The standard enthalpy of atomization of an element, $\Delta_{at}H^{\oplus}$, is the enthalpy change associated with the transfer of one mole of element from its standard state to that of gaseous atoms at 298 K. The $\Delta_{at}H^{\oplus}$ values therefore provide a more precise estimate of the strength of the interactions in the solid state. The trend reproduces that observed for the melting points of the elements.

The strength of the metal–metal bonding for these metals depends primarily on the ability of their s valence orbitals to overlap with the s orbitals of the neighbouring atoms in the infinite metallic lattice. The decreasing ns–ns orbital overlaps down the column is responsible for this trend. A similar trend is observed for the Group 2 elements. The metallic elements generally adopt close-packed structures—cubic close-packed and/or hexagonal close-packed—or the nearly close-packed structure, i.e. body-centred cubic. From Table 2.7 it is apparent that the metallic elements in many of the groups share a common structural type.

Table 2.7 Structures of elements of Groups 1–12

Group(s)	Elemental structures
1	All body-centred cubic
2	Be, Mg hexagonal close-packed; Ca, Sr cubic close-packed; Ba, Ra body-centred cubic
3	Sc, Y hexagonal close-packed; La non-standard close-packed
4	All hexagonal close-packed
5 and 6	All body-centred cubic
7	Mn unique structure, Tc, Re hexagonal close-packed
8	Fe body-centred cubic; Ru, Os hexagonal close-packed
9	Co hexagonal close-packed, Rh, Ir cubic close-packed
10 and 11	All cubic close-packed
12	Zn, Cd distorted hexagonal close-packed, Hg distorted cubic close-packed

Given that the energy differences between cubic close-packed, hexagonal close-packed and body-centred cubic structures are usually quite small it is not too surprising that for some of the groups a common structure is not observed. For the post-transition elements much larger variations in structure are observed as the elements become progressively more metallic down a column. The melting points and structures of the elements of Groups 13, 14, and 15 are given in Tables 2.8, 2.9, and 2.10, respectively.

Table 2.8 Melting points and structures of Group 13 elements

Element	Melting point/^{o}C	Structure
B	2573	Infinite structure based upon icosahedral B_{12} fragments; semiconductor
Al	660	Cubic close-packed metal
Ga	30	Distorted metallic structure with interacting dimers
In	157	Distorted cubic close-packed metallic structure
Tl	303	Hexagonal close-packed metallic structure

For the transition metals the trend in metal–metal bonding is just the opposite, as exemplified by the melting points of the Group 6 elements as shown in Table 2.6.

Table 2.6 Melting points and enthalpies of atomization of Group 6 elements

Element	m.p./^{o}C	$\Delta_{at}H^{\ominus}$ kJ mol^{-1}
Cr	2130	397
Mo	2890	658
W	3680	849

For the transition elements the d orbitals make the dominant contribution to metal–metal bonding and the efficiency of orbital overlaps increases as follows:

3d–3d < 4d–4d < 5d–5d.

The most important contributors to intermolecular forces are:
1. Dipole–dipole forces which depend on the magnitudes of the dipole moments associated with the molecules.
2. Dipole-induced dipole forces, the strength of which depend on the dipole moment of one molecule and the polarizability of the other.
3. London forces which result from the attractive interactions between temporary dipoles on neighbouring molecules. These occur even for molecules having zero dipole moments and single atoms. The interaction depends on the polarizabilities of the molecules.

All three types of molecular forces have a $1/r^6$ dependence (where r is the distance between the molecules).

For a more detailed discussion of intermolecular forces see N. W. Alcock, *Bonding and Structure*, Ellis Horwood, 1990.

Table 2.9 Melting points and structures of Group 14 elements

Element	Melting point/^{o}C	Structure
C(diamond)	3550	Infinite structure based upon tetrahedral coordination of carbon atoms
Si	1410	Infinite structure based upon tetrahedral coordination of silicon atoms
Ge	937	Infinite structure based upon tetrahedral coordination of germanium atoms
Sn	232	Tetragonal structure based on a distorted cubic close packing arrangement
Pb	328	Cubic close-packed

In the crystal structure of I_2 the molecules are layered in sheets with a separation of ca. 440 pm. Within layers the iodine molecules form infinite chains with contacts of 355 pm. This is 50 pm smaller than the sum of the van der Waals radii, confirming the increased importance of intermolecular interactions for the more polarizable molecules.

Table 2.10 Melting points and structures of Group 15 elements

Element	Melting point/^{o}C	Structure
N_2	−210	Diatomic molecules
P_4	44	P_4 molecules, also allotropes based on polymeric forms
As	814	Polymeric solid with hexagonal layers
Sb	631	Polymeric solid with hexagonal layers
Bi	271	Polymeric solid with hexagonal layers

At the far right side of the Periodic Table the Group 17 elements (the halogens) have solid state structures based on infinitely packed diatomic molecules, X_2, and they show a regular increase in their melting points; these are given in Table 2.11.

For this group of elements the intermolecular van der Waals interactions in the solid state between the X_2 molecules depend upon the polarizabilities of the molecules and since fluorine is the least polarizable and iodine the most polarizable the trend in the melting points is readily accounted for.

Where the structures of the elements in a column change dramatically this is reflected in their melting points. The Group 13, 14, 15, and 16 elements show such variations because the elements undergo structural changes from simple molecular, e.g. O_2, S_8, N_2, P_4, to infinite molecular, e.g. diamond, Si) and to metallic, e.g. Sb, Bi, and Tl. Also many of the elements occur in several modifications known as allotropes, which have significantly different chemical and physical properties.

Table 2.11 Melting points of Group 17 elements

Halogen element	m.p./^{o}C
F_2	−219
Cl_2	−101
Br_2	−7
I_2	114

Allotropy

Many elements occur in more than one form, known as allotropes, which have very different physical and chemical properties. The study of these has very important lessons for inorganic chemists, because although the atoms within the allotrope are identical, differences in their molecular and crystal structures may lead to dramatically different chemical properties. The physical and chemical properties of white and red phosphorus which are summarized in Table 2.12 underline some of these differences.

White phosphorus is less dense, it ignites in air, it is soluble in organic solvents, and is more poisonous than red phosphorus. Phosphorus has the electronic configuration $3s^2 3p^3$ and in white phosphorus it uses its valence electrons to form the tetrahedral molecules, P_4. Molecules of P_4 then pack in the solid state in a molecular fashion and the intermolecular forces are rather weak, because only van der Waals interactions are present. The atoms in the crystalline forms of red phosphorus also form three bonds but do so in such a way that an infinite cross-linked structure derived from one-dimensional chains is formed.

Table 2.12 Summary of physical differences between white and red phosphorus

White phosphorus	Red phosphorus
Wax-like, translucent solid, colourless, turns yellow when exposed to sunlight	Red opaque solid, powdery microcrystalline
Density 1830 kg m^{-3}, very soft, m.p. 40°C, b.p. 281°C. Brittle at 0°C	Density 2340 kg m^{-3}, m.p. 590°C, b.p. 725°C, sublimes at 416°C
Volatile, giving P_4 molecules in gas phase, may be steam distilled	Not volatile, cannot be steam distilled
Luminescent in air	Not luminescent
Ignites in air at 35°C, has to be stored under water.	Ignites only above 260°C
Soluble in CS_2, PCl_3, $POCl_3$, liq. SO_2, liq. NH_3 and hydrocarbons Et_2O	Insoluble in CS_2 and hydrocarbons
Only slightly soluble in water	Insoluble in H_2O
Soluble in NaOH	Insoluble in NaOH
Poisonous	Not poisonous. Tasteless and odourless
When heated at 230–300°C in absence of air converted into red allotrope	Thermodynamically more stable than white P
P_4 tetrahedral molecules	Amorphous generally, but various crystalline forms exist, e.g., Hittorf's violet and Schlenck's scarlet forms. These contain either one-dimensional chains based on linked P_8 and P_2 fragments or sheets of P atoms

Fig. 2.9 contrasts the structures of white phosphorus and the violet Hittorf form of red phosphorus. The latter consists of P_8 and P_9 wedge-like species linked by P_2 entities to give the one-dimensional chain which is cross-linked through P_2 units to give observed polymeric structure. Amorphous red phosphorus probably has similar fragments, but the lack of long range order leads to an amorphous material. The chemical and physical differences between white and red phosphorus highlight the following important differences between molecular and polymeric structures.

1.   Compounds with molecular structures generally are more volatile and have lower melting points (the m.ps of white and red phosphorus differ by more than 500°C). This arises because the individual P_4 molecules are held together by weak van der Waals forces, whereas the phosphorus atoms in red phosphorus are part of an infinite structure.

2.   The molecular compounds are more soluble in organic solvents. This is also related to the presence of weak van der Waals forces in the former.

Individual P₄ molecules
of white phosphorus

Chains in the structure of red phosphorus

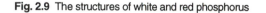

Fig. 2.9 The structures of white and red phosphorus

For some recent research on the structure of amorphous phosphorus see H. Hartl, *Angew. Chem. Int. Ed.*, 1995, **34**, 2637.

3.   The infinite structures are generally denser and harder. Smaller molecular entities generally pack less efficiently and form less rigid crystalline structures than an infinite structure, where all the atoms are chemically bonded to neighbouring atoms.

4.   Generally the molecular form is more reactive than the polymeric form, because the rates of reaction of the molecular allotrope with reactive molecules such as O_2 and Cl_2 are faster. If the rate determining step in the reaction with the gas is the departure from the surface of a small part of the solid state structure then clearly this has a lower activation energy for a molecular allotrope. The enhanced reactivity of white phosphorus is clearly illustrated in Table 2.12.

The red form of phosphorus is thermodynamically more stable than the white form. The red and white forms can be converted at 200°C and 12 000 atmospheres pressure to a black form. The black form is the thermodynamically most stable allotrope and is observed as three crystalline modifications and in an amorphous form.

Black phosphorus is more dense, semi-conducting, stable in air, and practically non-flammable. It should be stressed than the differences between molecular and infinite structures listed above become more pronounced if the dimensionality of the polymeric infinite structure is increased, i.e. a three-dimensional infinite cross-linked polymer is more dense, stable, harder, unreactive, and less soluble than a two-dimensional layer structure, which in turn displays these property differences more clearly than a one-dimensional chain structure.

The rhombic form of black phosphorus has a layer structure based on puckered hexagonal rings of 3-coordinate phosphorus atoms (density 3.56 kg m^{-3}), shown in Fig. 2.10, and the orthorhombic form (density 2.69 kg m^{-3}) has a related interlinked structure also shown in Fig. 2.10. Cubic black phosphorus (density 3.88 kg m^{-3}) has an infinite three-dimensional structure based on octahedral 6-coordinate phosphorus atoms.

The discussion above has centred on allotropes which are crystalline and whose structures have been clearly defined using single crystal X-ray crystallographic techniques. Other allotropes may exist which are amorphous, i.e. their structures do not have sufficient long range order to from crystalline solids. Such materials can also have distinctive chemical and physical properties. For example, there are many forms of amorphous carbon which are found in varying degrees in charcoal, coal, coke, carbon black, and lamp black. In essence they represent various forms of non-crystalline graphite and their properties depend critically on their surface areas. The finely divided forms absorb large volumes of gases. This property may be used to purify gases and if catalytic amounts of platinum and palladium are impregnated then they are very effective hydrogenation catalysts. Amorphous carbon is traditionally made by heating wood or sugar at 900°C in the absence of air; more recently this technology has been adapted to make high strength carbon fibres by thermally degrading oriented organic fibres (e.g. polyacrylonitrile) in the absence of air at 1500°C.

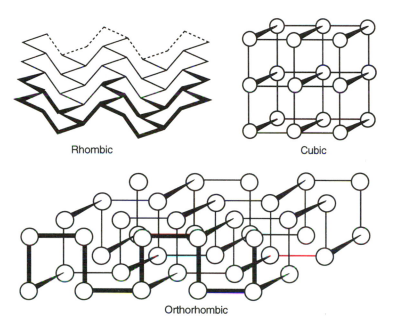

Rhombic   Cubic

Orthorhombic

Fig. 2.10 The structures of black phosphorus

The chemical and physical properties of the three crystalline allotropic forms of carbon, diamond, graphite, and C_{60}, are summarized in Table 2.13. The molecule C_{60} is one of a series of recently discovered allotropes of carbon which have molecular structures.

In general the properties of the allotropes reflect the differences associated with infinite (three-dimensional), layer (two-dimensional), and molecular structural types discussed previously. The structures of these allotropes are illustrated in Fig. 2.11.

The thermodynamically more stable form of carbon is graphite, which has a layer structure. Each layer has shared hexagons based on sp^2 carbon atoms and the layers are separated by 335 pm. Three of carbon's four valence electrons are used to form localized σ bonds and the remaining electron occupies a p orbital perpendicular to the hexagonal planes and the ensemble of p orbitals generates a delocalized band of π molecular orbitals which extend over the whole layer.

Table 2.13 Comparison of the physical and chemical properties of the allotropes of carbon

	Diamond	Graphite	C_{60}
Physical appearance	Transparent, crystalline, colour variable depending on impurities	Opaque, grey-black, crystalline flat plates	Thin crystals, mustard colour, large crystals very dark almost black
Density, hardness, conductivity	Very dense, 3510 kg m^{-3} Hardest substance known Poor conductor of electricity Very high thermal conductivity	Less dense, 2220 kg m^{-3} Soft, smooth Good conductor of heat and electricity	
Physical properties	Brittle, cleaves to give well defined facets, high refractive index	Brittle, lubricant because of softness Sublimes at 3850°C	Melting point 280°C Sublimes 400°C (in vacuo)
Reactions with O$_2$/F$_2$	Burns in O$_2$ at 800°C, and F$_2$ at 700°C	Burns in O$_2$ at 700°C, and F$_2$ at 500°C Forms intercalation compounds with halogens and metals, K, Rb, Cs	With F$_2$, gives $C_{60}F_6$, $C_{60}F_{42}$ and eventually $C_{60}F_{60}$ With other halogens has substituted derivatives
Reactions with acids and alkalis	Does not react with H$_2$SO$_4$	No reactions with H$_2$SO$_4$, but does react with HNO$_3$/KClO$_4$	
Solubility	Insoluble in all solvents	Insoluble in all solvents	Soluble in organic solvents
Reactions with metals	At high temperatures reacts with most metals to form carbides		Reduced by Li in liquid NH$_3$ to $C_{60}H_{36}$ K_3C_{60} is a superconductor at low temperatures
Stability	Transforms into graphite at 1500°C in absence of air	Thermodynamically stable form by 3 kJ mol^{-1} at 300 K and 1 atm Can be converted into diamonds only under very high pressures and temperatures	
Structure	Infinite three-dimensional structure based on tetrahedral carbon atoms	Layer structure, sp^2 carbons in hexagonal nets, inter-layer separation 335 pm	Molecular spherical cage structure with 5 and 6 membered rings, resembling a soccer ball

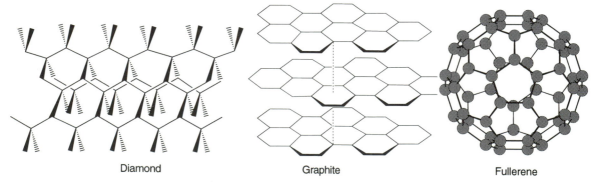

Diamond   Graphite   Fullerene

Fig. 2.11 The structures of diamond, graphite, and the C_{60} fullerene

The resultant bond order is 1.33 which is intermediate between that in C_2H_6 and that in benzene, where the bond order is 1.5. Therefore, the C–C distance of 141.5 pm is slightly longer than that in benzene (139 pm). The strong π bonding between carbon atoms makes the structure more stable than that of diamond. However, for the heavier Group 14 elements the decreasing importance of p_π–p_π bonding leads to a preference for the diamond structure. The delocalized π system in graphite causes it to be a zero band-gap conductor and to have a black colour (cf. diamond which is an insulator and colourless).

This two-dimensional layer structure leads to a density which is significantly less than that of diamond, and qualitatively can account for its flaky appearance, softness, and lubricating properties. Halogens and alkali metals intercalate between the hexagonal layers of graphite. The π electrons in the hexagonal layers form a delocalized system which extends over the whole net with a zero band gap between bonding and antibonding orbitals. Therefore graphite is conducting and detailed measurements have shown that it is 3 to 4 orders of magnitude more conducting along the layers than between the layers.

Diamond has an infinite solid state structure based on tetrahedrally linked carbon atoms. The saturated nature of the bonding leads to a transparent crystalline material, which is very dense. Although it is an insulator, it has a high thermal conductivity.

Other allotropic forms of carbon were discovered in the 1980s which are molecular and have spherical shell structures. The most stable of these, Buckminsterfullerene, has the formula C_{60}, but C_{70}, C_{76}, C_{78}, and others are also known. These molecular allotropes are thermodynamically less stable than diamond and graphite. The σ bonds in the hexagonal and pentagonal rings of the cage are based on sp^2 hybrids and additionally there is a delocalized π system perpendicular to the surface of the spherical shell.

There are instances where the allotropes of the elements do not have fundamentally different dimensionalities in the crystal structures, but still display significantly different properties. For example, many allotropes of sulfur are known which are all molecular but have different sizes of sulfur rings. For example, S_6, S_7, S_8, S_9, S_{10}, S_{11}, S_{12}, S_{18}, and S_{20} rings have been observed in the solid state. The thermodynamically most stable form is based on S_8 rings. All the allotropes have yellow to orange colours, are soluble in organic solvents, and are non-conductors. Their stabilities vary considerably because some interconvert more readily into S_8 than others. For example, S_{10} decomposes above 0°C. Their reactivities with other reagents also differ. Their densities, however, only span a relatively narrow range 1950–2200 kg m^{-3} because they are all molecular. Octasulfur, S_8, which occurs as crown-shaped rings, crystallizes in alternative lattices giving two monoclinic and an orthorhombic crystalline forms. However, the differences in their physical and chemical properties are very subtle because in each instance molecular S_8 rings are packed in the crystal and held together by weak van der Waals forces, although their precise packing modes within the crystals differ.

For elements whose allotropes adopt alternative infinite structures the change in packing modes may lead to very different properties. For example,

It is intuitively attractive to relate the lubricating properties of graphite to its layer structure; however, it has been found that in the absence of oxygen graphite is a poor lubricant. The formation of graphite oxides are necessary to reduce the interactions between the layers.

The 1996 Nobel Prize in Chemistry was awarded to Kroto, Smalley, and Curl for their discovery of C_{60}.

the stable allotrope of tin at room temperature is white tin which has a tetragonal metallic structure, but at low temperature it transforms into grey tin which has a diamond structure. This leads to the differences in physical properties given in Table 2.14.

The large difference in densities between the allotropes means that the phase change from white to grey tin requires large changes in the arrangement of atoms in the unit cells. Consequently the crystallinity of the metal is lost and a powdering of the metallic tin occurs. This transformation can have disastrous consequences to architectural structures built of tin when the temperature falls below the transition temperature of 14°C.

Table 2.14 Some physical properties of the allotropes of tin

	Density/kg m^{-3}	Band gap/eV	Coordination number
White (tetragonal)	7280	0	6
Grey (diamond)	5750	7.7	4
			(plus12 next nearest neighbours)

Summary of allotropes of Groups 13 to 16

Group 13

Boron $(2s^2 2p^1)$ has insufficient electrons to form a tetrahedrally based diamond-like structure with four localized two-electron bonds radiating from each atom. This electron deficiency is relieved by adopting structures based on B_{12} icosahedra, where each boron has at least five nearest neighbours and therefore can share the electron deficiency by delocalization. Elemental boron has several allotropic structures. The α-rhombohedral boron has a close-packed arrangement of B_{12} icosahedra and tetragonal boron has layers of B_{12} icosahedra linked by B–B bonds. The β-rhombohedral form of boron has a complicated structure based on a central B_{12} icosahedron with successive surrounding shells of boron atoms. The three-dimensional electron delocalization leads to semiconducting properties for elemental boron.

Aluminium has a regular cubic close-packed metallic structure. The structures of gallium and indium are no longer close-packed and that of gallium is particularly irregular and has pairs of gallium atoms only in close contact. Compared to aluminium, gallium, indium, and thallium are relatively soft. Aluminium, despite being hard, is light, malleable, and ductile and therefore is widely used as a structural material. The structure of gallium is closely related to that of iodine.

Group 14

The structures of the allotropes of carbon have been discussed in some detail above. Silicon, Ge, and Sn (grey allotrope) adopt the diamond structure which has tetrahedrally coordinated centres linked to form an infinite lattice. The ns^2np^2 electron configurations of these elements allow, after an s to p promotion to give ns^1np^3, the formation of localized bonds at each tetrahedral centre. Tin has a second allotrope (white tin) which has a distorted close-packed metallic structure. The properties of these allotropes were discussed above. Lead is only observed as a close-packed cubic metallic structure in the solid state.

Group 15

Nitrogen has a diatomic molecular structure based on a triple bond which utilizes the unpaired electrons of the atom in its $2s^2 2p^3$ ground state. Phosphorus has several allotropes. The structures and properties of the allotropes of phosphorus have been discussed above. Arsenic and antimony form the thermodynamically less stable yellow forms which are analogous to white phosphorus (P_4), but they transform into 'metallic' rhombohedral forms which are isomorphous with rhombohedral black phosphorus. The structures of rhombohedral As, Sb, and Bi are isomorphous, but the ratio of the primary and secondary element–element distances becomes larger down the column. The average coordination number is thereby increased and the elements become more metallic. A similar phenomenon is observed for Se and Te.

Group 16

Oxygen has two allotropic molecular forms—dioxygen O_2 and trioxygen (ozone) O_3. The more stable form, dioxygen, has a formal bond order of two which results from the utilization of the unpaired electrons in the atom ($2s^2 2p^4$).

Ozone has an angular geometry (bond angle of 117°) and a longer O–O bond length which is consistent with a lower formal bond order of 1.5. Sulfur forms many allotropes, the majority of which have rings of sulfur atoms, S_6, $S_8, \ldots S_{20}$. The thermodynamically more stable forms have S_8 rings. Plastic sulfur has chains of sulfur atoms. The stable allotrope of selenium (the grey form) also has infinite helical chains of atoms and tellurium has a closely related structure.

The weak interchain interactions between the Se atoms are not sufficiently strong for it to exhibit normal metallic conductivity, but it is an effective photoconductor, i.e. it conducts electricity when light shines on its surface. This property leads to its widespread use in photocopying machines.

Metastable forms of Se_6 and Se_8 with ring structures are known and the latter is a red crystalline solid which is soluble in CS_2.

Some general points

1. The elements of Groups 13–16 progressively become more metallic down each group. The transition from non-metal to metal character does not occur in a specific period, but lies along a diagonal of the Periodic Table. The reasons for this are discussed in more detail in Chapter 3.

2. For the lighter non-metallic elements the octet rule is obeyed and the structures arise from the full utilization of electrons in bond formation. This leads to the $(8 - N)$ rule which states that a p block element with N valence electrons forms structures with $8 - N$ bonds. Examples of this rule are given in Table 2.15.

3. The second row elements show a preference for structures where multiple bonding is involved. Examples include:

C	Graphite
N	$N \equiv N$
O	$O = O$

Table 2.15 Some examples of the $8 - N$ rule

Group	Electron configuration	Number of bonds	Examples
14	ns^2np^2	4	Diamond, graphite
15	ns^2np^3	3	N_2, P_4
16	ns^2np^4	2	O_2, S_8
17	ns^2np^5	1	F_2, Cl_2
18	ns^2np^6	0	He, Ne

The widespread occurrence of allotropy emphasizes how even when considering a single element the subtle balance of forces which occur between the atoms leads to a range of structures with remarkably different m.ps, densities, and chemical properties.

For the subsequent rows of elements structures based on σ bonded polyhedra (P_4), rings (S_8), chains (Se), and sheet (P, As) structures are preferred. This may be related to the fact that $2p_\pi - 2p_\pi$ overlaps are particularly favourable for multiple bond formation.

4. Within the framework of the $(8 - N)$ rule the elements can form alternative structures which lead to a widespread occurrence of allotropy.

5. The intermolecular interactions increase down the column and the intramolecular interactions decrease and consequently the ratio of primary to secondary distances in the solid state structures decrease.

In general the strengths of the covalent bonds for the elements decrease down a column, but the van der Waals interactions, which depend primarily on the polarizabilities of the atoms and reflect their atomic volumes, increase. Consequently, for a column of elements the intermolecular forces increase as the intramolecular forces decrease and therefore the difference between van der Waals and covalent radii decrease. Therefore, the molecular identity becomes less well defined and elements increasingly show metallic behaviour. The conduction of electrons from molecule to molecule is facilitated when the difference in strength between intra- and intermolecular attractive forces is reduced. If the element is put under pressure the intermolecular contacts are reduced and the conductivity increases.

Densities

The relative increase in atomic masses down a column usually outweighs the effect of larger atomic volumes and therefore densities generally get larger. The densities of the elements of Groups 1, 2, and 13 are summarized in Table 2.16.

A steady increase in densities is observed for Groups 1 and 13 and indeed for all other groups in the Periodic Table with the exception of Group 2. Clearly the effect of increasing relative atomic masses (RAM) outweighs the increasing atomic volumes ($\rho \propto RAM/r^3$). The anomalous behaviour of the Group 2 elements is shown in Fig. 2.12.

Table 2.16 Densities of Groups 1, 2, and 13 elements/kg m^{-3}

Li	Na	K	Rb	Cs
534	699	862	1532	1873
Be	Mg	Ca	Sr	Ba
1850	1740	1540	2620	3510
B	Al	Ga	In	Tl
2340	2698	5907	7310	11 720

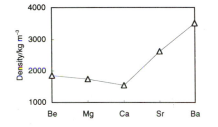

Fig. 2.12 Variation in density for Group 2 elements

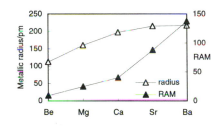

Fig. 2.13 Variations in metallic radius and RAM for the Group 2 elements

The variations in relative atomic mass (RAM) and metallic radius (in pm) for the Group 2 elements are shown in Fig. 2.13. When the relationship RAM/r^3 (proportional to the density) is plotted as in Fig. 2.14, based upon the RAM and r values shown in Fig. 2.13, the observed trend is reproduced.

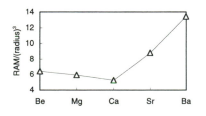

Fig. 2.14 Variation of the function RAM/r^3 for the Group 2 elements

The observed trend in densities occurs because the radii and the relative atomic masses increase parabolically, but the curvatures of the graphs differ at the ends of the series. The introduction of the d and f block elements leads to larger increases in RAM down the column, but smaller variations in *r*.

> This analysis emphasizes the very important general point that when an observable property depends on several variables, each of which varies regularly, the property itself may show a discontinuity because of the non-linear nature of the relationships.

As noted above the density of an element can change dramatically with structure, for example the densities of graphite and diamond are 2260 and 3513 kg m^{-3} respectively, i.e. the carbon atoms in diamond are much more tightly packed.

Resistivities and thermal conductivities

The thermal conductivities of the elements can show a regular progression down a column if the structure of the elements remain similar, as shown by the data given in Table 2.17. In a metal the delocalized electrons are able to conduct heat rapidly and they are also responsible for conducting electricity and therefore in general a good correlation is observed between thermal and electrical conductivities.

The electrical conductivities of metals vary greatly: the Group 11 metals show the highest conductivities (Cu > Ag > Au) followed by aluminium, beryllium, and magnesium. In general, for the s and p block metals the thermal and electrical conductivities decrease down a group and increase across a period as the number of valence electrons is increased. Table 2.17 summarizes the vertical trends for Group 1 metals and illustrates the anomalous behaviour of lithium.

Table 2.17 Thermal conductivities and resistivities of Group 1 elements

Element	Thermal conductivity / W m^{-1} K^{-1}	Resistivity /10^{-8} Ω m
Li	85	8.6
Na	141	4.2
K	102	6.2
Rb	58	12.5
Cs	36	20.0

For semiconductors the band gap (i.e. the energy difference between the highest occupied delocalized molecular orbitals in the solid state and the lowest unoccupied levels) is important in influencing their conductivity properties. The band gaps and resistivities of the Group 14 elements are given in Table 2.18.

The decreasing band gap from C to Ge, which all have diamond structures, is related to the decreasing overlap between the s and p orbitals of the elements. The bonding and antibonding orbitals are separated by a smaller energy gap because the orbitals do not overlap as effectively.

Table 2.18 Band gaps and resistivities of SiC and the Group 14 elements

Element/compound	Band gap/eV	Resistivity/Ω m
C (diamond)	5.47	10^{11}
SiC	3.0	2
Si	1.12	0.001
Ge	0.66	0.46
Sn (white)	0	11×10^{-8}
Pb	0	20.7×10^{-8}

2.3 Trends associated with the chemical properties of the elements

Standard reduction potentials

By definition an element undergoes a change in oxidation state when it is converted into its compounds and therefore it is instructive to establish whether the standard electrode potentials show systematic trends. It needs to be emphasized that the standard reduction potential, $E^{\ominus}$, only defines the thermodynamic favourability of a reaction through its relationship to the standard Gibbs energy change. It gives no information concerning the kinetics of the reaction, i.e. the rate of the reduction or oxidation process. For a metal the rate of reaction with acid may be influenced by the ease with which small particles of metal break off the surface, the presence and permeability of the oxide surface, etc.

For example, fine powders are much more reactive than the bulk metal, as are blocks of metals whose surfaces which have been treated in such a manner that the oxide layer has been removed.

The relationship between standard electrode potential and the change in standard Gibbs energy is:
$$\Delta G^{\ominus} = -nFE^{\ominus}$$
where F is the Faraday constant and n is the number of moles of electrons participating in the redox reaction

Table 2.19 Standard reduction potentials/V for the Group 1 [$E^{\ominus}$ (M^+/M)] and Group 2 [$E^{\ominus}$ (M^{2+}/M)] elements

Li	Na	K	Rb	Cs
–3.02	–2.71	–2.92	–2.99	–3.02
Be	Mg	Ca	Sr	Ba
–1.7	–2.37	–2.87	–2.89	–2.9

The $E^{\ominus}$ values for the elements of Groups 1 and 2 are given in Table 2.19. For the Group 1 elements there is a first row anomaly, but the subsequent $E^{\ominus}$ (M^+/M) values show a small but regular variation and the metals become progressively more reducing. The first row anomaly disappears for the Group 2 elements, but the standard potential for Be is much less negative than those of the heavier elements.

The Group 17 elements, in contrast, are oxidizing agents rather than reducing agents, but become progressively less oxidizing down the column of the Periodic Table. Their standard reduction potentials, for reduction of the elements to their uni-negative ions, are given in Table 2.20.

Table 2.20 Standard reduction potentials for the Group 17 elements ($\frac{1}{2}X_2/X^-$)

Halogen	$E^\ominus$/ V
F_2	2.87
Cl_2	1.36
Br_2	1.04
I_2	0.54
At_2	0.2

Gibbs energy, G, is sometimes referred to as free energy or as Gibbs free energy. Gibbs energy is defined as $G = H - TS$, where H = enthalpy, T = absolute temperature and S = entropy. Changes in Gibbs energy, ΔG, are more useful in chemistry because they determine whether reactions are energetically favourable. The second law of thermodynamics states that in a spontaneous process the change in *entropy* is equal to or greater than zero ($\Delta S \geq 0$) for the system and its surroundings. The total energy change associated with a chemical reaction, 'the change in Gibbs energy', is given by:

$\Delta G = \Delta H - T\Delta S$

If $\Delta G \leq 0$ the reaction is thermodynamically favourable, and will achieve equilibrium defined by the equilibrium constant, K_{eq}. The equilibrium constant is related to the standard change in Gibbs energy, $\Delta G^\ominus$, for the reaction by :

$\Delta G^\ominus = -RT \ln K_{eq}$

if $\Delta G^\ominus = 0$, $K_{eq} = 1$, and the reaction is in 50:50 equilibrium.

If $\Delta G^\ominus < 0$, $K_{eq} > 1$, and the reaction proceeds towards the products (providing there is no kinetic barrier), and the more negative the value of $\Delta G^\ominus$ is, the more favourable the equilibrium constant is towards the products, $K_{eq} \gg 1$.

It is noteworthy that the range of $E^\ominus$ values for the halogens (Group 17 elements) is much larger than that for the Groups 1 and 2 metals. Consequently the halogens show a much greater variation in chemical properties than the alkali metals. For example, although fluorine is the strongest oxidizing element and reacts with almost every element, iodine is a sufficiently weak oxidizing agent that it can be used as an antiseptic.

For the elements in Groups 13–16 the trends in standard reduction potentials are more complex because the structures of the elements change dramatically. Also, multiple oxidation states become available to elements; this makes it important to specify exactly which standard reduction potentials are being compared. For example, the heavier Group 13 elements form compounds in the +1 and +3 oxidation states. The $E^\ominus$ values for the M^{3+}/M and M^+/M reductions are given in Table 2.21.

Table 2.21 Standard reduction potentials/V for the Group 13 elements

	B	Al	Ga	In	Tl
$E^\ominus$ (M^{3+}(aq)/M(s))	−0.87	−1.66	−0.56	−0.34	0.72
$E^\ominus$ (M^+(aq)/M(s))	–	0.55	−0.79	−0.15	−0.33

The lower positive oxidation state becomes progressively more stable and the metal becomes more 'noble' down the group. The term 'noble' is used to indicate that the metal is becoming progressively less reducing and therefore does not dissolve in simple mineral acids. The metal may only be converted into its metal ions (M^+ or M^{3+}) by oxidizing acids. Similar generalizations apply to the metals of Groups 14 and 15.

Thermodynamic analysis

The trends discussed above may be interpreted using thermodynamic cycles which are based on Hess's Law. The aim of such analyses is to reinterpret the observed trend in chemical behaviour, in this case $E^\ominus$, in terms of the atom, its ions and the element, since the variations in these properties are reasonably well understood.

The standard reduction potential $E^\ominus$ is related to the standard Gibbs energy change $\Delta G^\ominus$ by the equation:

$$\Delta G^\ominus = -nFE^\ominus$$

(F is the Faraday constant, 96 485 C mol^{-1}, n is the number of electrons used to reduce the oxidized state).

However, in qualitative interpretations of trends $\Delta_{at}G^\ominus$, $\Delta_{ion}G^\ominus$, and $\Delta_{hyd}G^\ominus$ are replaced by $\Delta_{at}H^\ominus$, $\Delta_{ion}H^\ominus$, and $\Delta_{hyd}H^\ominus$. For example, the half-reaction associated with the standard reduction potential of a metal may be expressed by the thermochemical cycle shown in Fig. 2.15.

This analysis replaces the observed $E^\ominus$ values by a combination of three variables and it is important to establish their relative importance. Values of the variables for the Group 1 elements are given in Table 2.22. The ionization and hydration enthalpies are significantly larger than the enthalpy of atomization and therefore represent the dominant energy terms in the cycle.

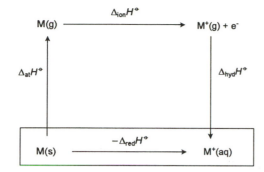

Fig. 2.15 A thermochemical cycle showing the contributions to the standard enthalpy change for the conversion of a metal, M, into its uni-positive ion, M^+

Table 2.22 $\Delta_{at}H^{\ominus}$, $\Delta_{ion}H^{\ominus}$, and $\Delta_{hyd}H^{\ominus}$ values/kJ mol^{-1} for Group 1 elements

	Li	Na	K	Rb	Cs
$\Delta_{at}H^{\ominus}$	159	107	89	81	76
$\Delta_{ion}H^{\ominus}$	520	496	419	403	376
$\Delta_{hyd}H^{\ominus}$	−520	−406	−322	−301	−277

The first two processes in the cycle leading to M(g) and M^+(g) require the input of energy, which is more than regained when the gaseous ion enters the solution as M^+(aq). Both $\Delta_{at}H^{\ominus}$ and $\Delta_{ion}H^{\ominus}$ decrease down the column and therefore the energy required to form M^+(g) decreases in the order:

$$Li > Na > K > Rb > Cs$$

These trends were discussed in Section 2.1. However, the energy released in the final step, the hydration enthalpy, is also largest for Li and least for Cs. Therefore, the two opposing energy effects approximately cancel out each other leading to the very small variation in $E^{\ominus}$ values. The anomalous $E^{\ominus}$ associated with Li may be related to its particularly favourable hydration enthalpy which results from the high charge-to-size ratio of the Li^+ ion (see Table 2.3). The hydration enthalpies of metal ions depend critically on their sizes and their formal charges. More energy is released when the water molecules surround and encapsulate a small highly charged ion, because the electrostatic forces involved are larger.

For the Group 17 elements, the thermodynamic cycle shown in Fig. 2.16 is relevant. Data for the Group 17 elements are given in Table 2.23.

The enthalpies of atomization are relatively small and the dominant energy terms are the electron attachment enthalpies and the hydration enthalpies. The sum of $\Delta_{at}H^{\ominus}$, $\Delta_{ea}H^{\ominus}$ favours the following order of oxidizing abilities F > Cl > >Br > I and $\Delta_{hyd}H^{\ominus}$ also favours F > Cl > >Br > I. Therefore, it is not surprising that F is the strongest oxidizing agent and iodine is the weakest. The reasons for the large variations in $E^{\ominus}$ for these elements compared to the alkali metals are also evident from the above data. It is noteworthy that even for these elements which form negatively charged ions the smallest has the most favourable $\Delta_{hyd}H^{\ominus}$.

For a more detailed account of free energies see *Inorganic Energetics* by W. E. Dassent, Cambridge University Press, 1988.

M(s) → M(g)
$\Delta_{at}G^{\ominus}$ = standard Gibbs energy change of atomization.
$\Delta_{at}H^{\ominus}$ = standard enthalpy of atomization.

M(g) → M^{n+}(g) + ne
$\Delta_{ion}G^{\ominus}$ = standard Gibbs energy change of ionization.
$\Delta_{ion}H^{\ominus}$ = standard enthalpy change of ionization.

M^{n+}(g) → M^{n+}(aq)
$\Delta_{hyd}G^{\ominus}$ = standard Gibbs energy change of hydration.
$\Delta_{hyd}H^{\ominus}$ = standard enthalpy change of hydration.

For an excellent exposition of these thermodynamic analyses see D. A. Johnson, *Some Thermodynamic Aspects of Inorganic Chemistry*, Cambridge University Press, 2nd Ed., 1982.

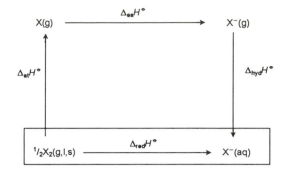

Fig. 2.16 A thermochemical cycle showing the contributions to the standard enthalpy change for the reduction of a Group 17 element, X, to its uni-negative anion, X⁻. The standard state for F_2 and Cl_2 is gas, that for Br_2 is liquid and that for I_2 is solid.

$\Delta_{ea}H^{\ominus}$ is the change in standard enthalpy when a gaseous ion accepts an electron to become a negative ion.

Table 2.23 $\Delta_{at}H^{\ominus}$, $\Delta_{ea}H^{\ominus}$, and $\Delta_{hyd}H^{\ominus}$ values//kJ mol⁻¹ for Group 17 elements

	F	Cl	Br	I
$\Delta_{at}H^{\ominus}$	79	122	112	107
$\Delta_{ea}H^{\ominus}$	−333	−348	−340	−297
$\Delta_{hyd}H^{\ominus}$	−461	−385	−351	−305

The importance of the methodology described above which re-expresses a chemical trend in forms of a Hess's law thermodynamic cycle cannot be over-stressed. It provides a much greater insight, not only into the overall trend, but also its magnitude and relates the trend to more fundamental atomic parameters. It can also provide a basis for explaining anomalies.

Standard enthalpies of formation of compounds

An alternative way of summarizing thermodynamic data which highlights periodic trends of the elements is based on the standard enthalpies of formation of their compounds. These data indicate the thermodynamic favourability for forming that compound from the elements (in their standard states) and therefore give a commentary on the stabilities of the compounds relative to the element. The data, of course, provide no information concerning the kinetics of such reactions. Table 2.24 provides typical data primarily for compounds of the Group 14 elements.

For the tetrahedral hydrides, the $\Delta_f H^{\ominus}$ becomes less favourable down a column of the Periodic Table in a systematic fashion and illustrates the Mendeleev methodology very well. However, the methodology may not be extrapolated to the corresponding fluorides and chlorides, both of which show a first row anomaly. The $\Delta_f H^{\ominus}$ values for CF_4 and CCl_4 are less exothermic than would be anticipated from a simple extrapolation and this arises because steric repulsions between the halogens around the small carbon atom make destabilizing contributions.

The standard enthalpies of formation for the hydrides are directly related to the observed relative stabilities and reactivities of the compounds. The hydrides become increasingly difficult to make and isolate down a column and they become more reactive. For example, SiH_4 is much more reactive than CH_4 and ignites spontaneously in air and reacts explosively with Cl_2 and Br_2. It should be noted that the ease of isolation and reactivity of a compound reflects not only thermodynamic factors, but also kinetic factors.

Table 2.24 Standard enthalpies of formation, $\Delta_f H^\ominus$ /kJ mol^{-1}, for some compounds of the Group 14 and 15 elements

$CH_4(g)$	$SiH_4(g)$	$GeH_4(g)$	$SnH_4(g)$	$PbH_4(g)$
−74.4	+34.3	+87	+162.8	+250
$CF_4(g)$	$SiF_4(g)$	$GeF_4(g)$	$SnF_4(s)$	$PbF_4(s)$
−933.6	−1615	−1090.2	—	−941.8
$CCl_4(l)$	$SiCl_4(l)$	$GeCl_4(l)$	$SnCl_4(l)$	$PbCl_4(l)$
−128.2	−687	−531.8	−511.3	−329.3
$C(CH_3)_4(g)$	$Si(CH_3)_4(g)$	$Ge(CH_3)_4(g)$	$Sn(CH_3)_4(g)$	$Pb(CH_3)_4(g)$
−167	−245	−71	−19	136
$CO_2(g)$	$SiO_2(s)$	$GeO_2(s)$	$SnO_2(s)$	$PbO_2(s)$
−394	−911	−551	−581	−277
$N_2O_5(s)$	$P_4O_{10}(s)$	$As_2O_5(s)$		
−43	−2984	−916		

Mean bond dissociation enthalpies, D(E–X)

In general the mean bond dissociation energies, D(E–X), decrease down the group if there are no lone pairs on X. Tables 2.25–2.28 summarize the mean bond dissociation enthalpies for E–H, E–C, E–Cl, E–F, and E–O bonds for the p block elements. However, when there is a lone pair on the element X the value of D(E–X) tends to be largest for the second third row element. The values of D(E–X) decrease in the order: F > Cl > Br > I.

Table 2.25 Typical values for some mean bond dissociation enthalpies for element–hydrogen bonds/kJ mol^{-1}

C–H	416	N–H	391	O–H	463	F–H	571
Si–H	322	P–H	322	S–H	367	Cl–H	432
Ge–H	288	As–H	297	Se–H	317	Br–H	366
Sn–H	253	Sb–H	257	Te–H	267	I–H	298

It is worth emphasizing that the value of D(E–X) depends on the valency of the element E. As the coordination number increases, the value of D(E–X) decreases, reflecting the greater steric crowding and the additional promotional energy terms. Some data are given in Table 2.28 which exemplify this point.

Table 2.26 Values of some mean bond dissociation enthalpies for Group 14 and 15 elements with carbon, oxygen, fluorine, and chlorine/kJ mol^{-1}

C–Cl	346	N–F	278	N–O	214	C–N	286	C–C	347
Si–Cl	400	P–F	490	P–O	360	Si–N	333	Si–C	307
Ge–Cl	340	As–F	487	As–O	326	Ge–N	257	Ge–C	245
Sn–Cl	315							Sn–C	212

Promotion energies

The Group 14 elements have a ground state electronic configuration ns^2np^2 in their atomic states, i.e. there are only two unpaired electrons. In order for the atoms to form four covalent bonds it is necessary to promote an electron from the s to the p shell, i.e. to generate a configuration ns^1np^3 with four unpaired electrons. The energy required for this process is the 'promotion energy'. The process is favourable because the energy released when the atom forms the additional covalent bonds exceeds the promotion energy.

For a discussion of the influence of electronegativity on mean bond enthalpies see K. B. Wiberg, *J. Chem. Educ.*, 1996, **73**, 1684.

Table 2.27 Values of some typical mean bond dissociation enthalpies for element-halogen bonds/kJ mol^{-1}

P–F	490	As–F	487	Si–F	597	Ge–F	471
P–Cl	322	As–Cl	309	Si–Cl	400	Ge–Cl	340
P–Br	263	As–Br	256	Si–Br	330	Ge–Br	281
						Ge–I	214

Table 2.28 Values of mean bond dissociation enthalpies for element-halogen bonds where the element has more than one oxidation state/kJ mol^{-1}

PF_3	490	PCl_3	323	SF_2	367	XeF_2	133
PF_5	461	PCl_5	257	SF_4	339	XeF_4	131
				SF_6	329	XeF_6	126

The trend noted above for standard enthalpies of formation reflect the decreasing bond enthalpies down a column of the Periodic Table. The element-carbon bond enthalpies follow a similar pattern to the hydrides and therefore the organometallic compounds of the p block elements would be expected to become less stable down the column. The data demonstrate that this trend is followed after carbon, but carbon itself has an anomalously unfavourable enthalpy of formation, presumably because of the steric repulsions between methyl groups in $C(CH_3)_4$. The endothermic nature of the reaction for tin and lead makes these compounds very susceptible to decomposition reactions which have as their initial reaction homolytic dissociation of the metal-carbon bond.

Organometallic compounds are those which contain metal–carbon bonds

Many of the organometallic compounds of the s and p block elements are pyrophoric, particularly those which can form monomers which are coordinatively unsaturated, e.g. $Al(CH_3)_3$ and $Be(CH_3)_2$, and which can coordinate an oxygen molecule readily and thereby initiate the oxidation reaction. Compounds such as $Si(CH_3)_4$ which are coordinatively saturated are much less reactive and only react with oxygen at elevated temperatures.

For the p block elements the carbanion character of the organic group increases down the group for the third and subsequent rows, i.e. $Al_2Me_6 >$ $GaMe_3 > InMe_3 > TlMe_3 > BMe_3$. This cannot be related directly to electronegativity differences which follow the order B < Ga < In < Al < Tl. The aluminium compounds react readily with water and this leads to the precipitation of $Al(OH)_3$, whereas Ga, In, and Tl form initially the cation MMe_2^+ which only hydrolyse under acidic conditions with the relative ease Ga > In > Tl.

The enthalpies of formation for the oxides and fluorides are particularly favourable for the third row elements (see Table 2.24). The bond enthalpies of the third row atoms with fluorine and oxygen are particularly strong and this has several important consequences. Firstly, Si, P, and S form a range of very stable compounds with oxygen. The silicates, phosphates, and sulfates are therefore important and widespread as minerals on Earth. Secondly, the relative stability of the oxides and instabilities of the hydrides make the hydrides of these elements more reducing than those of the second row elements and the oxides less oxidizing. For example, PH_3 is more reducing than NH_3 and nitric acid is a much stronger oxidizing acid than phosphoric acid. Thirdly, the stability of the fluorides makes the oxides soluble in HF, e.g. glass can be etched by aqueous HF, and the third row elements form a range of very stable complexes $MF_6{}^{n-}$, e.g. M = Al, Si, P, and S.

Differences in the standard enthalpies of formation for series of compounds may give an interesting insight into the relative stabilities of compounds. For example for the zinc, cadmium, and mercury (Group 12) compounds listed below, the fluorides and the oxides always have a more negative enthalpy of formation, but the difference becomes smaller as the group is descended.

Table 2.29 Standard enthalpies of formation, $\Delta_f H^{\ominus}/kJ\ mol^{-1}$, for some compounds of the Group 12 elements

	ZnF_2	CdF_2	HgF_2
	−764	−700	−418
	ZnI_2	CdI_2	HgI_2
	−240	−203	−105
Difference	**−524**	**−497**	**−313**
	ZnO	CdO	HgO
	−348	−255	−91
	ZnS	CdS	HgS
	−203	−144	−58
Difference	**−145**	**−111**	**−33**

The iodides and the sulfides are therefore stabilized relative to the fluorides and oxides as one descends the group. This observation is but a single example of a more general phenomenon whereby metals on the right-hand side of the Periodic Table show an increased tendency to form more stable compounds with more polarizable ligands such as I^- and S^{2-}. The so-called *Hard and Soft Acid and Base Principle* generalizes this phenomenon and states that hard metal ions, e.g. Ca^{2+} and Al^{3+}, which are small and not very polarizable prefer to form compounds with hard ligands, e.g. F^-, O^{2-}, Cl^-, which are small and also not very polarizable. In contrast, soft metal ions prefer to form compounds with soft ligands which are large and polarizable, e.g. I^-, S^{2-}, P^{3-}. This principle is discussed further in Chapter 5.

For a detailed account of the historical background and recent theoretical studies see *Chemical Hardness*, Ed. K. Sen, *Struct. Bond.*, 1993, **80**.

Reactions of elements with air

The majority of elements burn in air or oxygen when heated. The reaction occurs more rapidly and at times explosively if the element is in a finely divided form. On exposure to air the more electropositive elements form oxides, and the more electropositive the easier the reaction proceeds. The nature of the initial oxide layer formed strongly influences the reactivity of the metal. Beryllium, magnesium, and aluminium, as well as zinc and many of the transition metals, form coherent and impermeable oxide films which protect the metal from further attack. For other metals such as iron the oxide layer is permeable to air and moisture and the metal 'rusts away' when exposed to moist air. As the metals become less electropositive they have to be heated to higher temperatures in order to form their oxides. Mercury forms an oxide at about 350°C, but at higher temperatures the oxide decomposes into its elements. *Noble metals* such as gold and platinum do not react with oxygen even when heated strongly.

The differences in stabilities of oxides may also be used in the separation of metals from their ores. For example, when HgS is roasted in air the metal is liberated and SO_2 is released.

$$HgS + O_2 \rightarrow Hg + SO_2$$

In contrast, zinc and cadmium sulfides give a mixture of oxides and sulfates when roasted in air. When the resulting mixture is dissolved in H_2SO_4 and reduced, cadmium metal is formed whereas the zinc remains in solution as the sulfate. This behaviour reflects the standard reduction potentials of Zn, Cd, and Hg given below.

$Zn^{2+}(aq) + 2e$	$\rightarrow$	Zn	$E^{\ominus} = -0.76$ V
$Cd^{2+}(aq) + 2e$	$\rightarrow$	Cd	$E^{\ominus} = -0.40$ V
$Hg^{2+}(aq) + 2e$	$\rightarrow$	Hg	$E^{\ominus} = 0.85$ V

The alkali metals show interesting differences when burnt in air. Lithium forms Li_2O (oxide), sodium forms Na_2O_2 (peroxide), and potassium, rubidium, and caesium form MO_2 (superoxide).

The standard enthalpy of formation of the monoxide, M_2O, of a metal, M, has contributions from the processes shown in Fig. 2.17. The conversion of dioxygen into oxide requires the breaking of the O–O bond and reduction of the resultant atom, both of which require a considerable energy input. The process leading to the metal oxide is only energetically favourable if these energy terms are compensated for by a large lattice enthalpy for the metal oxide which is eventually formed in the solid state. Therefore, it is essential to understand in more depth the factors which influence the lattice enthalpies.

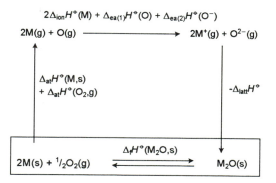

Fig. 2.17 A thermochemical cycle showing the contributions to the standard enthalpy of formation of the monoxides of the element, M

Lattice enthalpies

The lattice enthalpy $\Delta_{latt}H^{\ominus}$ of a crystal is a measure of the strength of the electrostatic forces which exist in an ionic solid. Formally, it is defined as the change of enthalpy when one mole of solid is transformed into one mole of each of the separated ions in the gas phase at 0 K, e.g.

$$NaCl(s) \rightarrow Na^+(g) + Cl^-(g) \qquad \Delta_{latt}H^{\ominus} = 787 \text{ kJ mol}^{-1}$$

The Born–Landé equation, which allows the calculation of lattice enthalpy, is based on the electrostatic interactions of ions in an ideal ionic solid coupled with a term that accounts for short distance repulsions caused by overlapping electron clouds, is given below:

$$\Delta_{latt}H^{\ominus} = \frac{N_A M z^+ z^- e^2}{4\pi\varepsilon_0 r_0}\left[1 - \frac{1}{n}\right]$$

where N_A is the Avogadro constant, M is the Madelung constant, e is the electronic charge, ε_0 is the permittivity of a vacuum, r_0 is the minimum interionic distance (r^+ and r^-), and n (the Born exponent) has values dependent upon the electronic configuration of the participating ions. The values of n appropriate to various electronic configurations are given in Table 2.30.

The lattice energy depends primarily on the ionic radii of the ions, r^+ and r^-, and the numerical positive values of their charges, z^+ and z^-, as expressed by the approximate relationship:

$$\Delta_{latt}H^{\ominus} \propto \frac{z^+ z^-}{r^+ + r^-}$$

In general lattice enthalpies decrease down a column of the Periodic Table as illustrated by the data given in Table 2.31.

Their are the two electron attachment enthalpy changes when the oxygen atom accepts two electrons, $\Delta_{ea}H_1^{\ominus} = -136$ kJ mol^{-1}; $\Delta_{ea}H_2^{\ominus} = 850$ kJ mol^{-1}

The Born exponent term defines the repulsion between the closed shells of electrons associated with the ions. The equilibrium distance in the salt represents the point where the resultant electrostatic attractive forces between the ions are balanced by the repulsive forces between the electron clouds on adjacent ions.

Table 2.30 The values of the Born exponent appropriate to the ion configuration. Mean values are used if the two ions have different values

Electronic configuration of ion	Value of n
He	5
Ne	7
Ar or Cu$^+$	9
Kr or Ag$^+$	10
Xe or Au$^+$	12

Table 2.31 Lattice enthalpies for the Group 1 halides

$\Delta_{latt}H^{\ominus}$ /kJ mol^{-1}	F	Cl	Br	I
Li	1037	862	785	729
Na	918	788	719	670
K	817	718	656	615
Rb	784	694	634	596
Cs	729	672	603	568

The trends are apparent for the alkali metal salts for a given halide and also for the halides of a given alkali metal cation. These variations in the lattice enthalpies are generally reflected in the melting and boiling points of these ionic compounds. For example, those for the sodium halides are given in Table 2.32. However, the m.ps and b.ps of lithium salts are anomalously low.

Table 2.32 Melting and boiling points of the sodium halides

	NaF	NaCl	NaBr	NaI
m.p. /$^\circ$C	996	801	747	660
b.p. /$^\circ$C	1695	1413	1390	1304

With regard to the reactions of Group 1 metals with dioxygen, since the lattice enthalpies for a given anion decrease down a column of the Periodic Table, only lithium oxide has a sufficiently large lattice enthalpy to drive the reaction to completion. The sodium ion provides sufficient lattice enthalpy to reduce dioxygen to the peroxide dianion, but not sufficient to break the O–O bond. Finally the potassium, rubidium, and caesium ions only give sufficiently large lattice enthalpies to reduce dioxygen to the superoxide mono-anion, O_2^-.

The decomposition of peroxide salts into oxides represents a single example of a more general class of decomposition reactions of oxoanion salts of the Group 1 and 2 metals. The alkali and alkaline earth metals form a wide range of salts with oxoanions, e.g. CO_3^{2-}, NO_3^-, and SO_4^{2-}, and there is a wealth of comparative data on their decomposition temperatures and enthalpies. The compounds are stable at room temperature, i.e. $\Delta_{decomp}H^{\ominus}$ is positive, but decompose above 800°C. The decomposition reactions result in the evolution of gases and consequently ΔS is positive and at the decomposition temperature $T\Delta S > \Delta H$ and ΔG is negative.

$$MCO_3(s) \longrightarrow MO(s) + CO_2(g)$$

$$M(NO_3)_2(s) \longrightarrow MO(s) + {}^1/_2O_2(g) + 2NO(g)$$

$$MSO_4(s) \longrightarrow MO(s) + SO_3(g)$$

The stability orders are: Li < Na < K < Rb < Cs; Mg < Ca < Sr < Ba

The outcome of chemical reactions depends on the relative stabilities of the reactants and products. Therefore, the interpretation of periodic trends must not be based solely on the properties of the reactants or products, but their difference. The discussion of the decomposition temperatures of oxo-anion salts provides an excellent illustration of the importance of considering the lattice enthalpies of the salt and the product oxide.

$$\Delta_{latt}H^{\ominus} \propto \frac{z^+z^-}{r^+ + r^-}$$

The decomposition of sulfates, nitrates, carbonates, etc. can be analysed using arguments similar to those developed above. Specifically, the thermochemical cycle shown in Fig. 2.18 indicates that the relative lattice enthalpies of the oxoanion and the oxide are the dominant factors.

The reaction is driven towards the right by the larger lattice energies of the metal oxides. As indicated previously, the lattice enthalpies, $\Delta_{latt}H^{\ominus}$, depend primarily on the sizes of the ions r^+ and r^-. Consequently $\Delta_{latt}H^{\ominus}(MO)$ is always larger than $\Delta_{latt}H^{\ominus}(MCO_3)$; however, the difference depends on the metal. For small cations such as Mg^{2+} the difference:

$$\Delta_{latt}H^{\ominus}(MO) - \Delta_{latt}H^{\ominus}(MCO_3)$$

is greater than for larger metal ions such as Ba^{2+}.

Table 2.33 provides estimates of the lattice enthalpies for the oxides and carbonates of the Group 2 metal ions. The lattice enthalpies are largest for the combination with smaller ions. Therefore, there is a greater thermodynamic driving force for forming the oxides when the cation is small. This procedure is particularly important because the thermodynamic stability can rarely be related solely to a property of the salt in isolation.

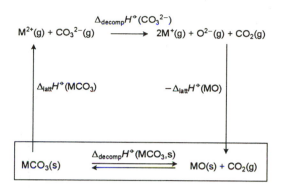

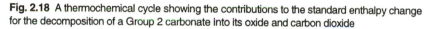

Fig. 2.18 A thermochemical cycle showing the contributions to the standard enthalpy change for the decomposition of a Group 2 carbonate into its oxide and carbon dioxide

Table 2.33 Estimated changes in lattice enthalpies for oxides and carbonates

	r^+/pm	$1000/(r^+ + r(O^{2-}))$ $r(O^{2-}) = 140$ pm	$1000/(r^+ + r(CO_3^{2-}))$ $r(CO_3^{2-}) = 185$ pm	Difference /%
Be^{2+}	45	5.41	4.35	20
Mg^{2+}	72	4.72	3.89	18
Ca^{2+}	100	4.17	3.51	16
Sr^{2+}	118	3.88	3.30	15
Ba^{2+}	135	3.64	3.13	14

Similar arguments may be developed for interpreting the decomposition temperatures for the following related reactions:

$$M(OH)_2(s) \longrightarrow MO(s) + H_2O(l)$$

$$MO_2(s) \longrightarrow MO(s) + \tfrac{1}{2}O_2(g)$$

$$MHF_2(s) \longrightarrow MF(s) + HF(g)$$

$$MI_3(s) \longrightarrow MI(s) + \tfrac{1}{2}I_2(s)$$

The hyperbolic nature of the relationship between $\Delta_{latt}H^\circ$ and $(r^+ + r^-)$ shown below suggests that for a given change in $r^+ + r^-$, Δr, the change in $\Delta_{latt}H^\circ$ is greater for smaller values of $r^+ + r^-$.

In each case the most stable salt occurs when the large anion is present with a large cation although this combination does not give the most favourable lattice enthalpy. The decomposition temperature cannot be interpreted in terms of a single parameter associated with the initial salt, but results from a difference between the lattice enthalpies of the initial salt and its decomposition product. Furthermore, alkali metal salts are generally more stable than the corresponding alkaline earth metal salts. The larger lattice enthalpies for the M^{2+} ions in combination with the smaller anions on the right hand side of the equations above favour the decomposition process.

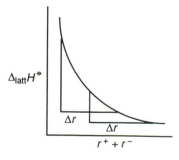

The re-expression of trends in the chemical properties of elements belonging to the same column of the Periodic Table using a thermodynamic cycle provides a powerful means of establishing the important atomic and molecular parameters underlying the trend. This procedure is particularly important because the thermodynamic stability can rarely be related solely to a property of the isolated compound. It is also useful for establishing whether a wide variation in properties occurs for the column of elements and rationalizing anomalies.

The relative stabilities of salts with different oxoanions reflect the acidities of the oxide being liberated, i.e. since the acidities follow the order

$$SO_3 > N_2O_5 > CO_2,$$

the stability order is:

$$CaSO_4 > Ca(NO_3)_2 > CaCO_3$$

2.4 Structural and bonding aspects of compounds

Ionic–covalent transition

The ionic and covalent bonding models are idealized theoretical concepts and in real molecules and solids the bonding is of an intermediate nature which involves some characteristics of both these extremes. In terms of the valence bond theory this intermediate character is represented by taking a linear combination of the wave functions which represent the extreme ionic and covalent contributions and this is represented schematically as follows:

$$M^+ \ X^- \quad \longleftrightarrow \quad M–X$$

ionic $\qquad\qquad\qquad$ covalent

The relative contributions of the ionic and covalent canonical forms is determined by the difference in electronegativities between M and X. If the electronegativity difference is large then the bonding is assumed to be ionic and if small then the bonding is covalent. In general the % ionic character in a bond may only be established by detailed spectroscopic measurements, and consequently, the effects of bond polarities are only observed indirectly through structural and chemical properties.

General trends

The discussion of allotropes (page 48) emphasized how the melting points and chemical properties depended on whether the element crystallized in a molecular form or formed a higher dimensional layer or infinite structure. Similar considerations apply to compounds and the following general classes of behaviour are observed.

1. Ionic compounds (large electronegativity difference $\chi_A-\chi_B$)

Infinite structures are always adopted in the solid state (molecular entities are only observed in gas phase at high temperatures) which may be interpreted in terms of close-packed A^{m+} and B^{n-} ions. There are no short A..A or B..B contacts and the ions occupy high symmetry coordination environments, e.g. cubic, octahedral, tetrahedral. The lattice energies may be accurately calculated from an electrostatic model (see Table 2.34), and the compounds generally have high melting points and are hard and brittle. As melts they are electrical conductors. Their solubilities vary greatly, and they dissolve only in solvents with high dielectric constants which reduce the interactions between oppositely charged ions. Their oxides and hydroxides are generally basic and their halides do not hydrolyse to liberate HX (H = Cl, Br).

2. Polarized ionic and polar covalent compounds (electronegativity difference 0.1–2.5). The structures adopted by these compounds may be infinite three-

dimensional, layers, chains or molecular structures. Precisely which type does not depend only on electronegativity differences. Some specific examples of these structural types are listed below.

(a) Symmetric infinite structures which have the cations and anions in high symmetry environments, e.g. ZnS, where both ions are tetrahedral, SiO_2, tetrahedral silicon, AlF_3, octahedral aluminium ions.

(b) High symmetry infinite structures which have short metal–metal contacts indicative of some metal–metal bonding, e.g. NiAs, CoS.

(c) Layer structures where the metal atoms are in relatively high symmetry environments, but the anions show significant van der Waals interactions with nearest neighbours which are also anions, e.g. $CdCl_2$ and CdI_2; $AlCl_3$.

(d) Layer structures where the metal is in a low symmetry or low coordination number environment, e.g. PbO.

(e) Chain structures, e.g. SeO_2, Sb_2O_3.

(f) Ring and polyhedral-molecular structures, where a distinctive molecular structure containing several identical atoms may be identified in the solid state, e.g. P_4O_{10}, P_4S_3, N_4S_4.

(g) Simple molecular structures containing only a single central atom, e.g. CO_2, PF_5, PF_3.

(h) Molecular salts, e.g. $[PCl_4]^+[PCl_6]^-$.

Table 2.34 Experimental and calculated lattice energies (kJ mol^{-1})

	Exp.	Calc.*
LiF	1033	1033
LiCl	845	840
LiBr	799	794
LiI	741	729
NaF	916	919
NaCl	778	763
NaBr	741	724
NaI	690	670
CsF	749	720
CsCl	653	626
CsBr	632	600
CsI	602	563
AgF	955	878
AgCl	905	734
AgBr	890	699
AgI	876	608

* calculated from the Born-Landé equation (see page 67)

The compounds in categories (a) to (e) have reasonably high melting and boiling points. They are softer and less brittle than ionic compounds. Their solubilities are variable but those that dissolve do so in solvents with either high dielectric constants or are polar donor organic solvents. The lattice energies of these compounds are not generally accurately calculated from an electrostatic model. As melts some are conductors and some are non-conductors. Oxides belonging to these categories are generally amphoteric, but tend to be acidic if the central atom has a high electronegativity.

The compounds in categories (f) to (h) have relatively low melting and boiling points, the precise melting and points may be interpreted in terms of the dipole moments and polarizabilities of the molecules.

Specific trends

In general the electronegativity of an atom decreases down a group and therefore if the atom is bonded to a more electronegative atom the electronegativity difference increases and the bonding becomes more ionic. For the alkali and alkaline earth metals the salts MX and MX_2 (X = halide) become progressively more ionic in the series.

$$LiX \quad < \quad NaX \quad < \quad KX \quad < \quad RbX \quad < \quad CsX$$
$$BeX_2 \quad < \quad MgX_2 \quad < \quad CaX_2 \quad < \quad SrX_2 \quad < \quad BaX_2$$

For variations in the anion the ionic character increases in the series:

$$LiI < LiBr < LiCl < LiF$$
$$Na_2Te < Na_2Se < Na_2S < Na_2O$$

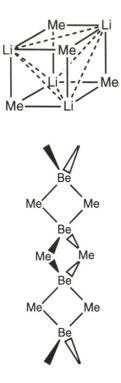

Fig. 2.19 The structures of Li_4Me_4 and $BeMe_2$

The high polarizing powers of Li^+ and Be^{2+} (see Table 2.3) arising from their high charge/size ratio make compounds of these metals significantly more covalent than those of the other members of the group. This has structural and reactivity consequences. For example, the salts of the Group 1 and 2 metals are invariably hydrated, and the maximum number of water molecules of crystallization is related to the polarizing power of the metal cation. For example, magnesium salts have 9–12 water molecules of crystallization: $MgCl_2.12H_2O$, $MgSO_4.12H_2O$, $Mg(NO_3)_2.9H_2O$; calcium and strontium salts have 2–6 water molecules of crystallization, e.g. $CaCl_2.6H_2O$, $SrCl_2.6H_2O$, $CaSO_4.2H_2O$, $Sr(NO_3)_2.4H_2O$; and barium salts are frequently anhydrous, e.g. $BaSO_4$. Therefore, it is not surprising that magnesium salts are commonly used as drying agents for removing traces of water from organic solvent mixtures.

Similar covalency effects are observed for the organometallic compounds of these metals. LiMe has a tetrahedral molecular structure, shown in Fig. 2.19, based on Li_4Me_4, whereas KMe has an infinite structure based on the NiAs lattice. Consequently lithium organometallics are more soluble in organic solvents and more volatile, whereas the organometallic compounds of K, Rb, and Cs are insoluble in organic solvents and have much higher melting points.

Similarly $BeMe_2$ has a chain polymeric structure, shown in Fig. 2.19, with three-centre two-electron bonds. The greater covalent contribution to the M–C bond in lithium and beryllium alkyls reduces the carbanion character of the organic anion and therefore they are less reactive.

$$M \longrightarrow R \quad \longleftrightarrow \quad M^+ \quad R^-$$

Similar trends are apparent for the Group 13 elements. Boron compounds do not have properties consistent with the presence of B^{3+} ions and the compounds of boron are essentially molecular and covalent. This is consistent with the very high polarizing power of B^{3+} in Table 2.3 on page 43. Aluminium compounds have more ionic character and gallium compounds reinforce the trend. With the exception of SnF_4 and PbF_4 the halides of the Group 14 elements are predominantly molecular covalent. They are volatile and many hydrolyse readily. For Group 15 elements compounds with +5 ions are not observed but Sb and Bi have compounds in the +3 oxidation state which are essentially ionic. Similarly although Bi_2O_3 may be described as ionic, the arsenic oxides are macro-molecular and the phosphorus oxides are molecular covalent. In the Group 16 few compounds can be described as ionic; PoF_4 represents a rare example.

Illustrations of some common infinite and layer structures are shown in Fig. 2.20.

The discussion above has centred on those elements which are more likely to form a cation. If it is now redirected to the anion, the most electronegative elements are found in the top right section of the Periodic Table and consequently the degree of ionic character in MX_n follows the orders:

F > Cl > Br > I,   O > S > Se > Te, and   N > P > As > Sb > Bi.

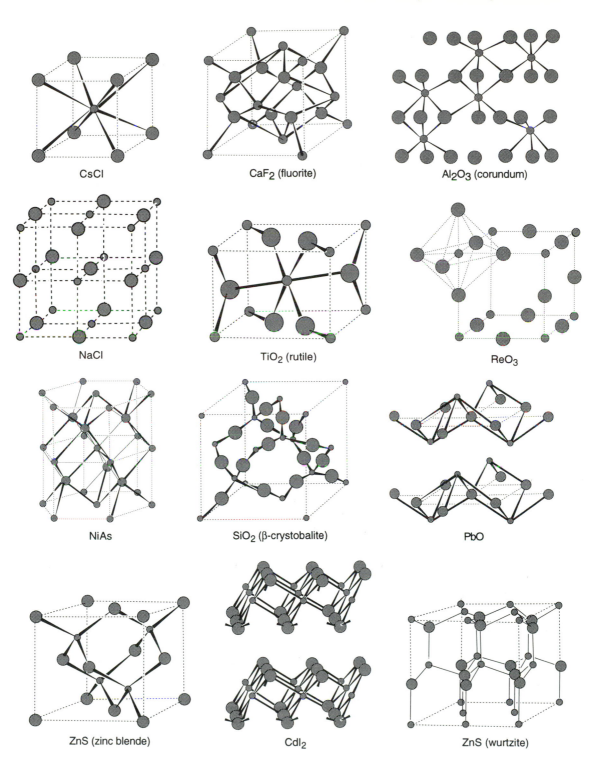

Fig. 2.20 Some common infinite and layer structures

BeF_2 ($\Delta\chi = 2.63$), m.p. 552°C, infinite lattice (4:2) quartz-like structure.

MgF_2 ($\Delta\chi = 2.87$), m.p. 1263°C, infinite lattice (6:3) rutile.

CaF_2 ($\Delta\chi = 3.06$), m.p. 1418°C, infinite lattice (8:4) fluorite.

SrF_2 ($\Delta\chi = 3.11$), m.p. 1477°C, infinite lattice (8:4) fluorite.

BaF_2 ($\Delta\chi = 3.13$), m.p. 1368°C, infinite lattice (8:4) fluorite.

BF_3 ($\Delta\chi = 2.09$), m.p. −128°C, molecular.

AlF_3 ($\Delta\chi = 2.63$), m.p. 1270°C, infinite lattice (6:3) octahedral Al^{3+}.

GaF_3 ($\Delta\chi = 2.28$), m.p. 980°C, infinite lattice, vertex sharing octahedra.

InF_3 ($\Delta\chi = 2.61$), m.p. >1000°C, infinite lattice, ReO_3.

TlF_3 ($\Delta\chi = 2.63$), m.p. 550°C, infinite (9:3) BiF_3 structure.

CF_4 ($\Delta\chi = 1.60$), m.p. −183°C, molecular.

SiF_4 ($\Delta\chi = 2.36$), m.p. −96°C, molecular.

GeF_4 ($\Delta\chi = 2.68$), m.p. −37°C, molecular.

SnF_4 ($\Delta\chi = 2.38$), m.p. 705°C, infinite layer structure (octahedral).

PbF_4 ($\Delta\chi = 2.55$), m.p. 600°C, infinite layer structure (octahedral).

For cations the compounds generally become more ionic in character down a column of the Periodic Table and for anions the compounds become less ionic and more covalent. The effects of changes in bond types should not be confused with changes in stoichiometry which can also influence whether the resultant compounds form infinite lattices, layer compounds or molecular compounds. This aspect is discussed further in Chapter 3.

The change in ionic-covalent character can also influence the type of lattice adopted by a compound in the solid state. The halides of Mg show a common transition from an infinite lattice to a layer lattice structure; MgF_2 has an infinite rutile 6:3 structure, $MgCl_2$ has a cadmium chloride 6:3 layer lattice, and $MgBr_2$ has a cadmium iodide 6:3 layer lattice.

A layer structure has anions in successive layers without an intervening layer of cations and therefore such structures are only observed when the anion–anion repulsion terms are not excessively large, i.e. the metal–halide bonds have significant covalent character and the anions are highly polarizable. The high polarizability leads to attractive van der Waals interactions between the anions which help to compensate for the electrostatic repulsions between them. Sulfides and selenides show a much greater tendency to form layer structures than oxides for analogous reasons.

Range of available oxidation states

The Mendeleev construction was initially based on regularities in the formulae of the commonly occurring compounds of the elements. Since his time, the number of known compounds for each element has multiplied by several orders of magnitude and therefore the generalizations developed by Mendeleev have to be amplified and some important caveats have to be added. The number of valence electrons of an element may set an upper limit to the oxidation state which the elements exhibits in its compounds, but the great majority of elements also form compounds in other oxidation states. In the following section the important trends in observed oxidation states are reviewed.

Group 1

The compounds of the alkali metals almost all have the metal in the +1 oxidation state corresponding to the loss of the single electron from the s shell and therefore the observed compounds have the formulae, MX (X^{-I}), M_2X (X^{-II}), M_3X (X^{-III}), etc. The only other oxidation state is −1, observed in the anions M^- (M = Li, Na, and K) and corresponds to the filling of the s shell of the metal atom. The alkali metal anions, M^-, are stabilized by shifting the following disproportionation reaction to the right-hand side:

$$2M(s) \quad \rightleftharpoons \quad M^+ \; + \; M^-$$

This reaction is thermodynamically unfavourable, but strong coordination of M^+ by a polydentate ligand shifts the equilibrium to the right. Ligands which are capable of achieving this are crown ethers and cryptands, typical structures of which are shown in Fig. 2.21. The resulting complexes are sufficiently stable to be crystallized as solids, whose structures can be determined.

Group 2

All the compounds of the alkaline earth metals have the metal in the +2 oxidation state corresponding to the ionization of both electrons from the s shell.

Group 13

For the Group 13 elements two oxidation states are commonly observed, +3 and +1 corresponding to the ionization of three electrons from the atom with ns^2np^1 configuration, or only the p valence electron leading to the M^+ ion with an ns^2 configuration. In an earlier section of this chapter it was noted that for the p block elements the s orbital is significantly more stable than the p orbital and the occurrence of these alternative oxidation states provides an important consequence of this energy separation.

The s–p energy separation therefore provides the possibility of a well defined subshell structure as well as the more commonly recognized noble gas shell structure associated with completely filled shells. The +1 oxidation state becomes progressively more stable than the +3 state down the group. Gallium dichloride, 'GaCl$_2$', does not have the metal in the +2 oxidation state, corresponding to a paramagnetic ns^1 ion, but is actually the mixed valence compound $Ga^IGa^{III}Cl_4$. There are some compounds where the metal has a +2 oxidation state and which are diamagnetic. These arise when the compound has a metal–metal bond. Examples are shown below. NB, in the definition of oxidation state a metal–metal bond does not contribute in the calculation of oxidation state.

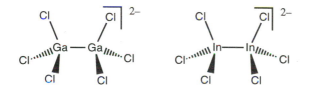

Group 14

The +4 and +2 formal oxidation states dominate the chemistry of the elements of this group, again corresponding either to the ionization of all the valence electrons or the p electrons exclusively. As with Group 13, the lower oxidation state becomes progressively more stable down the column. The +3 oxidation state is only observed in those cases where the element forms an element–element bond or the ligands around the metal are so bulky that the resulting radical is sterically prevented from forming a metal–metal bond. Examples of such compounds are illustrated as follows.

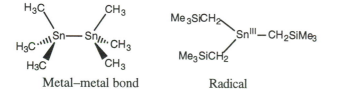

Metal–metal bond   Radical

Group 15

The compounds of the elements in Groups 1, 2, 13 and 14 are dominated by the metals in positive oxidation states. The Group 15 elements have a half-filled p shell and therefore it becomes possible to complete the p shell by the addition of three electrons. The resulting –3 oxidation state is observed in nitrides and phosphides of the electropositive metals. At the other extreme,

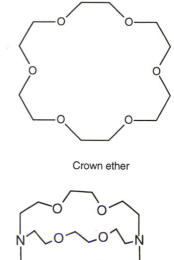

Crown ether

Cryptand

Fig. 2.21 Structures of the crown ether 18-C-6 and the cryptand -1,1,1

The presence of element–element bonds causes problems for the oxidation state formalism which partitions the bond into positively and negatively charged fragments, because the metal–metal bond does not contribute in the calculation of the oxidation state. However, since it involves sharing an electron pair it does contribute to the valency of an atom. Indeed in terms of number and type of bonds which they form $Ga_2Cl_6^{2-}$ is more closely related to $GaCl_4^-$ than the hypothetical radical species $GaCl_2$. In both cases four two-centre two-electron bonds, tetrahedrally arranged, are formed around the gallium atom. Therefore, for metal–metal bonded compounds it is usually more helpful to emphasize the valency than the oxidation state. For some recent examples see N. Wiberg, *Angew. Chem. Int. Ed.*, 1996, **35**, 65.

Steric effects are discussed on page 113.

the elements form compounds with a formal oxidation state of +5 corresponding to the removal of all the valence electrons. The compounds shown in Table 2.35 illustrate the wide range of oxidation states available to these Group 15 elements.

Table 2.35 Examples of compounds showing a range of oxidation states of the phosphorus atom

−3	−2	−1	1	3	5
P^{3-}					
PH_3	P_2H_4	H_2POH	H_3PO_2	H_3PO_3	H_3PO_4

The oxidation states change in steps of two, unless element–element bonds are formed, e.g. P_2H_4; consequently all the compounds are diamagnetic. The compounds H_3PO_2, H_3PO_3, and H_3PO_4 are all based on tetrahedral geometries as shown below and share a common phosphorus valency of 5.

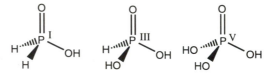

Since oxygen is more electronegative and hydrogen is less electronegative than phosphorus, the formal replacement of OH⁻ by H⁺ in this series of compounds leads to a regular decrease in formal oxidation state by 2.

The above series of compounds illustrates the limitations of the oxidation state concept. In the series the bond polarities are not great and the bonding is better described as covalent than ionic. Therefore, the formal assignment of a +5 oxidation state in $P^{+5}(O^{2-})(HO^-)_3$ is not very meaningful. The compounds shown in the diagram do exhibit a common valency and geometry. Each compound is tetrahedral and the phosphorus atoms form five covalent bonds; one double and three single. Therefore, in each of the compounds the phosphorus atom is utilizing all five of its valency electrons in forming the five covalent bonds; it is five-valent. The different formal oxidation states originate because the electronegativity order; O > P > H, changes the polarities of the P–H and P–OH bonds:

$$P^{\delta-}-H^{\delta+} \qquad P^{\delta+}-O^{\delta-}H$$

For elements such as phosphorus with intermediate electronegativities the residual charge on the atom may be either positive or negative depending on the substituent. For example, PF_3 and PH_3 have similar geometries, but the former has a formal oxidation state of (+3) and the latter (−3) because of electronegativity differences: F > P > H. Therefore, for compounds with intermediate electronegativities, it is more useful to define the valency of the atom rather than its oxidation state.

Group 16

Once again the lowest oxidation state (−2) is decided by the capture of two electrons to complete the noble gas configuration and the highest by the loss of all the valence electrons (+6). The compounds in Table 2.36 exemplify the commonly observed oxidation states for sulfur.

Table 2.36 Examples of compounds showing a range of oxidation states for sulfur in the compounds shown

-2	-1	2	4	6
H_2S	H_2S_2		SO_3^{2-}	SO_4^{2-}
		SF_2	SF_4	SF_6

The +4 and +6 states are not observed for oxygen. The +4 oxidation state becomes progressively more stable down the group and the +6 oxidation state exhibits the following order of oxidizing ability; Po >> Se > Te > S.

Group 17

Compounds of the halogens exhibit oxidation states which range from −1 to +7; examples are given in Table 2.37.

Table 2.37 Compounds illustrating the oxidation states of the halogens

−1	1	3	5	7
F^-				
Cl^-	Cl–F	ClF_3	ClF_5	
Br^-	Br–F*	BrF_3	BrF_5	
I^-	I–F*	$(IF_3)_n$	IF_5	IF_7

*Very unstable

The highest oxidation state is only observed for I in IF_7. Therefore, in contrast to Groups 13, 14, and 15, the highest oxidation compounds become progressively more stable for the heavier elements of the column.

Group 18

Only xenon forms compounds readily, in the +2, +4, and +6 oxidation states, again showing the trend noted for Group 17 that the heavier elements form compounds in the higher oxidation states more readily. Additionally, the +8 oxidation state is observed in XeO_6^{4-}. Radon has a lower ionization energy than Xe and therefore a wide range of compounds is anticipated. However, its radioactivity has limited the investigation of its compounds to RaF_2 and RaF^+. Krypton has higher ionization energies than xenon and therefore KrF_2 has only been isolated at low temperatures. The highest oxidation states of the elements are invariably displayed for the fluorides and oxides. Since the compounds in their highest oxidation states are usually molecular, this observation may be discussed in terms of the following Hess's law analysis, based upon the thermochemical cycle shown in Fig. 2.22 for the formation of the molecular fluorides, MF_n. Fluorine leads to a favourable value of $\Delta G^{\ominus}$ for this process because the dissociation energy of the fluorine molecule is small and the mean bond enthalpies, $E(M–F)$, are generally the strongest, i.e. $E(M–F) > E(M–Cl) > E(M–Br) > E(M–I)$.

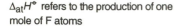

$\Delta_{at}H^{\circ}$ refers to the production of one mole of F atoms

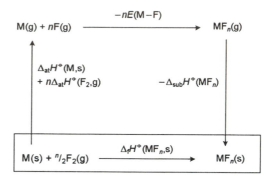

Fig. 2.22 A thermochemical cycle showing the contributions to the standard enthalpy of formation of molecular fluorides, MF_n

The range of oxidation states for the Group 13–15 elements illustrate clearly the shell and subshell structures of the Periodic Table. Elements in these groups form compounds which formally are associated either with the complete loss of all the valence electrons, N, i.e. have a formal oxidation state of $+N$ or $-(8-N)$, and also compounds with the formal oxidation state $+(N-2)$ resulting from the retention of a s^2 subshell. For these elements the increased covalency of the bonds leads to compounds with element–element bonds where the value of the formal oxidation state procedures becomes limited. For Groups 16–18 intermediate oxidation state compounds are observed which correspond to electronic configurations with partially filled p shells with even numbers of electrons.

'Inert pair' and alternation effects

The account given above of the range of oxidation states has not included any detailed discussion of the relative stabilities of the oxidation states. The Group 13, 14, and 15 elements exhibit the oxidation states N ($N = 3$, 4, and 5) and $N-2$, ($N-2 = 1$, 2, and 3). The compounds with lower oxidation states become progressively more stable down the column. The data given in Table 2.38 suggest that the relative stabilities of the two oxidation states is thermodynamically controlled.

Table 2.38 Enthalpies of formation, $\Delta_f H^{\circ}$ /kJ mol^{-1} for MX_4 and MX_2

	MF_2	MF_4	MCl_2	MCl_4	MBr_2	MBr_4	MI_2	MI_4
Si		−565		−381		−310		−234
Ge	−481	−452	−385	−349	−326	−276	−264	−212
Sn	−481	−414	−386	−323	−329	−273	−262	−205
Pb	−394	−313	−304	−243	−260	−201	−205	−142

These data suggest that the enthalpy change for the reaction:

$$MX_4 \quad \rightarrow \quad MX_2 \quad + \quad X_2$$

becomes more favourable in the order: Pb > Sn > Ge > Si. The factors responsible for the trend may be analysed in more depth using the thermodynamic cycle shown in Fig. 2.23.

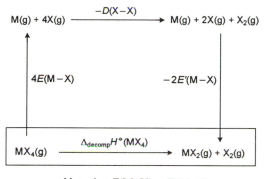

Note that $E(M–X) < E'(M–X)$

Fig. 2.23 A thermochemical cycle showing the contributions to the standard enthalpy change for the decomposition of MX_4 into MX_2 and X_2

$D(X–X)$ is the dissociation enthalpy of the X_2 molecule and $E(M–X)$ is the mean bond enthalpy term for the M–X bond.

The value of $\Delta_{decomp}H^{\ominus}$ becomes more negative as the value of $E(M^{IV}–X)$ decreases, but more negative as $D(X–X)$ increases.

$$\Delta_{decomp}H^{\ominus} = 4E(M^{IV}–X) - D(X–X) - 2E(M^{II}–X)$$

Down the column, the mean bond enthalpies for MX_4, $E(M–X)$, decrease, i.e. $E(Si–Cl) > E(Ge–Cl) > E(Sn–Cl) > E(Pb–Cl)$ and consequently it becomes progressively less attractive to form additional M–X bonds and attain the higher oxidation state.

In order to form MX_4 the additional energy required to promote the atom from its ground state ns^2np^2 to its valence state ns^1np^3 has to be compensated for by the formation of the two additional bonds. If $E(M–X)$ decreases this becomes progressively more difficult. Such trends in $E(M–X)$ are illustrated for Group 13 and 14 elements in combination with carbon, hydrogen, and chlorine atoms in Table 2.39.

Unfortunately only limited data are available for $E(M^{II}–X)$, but for a column of elements the values of $E(M^{II}–X)$ generally mirror those of $E(M^{IV}–X)$. Also the mean bond enthalpies for compounds in the lower oxidation state are significantly greater than those in the high oxidation state (see Table 2.39), thereby favouring the formation of the lower oxidation state compound.

The Group 15 and 16 elements do not show such straightforward trends and the relative stabilities of the oxidation states alternate. The occurrence (or non-occurrence) of the following compounds illustrates the phenomenon.

1. The Periodic Table provides an excellent basis for defining the highest and lowest oxidation states of an element in its compounds, because they correspond formally either to the loss of all the valence electrons of the atom or the formal gain of electrons to complete the valence shell, e.g. P with an atomic ground state of $[Ne]3s^23p^3$ gives rise to PO_4^{3-}, P^V $[Ne]3s^03p^0$, and PH_3, P^{-III} $[Ne]3s^23p^6$.

2. The s and p block elements also form compounds where the atoms have intermediate oxidation states. In general these oxidation states differ from the maximum in increments of two, e.g. $SF_6 – S^{VI}$, $SF_4 – S^{IV}$, and $SF_2 – S^{II}$ and the resultant compounds are diamagnetic.

3. Compounds with element–element bonds and compounds which are kinetically stabilized radicals have oxidation states which differ from those defined by 1 and 2 above, e.g. $S_2F_{10} – S^V$ and $Sn(CH_2SiMe_3)_3 – Sn^{III}$. The former class are diamagnetic and the latter paramagnetic.

4. The Periodic Classification does not provide a transparent explanation for the relative stabilities of oxidation states for a column of elements. The **inert pair effect** and **alternation effect** represent portmanteau terms for summarizing the important trends, but do not provide an interpretation. The more detailed analysis of the variations of the atoms, the promotion energies, and bond enthalpies provides some insight into the observed trends.

NF$_5$[*]	PF$_5$	AsF$_5$	SbF$_5$	BiF$_5$
NCl$_5$[*]	PCl$_5$	AsCl$_5$[†]	SbCl$_5$	BiCl$_5$[*]
NBr$_5$[*]	PBr$_5$	AsBr$_5$[*]	SbBr$_5$	BiBr$_5$[*]

[*] Unknown compound
[†] Unstable above –50°C

Table 2.39 Mean bond enthalpies for some Group 13 and 14 elements in combination with carbon, hydrogen, and chlorine atoms (kJ mol^{-1}) in MX$_4$ and MX$_3$ compounds. The figures in brackets refer to the bond enthalpies for MX$_2$ and MX compounds.

Element	E(M–C)	E(M–H)	E(M–Cl)
C	356	416	338
Si	301	326	391
Ge	255	289	349 (385)
Sn	225	251	323 (386)
Pb	130		243 (304)
B	360	373	440
Al	280	287	425
Ga	260	260	360
In	160	225	328 (435)
Tl	120		260 (364)

For a more sophisticated discussion see A. J. Downs, *Chemistry of aluminium, gallium, indium, and thallium*, Blackie, Glasgow, 1993.

This alternation in properties is described as the 'alternation' effect and may be related to the discontinuities in the ionization energies and electronegativities of these elements (see for example Fig. 2.7 on page 40).

The standard enthalpies of formation for the Group 16 fluorides given below confirms the thermodynamic origins of the phenomenon, as do the standard reduction potentials given in the margin.

	SF$_6$	SeF$_6$	TeF$_6$
$\Delta_f H^{\ominus}$/kJ mol^{-1}	–1209	–1029	–1318

	$E^{\ominus}$/V
ClO$_4^-$/ClO$_3^-$	1.23
BrO$_4^-$/BrO$_3^-$	1.74
IO$_4^-$/IO$_3^-$	1.64

For the Group 17 and 18 elements the higher oxidation state compounds become progressively more stable down the column. For these elements the mean bond energies follow the opposite trend, e.g. E(I–F) > E(Br–F) > E(Cl–F) in AF$_5$ (A = I, Br, or Cl). A similar trend occurs for the noble gases and it is not surprising that the first noble gas compounds were discovered for xenon. The very high ionization energies of the lighter Group 17 and 18 elements make it unfavourable for them to form a wide range of compounds where the central atom has a high positive formal oxidation state. Therefore, for Groups 13 and 14 the higher oxidation state is more stable for the lighter elements and for Groups 17 and 18 the higher oxidation states are more stable for the heavier elements. In between in Groups 15 and 16 the trends are less clear cut and alternation effects are more commonly observed.

Second row anomaly

One important aspect of the alternation effect which requires particular emphasis is the marked difference in properties between the second row elements (Li–Ne) and the third row elements (Na–Ar). This difference relates to the elements themselves and their compounds. Specific manifestations of the phenomenon include the following.

1. The anomalous properties of lithium and beryllium compared to the other remaining alkali and alkaline earth metals, e.g. the standard electrode potential of Li and its unique ability to form a nitride from N_2, and the greater covalency of lithium and beryllium compounds.

2. The greater thermodynamic stability of allotropes of the p block metals which have multiple bonds between the atoms, e.g. O_2 vs. S_8, N_2 vs. P_4, etc. This characteristic carries over to compounds of these elements, e.g. CO_2 (molecular) vs. $\{SiO_2\}_n$ infinite lattice; etc.

3. A reduced tendency of compounds of the second row p block elements to exhibit the maximum group valency, e.g. NF_3 vs. PF_5; OF_2 vs. SF_6, etc.

Multiple bonding is discussed further on page 89

4. The formation of higher coordination numbers for the third row elements and the ability to behave as Lewis acids even when the octet rule is obeyed.

5. The greater lability towards nucleophilic substitution for the third row halides, which makes them more susceptible to hydrolysis.

The second row elements are unique in the sense that they are the only row which have a core consisting only of a filled s shell ($1s^2$) and this leads to their valence electrons being exposed to a high effective nuclear charge. Their atomic properties follow from this unique situation and exhibit relatively high orbital ionization energies and high electronegativity coefficients (see page 40), small radii, and contracted atomic orbitals. These properties enable the orbitals to overlap very effectively leading to strong covalent bonds and particularly multiple bonds because the 2s and 2p orbitals are more similar in their energies and r_{max} than the 3s and 3p orbitals (see page 44). The additional filled shells which make up the core levels for the heavier p block elements result in greater inner shell repulsions between the atoms and prevents their close approach. This may reduce the extent of inter-orbital overlap and particularly p_π–p_π overlaps. In addition the third row elements have higher lying 3d valence orbitals which may contribute to bonding and which are not available for the second row elements.

There are clearly many reasons why the chemistry of the second row elements should differ from those of the subsequent elements, but the factors which currently appear to be most important are the large increase in radius (approximately 30%), the significant decrease in electronegativity, and the strength of the multiple bonding.

Effective atomic number (EAN) rule (octet rule)

In molecular compounds where the electronegativity of the central atom has an intermediate value the valency of the central atom is more important than its oxidation state and within the Lewis classification the valency of the atom is that which leads to the atom achieving a noble gas configuration (the effective atomic number—EAN—rule). The following molecules illustrate this principle since they each have the same number of valence electrons as the inert gas Ne.

	CH_4	NH_3	OH_2	FH	Ne
Total number of valency electrons:	8	8	8	8	8

Down a group of the Periodic Table the compounds of the Group 14–18 elements also exhibit higher valencies leading to compounds which do not obey the EAN rule; such compounds are frequently described as *hypervalent*. The valency does not exceed the number of electrons the central atom possesses in its valency orbitals. Some examples are given in Table 2.40.

Table 2.40 Examples of compounds and ions with different numbers of valency electrons surrounding the central atom

Number of valency electrons	Group 14	Group 15	Group 16	Group 17	Group 18
8	SiF_4	PF_3	SF_2	ClF	
10	SiF_5^-	PF_5	SF_4	ClF_3	
		AsF_5	SeF_4	BrF_3	KrF_2
12	SiF_6^{2-}	PF_6^-	SF_6		
			TeF_6	IF_5	XeF_4
14				IF_7	XeF_6
16	$[Sn(NO_3)_4]$		TeF_8^{2-}		XeO_4
18					XeF_8^{2-}

When a compound XF_n is oxidized by F_2 to XF_{n+2} the process is thermodynamically favourable. This suggests that in forming the two additional X–F bonds the gain in bond enthalpies must exceed the dissociation energy of F_2 and the promotion energy required to produce the appropriate spin state on the central atom.

The mean bond enthalpies for a series of compounds of the same elements decrease as the valency increases, as exemplified by the data given in Table 2.41 for series of sulfur, chlorine, and xenon fluorides. However, for these fluorides, the decrease is sufficiently small that it remains favourable to form the additional X–F bonds. The lighter elements do not form such compounds because $E(X–F)$ falls by a much greater amount as the valency is increased.

The second row elements of the Periodic Table (C to F) generally conform to the EAN rule, but the above data indicate that the subsequent series of the Periodic Table form a wide range of compounds, which exceed the octet rule. Two factors appear to be responsible for this trend. Firstly, the decrease in bond enthalpy with increasing valency is smaller for the third and subsequent row elements and, secondly, the increased size of the atom allows them to adopt higher coordination numbers. The formation of compounds which exceed the effective atomic number rule is favoured by central atoms with low ionization energies and peripheral atoms which have high electron affinities. This combination permits the transfer of excess electron density from the central atom to the outer atoms.

Table 2.41 Mean bond enthalpies, $E(X–F)/kJ\ mol^{-1}$

SF_2	350
SF_4	343
SF_6	328
ClF	255
ClF_3	174
ClF_5	151
XeF_2	133
XeF_4	131
XeF_6	125

The term 'hypervalency' suggests that the EAN rule is so ubiquitous that those compounds which have more electrons are very unusual and are adopting an unconventional bonding mode. In fact, there are so many examples of 'hypervalent' compounds for the elements in rows 3 to 6 that this view cannot be retained. In Chapter 4 the bonding in such compounds is discussed in more detail and it is demonstrated that the bonding in 'hypervalent' compounds can be described satisfactorily using modern valence theory.

Apart from beryllium and to a lesser extent lithium the majority of compounds of the alkali and alkaline earth metals do not conform to the effective atomic number rule. In the solid state the majority of these compounds have infinite structures and the bonding is primarily ionic. The coordination numbers are governed primarily by the sizes of the ions and the **electroneutrality principle**. The higher coordination numbers, usually 6 and 8, lead to more than eight electrons around the metal ions.

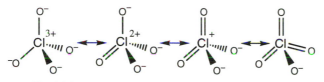

Fig. 2.24 Resonance forms of the tetroxochlorate(VII) ion

Coordination numbers

The radii of the atoms increase down a column of the Periodic Table and therefore both ionic and covalent compounds tend to adopt higher coordination numbers. The following molecular compounds illustrate the general trend:

Fluorine	F–F		
Chlorine	Cl–F	ClF_3	ClF_5
Bromine		BrF_3	BrF_5
Iodine		IF_3	IF_5	IF_7

Similar effects are observed in oxoanions:

Nitrogen	NO_2^- (angular)	NO_3^- (trigonal planar)
Phosphorus	PO_4^{3-} (tetrahedral)	
Arsenic	AsO_4^{3-} (tetrahedral)	
Antimony	$Sb(OH)_6^-$ (octahedral)	

In general the m.ps and b.ps of the compounds increase when they aggregate because the resultant molecular species has a larger size and the intermolecular forces which depend on the polarizabilities of the molecules increase. The tendency to adopt higher coordination numbers is also observed in the oxides of the heavier elements which increasingly adopt polymeric structures.

Electroneutrality principle

In a molecular compound the central ion attempts to attain a net charge lying between (–1) and (+1) by electron donation from the ligands and back donation from the central atom to the ligands. Unfortunately there is no simple way of calculating the extent of electron transfer in a specific complex, because it depends on the oxidation state of the central atom and the nature of the ligands present. Multiple bond formation can play an important role in this charge equalization process.

The bonding in the tetroxochlorate(VII) ion, ClO_4^-, may be regarded as being between a neutral central chlorine atom which participates in three Cl=O double bonds and one single bond to O^-. This and the bonding of the central chlorine atoms with charges of +1, +2, and +3 are shown in Fig. 2.24. According to the electroneutrality principle the canonical forms with +2 and +3 charged central atoms are less likely to contribute to the bonding of the ion.

It has to be recognized that the relative importance of these resonance forms also depends on the availablity of the chlorine 3d orbitals. Unfortunately, for the p block elements the nd orbitals lie at much higher energies and this limits the contribution which they can make. The role of these orbitals will be discussed further in Chapter 4.

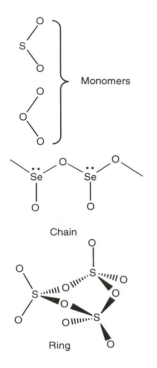

Monomers

Chain

Ring

Fig. 2.25 Examples of monomers, chain and ring structures. The trimeric structure is that of the γ-form of sulfur trioxide

In such structures the formation of additional bonds through bridging groups increases the coordination number. Such polymerization processes have a dramatic effect on melting and boiling points. Some relevant data are given in Table 2.42. The polymerization may occur in different ways leading to either rings or chains, some examples being shown in Fig. 2.25.

Solid SO_3 has three modifications, the simplest of which is the trimeric γ-form which is metastable, m.p. 17°C. The β-form consists of helical chains and is also metastable, depolymerizing at temperatures above 33°C. The most stable α-form consists of cross-linked helical chains and depolymerizes at temperatures above 62°C.

The tendency of a molecular compound to increase its coordination number influences its ability to function as a Lewis acid. For example, SiF_4 is a Lewis acid because it is able to expand in coordination number to either 5 or 6 whereas CF_4 is not. Similarly for the Group 16 elements SF_6, SeF_6, and TeF_6 are all known but they become progressively stronger Lewis acids and for example TeF_6 reacts readily with F^- to form TeF_8^{2-}. The ability of the molecules to functions as Lewis acids also influences their ease of hydrolysis. For example, SF_6 is completely inert to water, but TeF_6 is readily hydrolysed. Similarly CCl_4 is water stable, but $SiCl_4$, $GeCl_4$, and $SnCl_4$ fume in air because of hydrolysis.

The following main-group fluorides are able to function as Lewis acids.

PF_5, AsF_5, SbF_5
SF_4, SeF_2, SeF_4
TeX_2, TeX_4, TeF_6
(X = Cl, Br)

SnF_2, PbF_2
SiF_4, GeF_4, SnF_4, PbF_4
ClF, ClF_3
BrF, BrF_3
IF, IF_3, IF_5

Table 2.42 Structures and boiling points for some compounds

	Compound	b.p. (°C)	Compound	b.p. (°C)
Monomeric	CO	−192	CS_2	46
	CO_2	−79		(Sublimes)
			F_2O	−145
Weakly polymeric	SO_3	45 (γ-form)	SeO_2	340 (sublimes)
	P_4O_6 (cage)	175		
	As_2O_3 (layer structure)	457		
	Sb_2O_3 (double chains)	656		
Infinite polymeric structures	SiO_2 (4-coord.)	2230	GeO_2 (6-coord.)	2100

For the lighter elements Lewis acid behaviour is usually observed for compounds with coordination numbers less than 4, e.g. BF_3, NO_2^+.

The increasing sizes of atoms down the periodic columns and the associated changes in coordination numbers are very important for understanding many features of the chemistry of the p block elements. The following generalizations are helpful:

1. The second row elements generally favour a maximum coordination number of four. Although, of course, in transition states, and especially designed compounds higher coordination numbers may be achieved.
2. The third and fourth row elements favour maximum coordination numbers of six and octahedral geometries, e.g. SF_6 and SeF_6. The formation of such compounds negates the usefulness of the effective atomic number rule for such elements.
3. The heavier elements can form compounds with even higher coordination numbers, e.g. IF_7 and TeF_8^{2-}.

Secondary bonding

Down a column the van der Waals forces between molecules generally increase and the differences between covalent and van der Waals radii decrease. Consequently, it becomes less easy to define the molecular entity and the coordination number of the atoms in the crystal structure of a compound. For example, although ClO_3^- has a well defined trigonal pyramidal geometry, IO_3^- has three I–O bonds at 180 pm and three long I–O contacts at 280 pm. The six interactions when taken together generate a trigonally distorted octahedron around I. Similarly, secondary interactions are commonly observed in the oxides and halides of the heavier p block elements and lead to ambiguities in the designation of coordination numbers. This type of interaction resembles hydrogen bonding in terms of its strength and directionality. (See N. W. Alcock, *Bonding and Structure*, Ellis Horwood, 1990; *Adv. Inorg. Radiochem.*, 1972, **15**, 1)

The solid state infinite structures of ionic compounds of the pre-transition metals belonging to the same column of the Periodic Table exhibit similar trends in coordination numbers: (Structures at 0°C and 1 atm pressure)

BeO — Zinc blende, 4:4

MgO, CaO, SrO, BaO — Sodium chloride, 6:6

LiCl, NaCl, KCl, RbCl — Sodium chloride, 6:6

CsCl — Caesium chloride, 8:8

Although the size generalizations discussed above are almost universally valid for elements of the Periodic Table, there is one region where they must be used with caution. These are the Group 11 and 12 elements—Cu, Ag, Au and Zn, Cd, Hg. Some structural properties of these elements are summarized in Table 2.43.

Table 2.43 Structural properties of some Group 11 and 12 halides

Compound	Melting point/$^\circ$C	Structure	Metal coordination number
CuCl	430	Zinc blende	4-tetrahedral
AgCl	455	Sodium chloride	6-octahedral
AuCl	200 (dec.)	Saw-tooth linear chains	2-linear
$ZnCl_2$	318	Hexagonal close-packed	4-tetrahedral
$CdCl_2$	568	Hexagonal close-packed layer structure	6-octahedral
$HgCl_2$	280	Molecular	2-linear
$ZnBr_2$	394	Face-centred cubic	4-tetrahedral
$CdBr_2$	520	Cubic close-packed CdI_2 layer structure	6-octahedral
$HgBr_2$	289	Molecular or very distorted $CdBr_2$ layer structure	2-linear with four very long additional Hg–Br contacts

Although the atoms increase in size down these groups the compounds of the heaviest element of the group frequently exhibits a lower coordination number. The coordination number does often increase from 4 to 6 for the first two members of the group and then falls to 2 for the last member. The lower coordination number leads either to a molecular compound, e.g. $HgCl_2$, or a one-dimensional chain structure, e.g. AuCl, and the lower structural dimensionality is reflected in the lower melting points.

The compounds with a complete d^{10} subshell structure show a marked preference for linear coordination as the groups are descended. For example, the metal ions in SrO, BaO, and HgO have similar radii and SrO and BaO have sodium chloride structures (6:6 octahedral coordination) but HgO has linearly coordinated mercury ions. The s and d orbitals are relatively close in energy for these metals and it has been proposed that they may hybridize effectively in such cases. The 6p orbitals of Au and Hg are so high in energy that they are relatively unavailable for bonding and therefore the conventional sp^3 (tetrahedral) and d^2sp^3 (octahedral) hybridization schemes are energetically unfavourable compared to sd hybridization.

The relative promotion energies for the process $d^{10} \rightarrow d^9s^1$ are:

$$Cd^{2+} \; > \; Zn^{2+} \; > \; Hg^{2+}$$

$$Ag^+ \; > \; Cu^+ \; > \; Au^+$$

and for the process $d^{10} \rightarrow d^9p^1$ are:

$$Au^+ \; > \; Cu^+ \; > \; Ag^+$$

$$Hg^{2+} \; > \; Zn^{2+} \; > \; Cd^{2+}$$

Therefore, ions such as Au^+ and Hg^{2+} have relatively low d–s promotion energies and high d–p promotion energies favouring linear sd hybridization.

Molecular geometries

For a particular column of elements the molecular geometries for a series of compounds, EX_n, remain remarkably similar, particularly in the gas phase. If the molecule has no lone pairs, symmetry dictates that the molecules have identical geometries. For example, CF_4, SiF_4, GeF_4, and SnF_4 have tetrahedral geometries. When lone pairs are present the symmetry of the molecule is maintained, but subtle changes in the bond angles are generally observed. The data given in Tables 2.44 and 2.45 illustrate the general trends in bond angles for EX_3 and EX_2 molecules.

It is significant that when E is a second row element (N or O) the bond angle is much larger than for the other elements in the column. The subsequent variation is almost within the errors associated with the measurement of the bond angles. This second row anomaly is probably associated with the effective hybridization of the 2s and 2p valence orbitals which introduces more s character into the E–X bonds.

In Chapter 1 it was noted that the s and p orbitals of the 2nd row elements have similar maxima in their radial distribution functions and therefore hybridize effectively to form sp, sp^2, and sp^3 hybrid orbitals for chemical bond formation. For the subsequent rows of elements the bonding overlap with the p orbitals is concentrated into the bonds and the s electron density is concentrated into the lone pair regions. Therefore, the X–E–X bond angles approach 90°—the ideal angle for p orbital overlaps and the angle between the lone pair and the X atoms increases because of the increased s character in the lone pair hybrid orbital.

Table 2.44 Values of bond angles (°) for some EX_3 compounds

E	N	P	As	Sb	Bi
X					
H	107	94	92	91	
F	102	100	102		
Cl	107	100	98	100	100
Br		102	101	97	98
I	108	99	96		

Table 2.45 Values of bond angles (°) for some EX_2 compounds

E	O	S	Se	Te
X				
H	105	92	91	91
F	104	98		
Cl	111	103		
Br	111			98
CH_2	112	99	98	

The EMe_3 molecules undergo inversions by a mechanism which is shown in Fig. 2.26. The inversion barriers in the molecules EMe_3 (E = N to Sb) are given in Table 2.46 and do not follow a simple trend, but reach a maximum at phosphorus.

Table 2.46 Inversion barriers/kJ mol^{-1} for some Group 15 trimethyls

	NMe_3	PMe_3	$AsMe_3$	$SbMe_3$
Inversion barrier	34	133	122	112

The barriers in ERR'R" (E = P, As, or Sb) are sufficiently large for optically active compounds to be separated.

The stereochemical activity of lone pairs is also evident in sandwich and half-sandwich compounds of the p block elements and a specific example is shown in the margin.

Some of the sandwich compounds adopt polymeric structures in the solid state, e.g. In(η-C_5H_5) has a saw-tooth chain geometry and In(η-C_5Me_5) exists as an octahedron.

Occasionally a vertical comparison can reveal some very surprising geometric variations. For example, the alkaline earth halides, MX_2, when studied in the gas phase as molecules have the bond angles summarized in Table 2.47.

In this series of molecules the molecular symmetry changes from $D_{\infty h}$ (linear) to C_{2v} (angular). This difference may be related to the fundamental

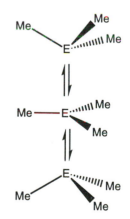

Fig. 2.26 An inversion mechanism for a pyramidal EMe_3 molecule

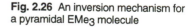

structure of the Periodic Table and the way in which the valence orbitals of the Group 1 and 2 elements change down a column of the table. For the second and third row elements the valence orbitals are ns and np. For the subsequent rows the inter-positioning of the d and f orbitals between the s and p shells alters the nature of the valence orbitals. For rows four and five the valence orbitals are ns, $(n-1)$d and np and for rows six and seven ns, $(n-2)$f, $(n-1)$d and np.

Table 2.47 Values of bond angles (^{o}) for some gaseous MX_2 compounds

M		F	Cl	Br	I
Be	$2s^2 2p^0$	180	180	180	180
Mg	$3s^2 3p^0$	155–180	180	180	180
Ca	$4s^2 3d^0$	133–155	180	173–180	180
Sr	$5s^2 4d^0$	108–135	120–143	133–180	161–180
Ba	$6s^2 5d^0 4f^0$	100–115	100–127	95–135	102–108

For the lighter elements the valence ns and np orbitals can form sp hybrids, which make a bond angle of 180°. For the heavier elements the valence orbitals in order of decreasing stability are ns $> (n-1)$d $> n$p and an alternative sd hybridization becomes energetically favourable. The latter achieves maximum overlap for bond angles of 90°. These differences are illustrated schematically in Fig. 2.27.

The linear geometries observed in d^{10} complexes were also attributed to sd hybridization effects on page 86 and this difference requires some clarification. In the Group 2 compounds the d^0 configuration of the M^{2+} ions results in two sd hybrids each of which can accept a lone pair from the ligands and a pair of localized two-centre two-electron bonds at 90° result. In the d^{10} ions one of the hybrids is occupied by an electron pair from the d shell and only one sd hybrid remains available for donation from the ligand lone pairs. Therefore, the ligands are forced to donate their electron pairs into the same sd hybrid and a bond angle of 180° results.

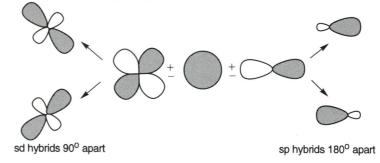

sd hybrids 90° apart   sp hybrids 180° apart

Fig. 2.27 The formation of sd and sp hybrid orbitals

For main group compounds which crystallize to form infinite structures the stereochemical role of lone pairs of electrons depends critically on the metal and the counter anions. For example, PbO has a layer structure and the coordination environment around each metal atom may be described as square-pyramidal with the lone pair occupying an apical site and the oxygen atoms in the four equatorial sites. The related compound PbS has a NaCl structure with both the Pb and S atoms in regular octahedral environments. As the ligand atoms become larger and the bond strengths become weaker it becomes less favourable to distort a symmetrical coordination polyhedron in order to locate a lone pair in a stereochemically active position.

In the gas phase the Group 15 halides are molecular and this structure persists in the solid state for the majority of P and As compounds. However, in AsI_3 the packing is such that the AsI_3 pyramid points towards three more distant I atoms to generate a distorted octahedron around As (AsI_3 has three As–I bonds at 256 pm and three at 350 pm).

In SbF_3, $SbCl_3$, $SbBr_3$, and SbI_3 similar structures are observed but the octahedral environment is more regular, and in BiI_3 the geometry is a regular octahedron. BiF_3 has what is described as a YF_3 structure with regular 9-coordinate trigonal prismatic environments around the Bi atom.

> Molecular geometries are generally characteristic of the periodic group and relatively minor angular changes are observed when a series of compounds belonging to the same column of the Periodic Table are compared. Exceptions occur either when the valence orbitals available to the central atom change dramatically or in the solid state when the lone pairs no longer exert a stereochemical role.

Multiple bonds

The second row elements are unique in having a core of electrons which involves only the filled 1s shell. This has two important consequences—the 2s and 2p orbitals of these elements are very contracted and secondly the radial distribution functions of the s and p orbitals leads to similar r_{max} values. Therefore, the s and p orbitals are able to hybridize effectively. The 2p orbitals of the second row elements (C–F) are capable of overlapping very effectively to form multiple bonds. Therefore, the compounds of these elements possess some of the strongest bonds in the Periodic Table, e.g. the triple bonds in N_2 (945 kJ mol^{-1}) and CO (1076 kJ mol^{-1}).

Tables 2.48 and 2.49 summarize the mean bond dissociation energies for multiple and single bonds of the p block elements. The data in the tables reinforce the view that the $2p_\pi$–$2p_\pi$ overlaps leading to multiple bonds are particularly favourable for C, N, and O.

The overlap between the np_π–np_π ($n > 3$) orbitals diminishes rapidly with the increasing sizes of the atoms and consequently the multiple bonding becomes progressively weaker.

The single bond enthalpies in Table 2.49 follow a similar trend for the Group 14 elements, but the Group 15, 16, and 17 elements show a distinctive first row anomaly. The values of E(N–N) and E(O–O) are much smaller than that anticipated by a simple extrapolation. These elements differ from C, Si, Ge, Sn, and Pb in also having lone pairs on the atoms. It appears that for the second row elements from N to F the presence of the lone pairs adjacent to the single bond, indicated in the structures of Fig. 2.28, exert a significant repulsive effect and destabilize the single bonds.

It is convenient for a chemist to consider that in a molecule the bonding occurs exclusively through the electrons which have been assigned to the bond and that the lone pairs of electrons are there as impartial spectators which do not influence the bond and its strength. This is too naive. The total

Down a group the halides generally become stronger Lewis acids. In part this results from their larger radii which allows them to expand their coordination numbers, and in part because the increasing electropositive character makes the central atom more positively charged. Therefore, in the solid state the EX_3 molecules are able to act as acceptors towards the lone pairs on the halides of adjacent molecules.
(See G. A. Fish and N. C. Norman, *Adv. Inorg. Chem.*, 1994, **41**, 233)

Table 2.48 Mean bond dissociation enthalpies of some multiple bonds /kJ mol^{-1}

C=C	N≡N	O=O
602	945	513
Si=Si	P≡P	S=S
310	493	430
Ge=Ge	As≡As	Se=Se
270	380	290
Sn=Sn	Sb≡Sb	Te=Te
190	293	218

Table 2.49 Mean bond dissociation enthalpies for some single bonds /kJ mol^{-1}

C–C	N–N	O–O	F–F
356	167	144	158
Si–Si	P–P	S–S	Cl–Cl
226	209	226	242
Ge–Ge	As–As	Se–Se	Br–Br
188	180	172	193
Sn–Sn	Sb–Sb	Te–Te	H
151	142	149	151

energy in a molecule reflects the interactions between all of the electrons. The contracted nature of the 2s and 2p valence orbitals brings the atoms together sufficiently closely to form strong bonds, but also it brings the lone pairs on adjacent atoms close enough together to exert a significant repulsive influence. For the subsequent rows of elements the increased size of the atoms leads to a longer single bond and consequently the repulsive effects between the lone pairs are less significant.

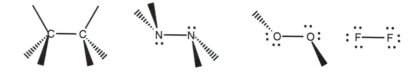

Fig. 2.28 Molecules with element–element single bonds and with lone pairs of electrons indicated

CO $\quad$ N$_2$ $\quad$ O$_2$
CO$_2$ $\quad$ NO $\quad$ O$_3$
C$_3$O$_2$ $\quad$ N$_2$O$_3$
CS$_2$ $\quad$ NO$_2$
$\quad\quad\quad$ N$_2$O

Since the multiple bonds of the second row elements (C to F) are particularly strong and the single bonds relatively weak for the elements N to F, it is not surprising that the second row elements provide the largest number of examples of small stable molecules with multiple bonds, some of which are given in the margin. The weakness of N–N and O–O single bonds leads also to the occurrence of radical molecules with unpaired electrons. For example, NO$_2$ forms a dimer with a N–N bond at lower temperatures, but it is not sufficiently strong to be maintained at higher temperatures.

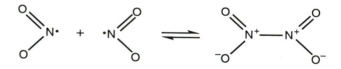

Dissociation is complete at 140°C and the dissociation enthalpy is 57 kJ mol^{-1}. Other analogous radical molecules are NO and O$_2$. The weak E–E single bond resulting from the dimerization of the radicals does not always compensate for the loss in multiple bond energies and therefore the driving force for dimerization is diminished.

The corresponding molecules for the third, fourth, and fifth row elements are generally polymeric and adopt structures with single bonds between the atoms, in preference to diatomic molecules with multiple bonds; examples are shown in Fig. 2.29. The bond enthalpy difference between P≡P and P–P is 284 kJ mol^{-1}, which may be contrasted with that between N≡N and N–N which is 778 kJ mol^{-1}.

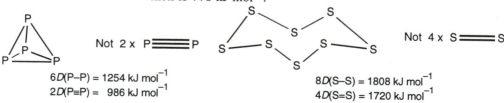

Not 2 x P≡P

$6D(\text{P–P}) = 1254$ kJ mol^{-1}
$2D(\text{P≡P}) = 986$ kJ mol^{-1}

Not 4 x S≡S

$8D(\text{S–S}) = 1808$ kJ mol^{-1}
$4D(\text{S=S}) = 1720$ kJ mol^{-1}

Fig. 2.29 Examples of molecules which have single rather than multiple bonds

Table 2.50 summarizes the bond lengths and bond enthalpies for oxygen and sulfur compounds which possess formal triple bonds.

Table 2.50 The bond lengths and enthalpies of dissociation of some triply bonded molecules

Molecule	Bond length/pm	Enthalpy of dissociation/kJ mol^{-1}
CO	113	1072
CS	153	728
SiO	151	760
SiS	193	620
SnO	184	532
SnS	221	460

It is noteworthy that for the same element the oxide always forms a stronger and shorter bond and the oxides and sulfides form weaker and longer bonds down the periodic group.

The differences in the strengths of multiple bonds for the different rows of the Periodic Table may also be used to rationalize many features of the chemistry of the p block elements, as the following examples show.

1. Elemental Si—diamond structure (no analogue of graphite is observed).
2. SiO$_2$—infinite structure based on tetrahedral SiO$_4$ units (cf. CO$_2$).
3. P$_4$O$_6$—tetrahedron of P atoms with bridging oxygen atoms (cf. N$_2$O$_3$).
4. N$_4$S$_4$—tetrahedron of S atoms with bridging nitrogen atoms (cf. NO).

Although P$_2$ and S$_2$ are not observed as simple molecules at room temperature they are observed in the vapour phase when the solids are heated strongly. Their presence can be detected using mass spectrometry and their geometries established using spectroscopic and electron diffraction techniques.

Although the heavier element analogues of multiply bonded compounds may not be readily isolated as simple hydrides, e.g. Si$_2$H$_2$, Si$_2$H$_4$, P$_2$H$_2$, the multiple bonds may be kinetically stabilized by replacing the hydrogen atoms with organic groups which are sterically demanding. The presence of these groups discourages the polymerization of the molecules which replace the multiple bonds by single bonds and also protects the double bond from small electrophilic reagents as indicated in Fig. 2.30. These steric effects are discussed in more detail in Section 2.6.

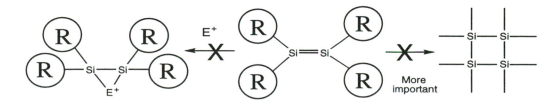

Fig. 2.30 Multiply bonded compounds tend not to polymerize or to react with electrophiles if the R groups are large

Table 2.51 compares the bond lengths in a series of Group 14 dimeric compounds with single and multiple bonds. It is significant that the bond contraction associated with multiple bond formation decreases down the column. This reinforces the view that the multiple bond is becoming progressively weaker going down the column. Although there are bond lengths quoted for SiO in Table 2.51, it is not a stable gas at room temperature like CO and is only observed at high temperatures in the gas phase.

Besides the bond length differences noted above, the geometries of these multiply bonded molecules may also change down the column. For example, the double bonded molecules $R_2X=XR_2$ becomes progressively less planar. The fold angle, θ, defined in Fig. 2.31, is 0° for X = C and increases to 32° for Ge and 41° for Sn. The heavier members of the series behave chemically like a pair of weakly interacting XR_2 'carbene-like molecules' rather than olefinic multiply bonded dimers.

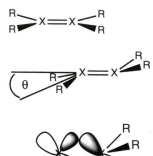

Fig. 2.31 Diagrams of planar and non-planar doubly bonded molecules showing the fold angle, θ, in the latter. The lowest diagram shows the overlapping orbitals in the non-planar case.

Volume 39 of *Advances in Organometallic Chemistry* is devoted to a discussion of the multiply bonded compounds of Group 13 and 14.

Table 2.51 Bond lengths in some carbon, silicon, germanium, and tin compounds with single and multiple bonds (data from Prof. A. J. Downs, Oxford)

	d(X–X)/pm	% contraction
H_3C–CH_3	154	
$H_2C=CH_2$	133	13.6
R_2HSi–$SiHR_2$	235	(R = mesityl)
$R_2Si=SiR_2$	214	9.0
R_3Ge–GeR_3	245	(R = Me, CH(SiMe_3)_2)
$R_2Ge=GeR_2$	235	4.0
R_3Sn–SnR_3	282	(R = Me, CH(SiMe_3)_2)
$R_2Sn=SnR_2$	276	2.0

	d(X–O)/pm	% contraction
$(H_3Si)_2O$	163	
$Si\equiv O$	151	7.7
$(H_3C)_2O$	142	
$C\equiv O$	113	20.0

In the delocalized ring compounds, shown in Fig. 2.32, the angle at the Group 15 atom decreases down the column. The largest change in bond angle again occurs between the elements of rows 2 and 3. The stabilities of the compounds also decrease down the column.

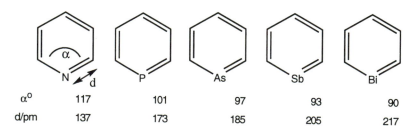

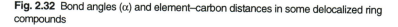

	N	P	As	Sb	Bi
$\alpha°$	117	101	97	93	90
d/pm	137	173	185	205	217

Fig. 2.32 Bond angles (α) and element–carbon distances in some delocalized ring compounds

Strong p_π–p_π multiple bonding is a characteristic of the second row elements. The strength of multiple bonding decreases down the columns of the Periodic Table and compounds of the heavier elements with multiple bonds may only be isolated if sterically demanding substituents are introduced. The reduced strength of the multiple bonds also leads to non-planar structures.

Catenation

The ability of an element to form compounds with element–element bonds (E–E) is described as *catenation*. If these compounds have single element–element bonds then this property is clearly related to the mean bond dissociation enthalpy for the E–E bond D(E–E). For example, for Group 14 elements the D(E–E) values are shown in Table 2.52.

Their values are consistent with the qualitative view that the catenating abilities of the elements are in the order:

$$C > Si \approx Ge > Sn \approx Pb$$

There are literally hundreds of hydrocarbons C_nH_{2n+2}, but a limited number of silanes Si_nH_{2n+2} and germanes Ge_nH_{2n+2} ($n = 1$–10), and very few polynuclear stannanes and plumbanes.

A similar ordering is observed for the ability of Group 13 compounds to catenate, i.e. B > Al > In > Ga > Tl. Boron forms a wide range of polyhedral boranes, B_nH_{n+4}, B_nH_{n+6}, and $B_nH_n^{2-}$ ions which in general are not emulated by the heavier elements.

In the Group 15 elements the order of ability to *catenate* is:

$$N < P > As > Sb > Bi$$

The difference from Group 14 can be rationalized in terms of the very low D(E–E) for the nitrogen–nitrogen single bond. Similar arguments can be proposed for Group 16 where the order of ability to form catenated compounds is:

$$O < S > Se > Te > Po$$

In Group 17, iodine is the only element which shows a significant number of compounds with element–element bonds and some examples are illustrated in Fig. 2.33.

2.5 Trends associated with physical properties of compounds

Dipole moments

An electric dipole consists of two equal and opposite charges q and $-q$ separated by a distance r. The dipole moment, μ, is a vector of magnitude, $\mu = qr$, which is directed from the positive charge to the negative charge. In a heteronuclear diatomic molecule the electronegativity difference leads to an

Table 2.52 Mean bond dissociation enthalpies for element–element bonds of Group 14 elements

	D (E–E) /kJ mol^{-1}
C–C	356
Si–Si	226
Ge–Ge	188
Sn–Sn	151
Pb–Pb	<100

A more detailed description of cage molecules of the Group 14 elements is given in A. Sekiguchi and H. Sakwai, *Adv. Organometal. Chem.*, 1995, **37**, 1.

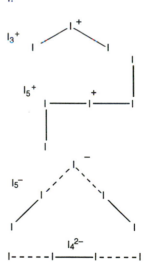

Fig. 2.33 Some examples of iodine compounds with I–I bonds

asymmetric charge distribution and a resultant dipole moment, as shown for example by the diatomic molecules of Table 2.53.

Table 2.53 Dipole moments (μ) and electronegativity differences ($\Delta\chi$) for dibromine and two bromine halides

	μ /D	$\Delta\chi$
Br_2	0	0
BrCl	0.52	0.1
BrF	1.42	1.4

The dipole moment increases as the electronegativity difference $\Delta\chi$ increases. In the hydrogen halides and the alkali metal chlorides, which exist as simple diatomic molecules in the gas phase, the dipole moments (some values are given in Table 2.54) also reflect the electronegativity differences. The larger dipole moments in the more ionic alkali metal chlorides are also noteworthy.

Table 2.54 Dipole moments of gas phase hydrogen halides and alkali metal chlorides

	μ/D		μ/D
HF	1.83	LiCl	7.13
HCl	1.10	NaCl	9.01
HBr	0.83	KCl	10.3
HI	0.45	RbCl	10.5

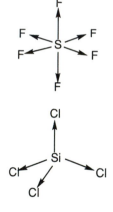

In molecules the resultant molecular dipole moment is approximately equal to the vector sum of the individual bond dipole moments. In high symmetry molecules such as SF_6 and $SiCl_4$ this can result in a net vector sum of zero, because the dipoles cancel out each other as shown in Fig. 2.34.

In polar molecules the vector sum is not zero and the molecule has a resultant dipole moment. The O=C=O molecule has a zero dipole, but O=C=S has a dipole moment of 0.71, the sulfur atom being less negative than the oxygen. Examples of other polar molecules are shown in Fig. 2.35.

Fig. 2.34 The bond dipoles of SF_6 and $SiCl_4$ molecules represented by arrows directed towards the atoms which possess partial negative charges. In both cases the vector sum is zero.

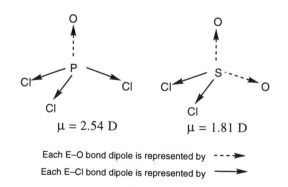

$$\mu = 2.54 \text{ D} \qquad \mu = 1.81 \text{ D}$$

Each E–O bond dipole is represented by `--->`
Each E–Cl bond dipole is represented by `——>`

Fig. 2.35 The bond dipoles of the polar molecules $POCl_3$ and SO_2Cl_2

In these examples the resultant dipole moment originates from the differences in the types of bonds within the molecule. In other cases it is the geometric location of bonds which influences the resultant dipole moment. For example, *cis-* and *trans-*isomers may be distinguished by differences in their dipole moments, as exemplified by the isomers of [SnCl$_4$(py)$_2$] shown in Fig. 2.36.

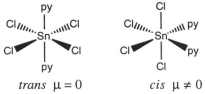

trans $\mu = 0$ *cis* $\mu \neq 0$

Fig. 2.36 The structures of the *trans-* and *cis-*forms of [SnCl$_4$(py)$_2$]

In addition to the dipoles associated with specific bonds, lone pairs can contribute to the resultant molecular dipole moment. In the series EH$_3$ (E = N, P, As, or Sb) the bond dipole and the lone pair dipole reinforce each other (see Table 2.55 and Fig. 2.37).

Table 2.55 Dipole moments of some Group 15 hydrides, chlorides, and fluorides

	μ/D		μ/D		μ/D
NH$_3$	1.47	NCl$_3$	0.39	NF$_3$	0.24
PH$_3$	0.57	PCl$_3$	0.56	PF$_3$	1.03
AsH$_3$	0.20	AsCl$_3$	1.59	AsF$_3$	2.59
SbH$_3$	0.12				

For the Group 15 elements the decreasing electronegativity leads to a smaller E–H bond dipole and an overall reduction in the resultant molecular dipole moment. Indeed, the bond dipole moments for the P–H, As–H, and Sb–H bonds are possibly negligible because these elements have electronegativities which are almost identical to those of hydrogen. For EF$_3$ (E = N, P, As) the E–F bond dipoles oppose the lone pair dipole and consequently the resultant molecular dipoles in NF$_3$ (see Fig. 2.37) and NCl$_3$ are very small. On descending the group the bond dipoles increase because of the increasing electronegativity difference for the E–Cl and E–F bonds. The total molecular dipole moment therefore increases.

When an isoelectronic series of multiply bonded molecules is compared, the electronegativity differences are suppressed to a large extent because the π-covalent bonding reduces the charge asymmetry as shown, for example, by the dipole moments of the molecules:

	N$_2$	CO	BF
μ/D	0	0.11	0.50

Molecular dipole moments have an important influence in the Lewis base properties of molecules, the intermolecular forces in compounds, and the properties of molecules as solvents.

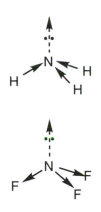

Fig. 2.37 The trigonal pyramidal NH$_3$ and NF$_3$ molecules showing the bond dipoles (full arrows) and the lone pair dipoles (dotted arrows). The arrows are directed towards the negative end of the dipoles

See page 47 for a more detailed analysis of Van der Waals forces.

Volatilities

The majority of molecular compounds of the Groups 14–18 elements crystallize as discrete molecules held together by van der Waals forces. The melting and boiling points consequently depend on the polarizabilities and dipole moments of the molecules. Down a column, a series of related compounds generally have increasing b.ps because of the increased atomic size and polarizabilities; some examples are given in Table 2.56.

Table 2.56 Melting and boiling point data ($^{\circ}$C) for some Group 14 and 15 hydrides and halides

	NF_3	PF_3	AsF_3	SbF_3
b.p.	−129	−101	63	376
	CH_4	SiH_4	GeH_4	SnH_4
b.p.	−161	−112	−88	−52
	PF_3	PCl_3	PBr_3	PI_3
m.p.	−152	−94	−42	61
b.p.	−101	76	173	>200 (decomp.)
	$SiCl_4$	$GeCl_4$	$SnCl_4$	$PbCl_4$
m.p.	−70	−50	−33	−15
b.p.	58	84	114	Explosive decomposition to $PbCl_2 + Cl_2$

The polarizabilities follow the order F < Cl < Br < I leading to higher melting and boiling points for the compounds containing heavier halogens.

Where there is a sudden dramatic change in melting or boiling point it generally suggests that the molecules have undergone an additional polymerization process involving the formation of additional halide bridges between molecules. Examples are given in Table 2.57.

F ─┊─ H ─┊─ F⁻   3-centre 4-electron bond

$\overset{\delta^+}{H}$──$\overset{\delta^-}{F}$----$\overset{\delta^+}{H}$──$\overset{\delta^-}{F}$ Electrostatic interaction

Table 2.57 Melting points ($^{\circ}$C) and molecular natures of some fluorides of Group 13, 14, and 16 elements

	m.p.	Nature of compound
BF_3	−101	Molecular
AlF_3	1291 (sublimes)	Infinite solid
GeF_4	−30	Molecular
SnF_4	705 (sublimes)	Infinite solid
SF_4	−121	Molecular
SeF_4	10	Molecular
TeF_4	130	Polymeric—fluorine bridged

The hydrides of the second row elements show anomalous trends if the molecules have lone pairs of electrons because of intermolecular hydrogen bonding interactions. Hydrogen bonding may be described either as a four-electron three-centre bond or as a strong electrostatic interaction between the residual charges on hydrogen and the electronegative atom on an adjacent molecule as shown above.

The theoretical aspects of hydrogen bonding are discussed further in Chapter 4. The important point to note for the present is that for both of these bonding descriptions the maximum stabilization is achieved when the hydrogen is bonded to an atom with a high electronegativity. Therefore, hydrogen bonding interactions are only energetically significant when they involve N, O, or F, and within any series the hydrogen bond strengths are in the order:

$$H\text{----}F \quad > \quad H\text{----}O \quad > \quad H\text{----}N$$

The importance of hydrogen bonding for the second row elements is reflected in the boiling points of the Group 14–17 hydrides given in Table 2.58.

The boiling points of NH_3, OH_2, and HF are anomalously high because of the strong hydrogen bonding interactions which result when the non-hydrogen atoms have a high electronegativity.

Table 2.58 Boiling points ($^{\circ}$C) of some Group 14–17 hydrides

HF	HCl	HBr	HI
19	−90	−67	−35
H_2O	H_2S	H_2Se	H_2Te
100	−61	−42	−2
H_3N	H_3P	H_3As	H_3Sb
−33	−87	−55	−17
H_4C	H_4Si	H_4Ge	H_4Sn
−162	−112	−98	−52

In general, compounds have higher volatilities if they are molecular and the van der Waals forces between the molecules are relatively weak. Therefore, the molecules must not have a large dipole moment, groups which are capable of either forming strong hydrogen bonds or are very polarizable. Simple hydrocarbon and fluorocarbon groups, e.g. CH_3 and CF_3, on the surface of the molecule generally promote higher volatilities.

The chelate complexes illustrated in Fig. 2.38 incorporate the required features and their relative volatilities are:

$Al(hfac)_3 > Al(hfac)_2(acac) > Al(hfac)(acac)_2 > Al(tfac)_3 > Al(fod)_3 > Al(acac)_3 > Al(thd)_3$.

Solubilities of salts

The solubilities of compounds at a specified temperature may be expressed either as their solubilities in 100 grams of water or, if very insoluble, as solubility products, K_{sp} where $K_{sp} = [M][X]^n$ for the ionic compound MX_n. Some typical values of solubilities and solubility products are summarized in

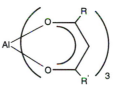

R = CH_3	R' = CH_3	= acac
R = CF_3	R' = CF_3	= hfac
R = CH_3	R' = CF_3	= tfac
R = C_3F_7	R' = CH_3	= fod
R = Bu^t	R' = Bu^t	= thd

Fig. 2.38 Some chelate complexes of Al^{III}. In general, fluorinated complexes are more volatile than their hydrocarbon analogues.

Table 2.59. The large variations, even within a related series of compounds, are particularly noteworthy.

Table 2.59 Some solubilities and solubility products of Group 2 compounds

	Solubility at 10°C /g per 100 g water		Solubility product
$MgSO_4$	31		
$CaSO_4$	0.2	$Ca(OH)_2$	5.5×10^{-9}
$SrSO_4$	0.1	$Sr(OH)_2$	3.2×10^{-4}
$BaSO_4$	2.7×10^{-3}	$Ba(OH)_2$	5.0×10^{-3}

Table 2.60 Some typical values of $\Delta_{sol}H^{\ominus}$ /kJ mol^{-1}

	$\Delta_{sol}H^{\ominus}$
NaF	1
NaCl	4
NaBr	−1
NaI	−9
NH_4Cl	14.7
AgF	−22
AgCl	66
AgBr	85
AgI	111

NH_4NO_3 has a quite positive $\Delta_{sol}H^{\ominus}$ (26 kJ mol^{-1}), but has a favourable entropy of solvation (110 kJ mol^{-1}K^{-1}) and therefore, it dissolves readily in water, but a noticeable cooling of the solution is observed because of the endothermic nature of the reaction.

The solubility product of a salt K_{sp} is related to the standard Gibbs energy of solution $\Delta_{sol}G^{\ominus}$ by the logarithmic relationship:

$$-\Delta_{sol}G^{\ominus} = RT\ln K_{sp}$$

This makes the observed solubility of salts very sensitive to small changes in $\Delta_{sol}G^{\ominus}$. A change of only 5.7 kJ mol^{-1} in $\Delta_{sol}G^{\ominus}$ causes a ten-fold increase in solubility and therefore a difference of only 10–20 kJ mol^{-1} can separate very soluble and very insoluble salts. Since, for an ionic solid, $\Delta_{sol}G^{\ominus}$ results from the difference of two rather large energy terms associated with the lattice energy of the salt, $\Delta_{latt}G^{\ominus}$, and hydration energies of the ions, discussions of trends must proceed carefully. Indeed, the entropy of solution contribution to the Gibbs energy change, $T\Delta_{sol}S^{\ominus}$, must also be considered because it is comparable in size to $\Delta_{sol}H^{\ominus}$ and its relative sign decides whether salts become more or less soluble as the temperature is increased. For the majority of salts $\Delta_{sol}S^{\ominus}$ is +ve and therefore the salts become more soluble as the temperature is raised. However, $\Delta_{sol}H^{\ominus}$ may be either negative (exothermic) or positive (endothermic) and some typical values are given in Table 2.60.

The Hess's Law thermodynamic cycle representing the components of $\Delta_{sol}G^{\ominus}$ is illustrated in Fig. 2.39 and must be analysed with some caution.

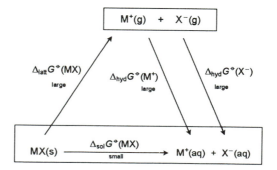

Fig. 2.39 A thermodynamic cycle showing the contributions to the Gibbs energy of solution of the compound, MX

The $\Delta_{latt}G^{\ominus}$ and $\Delta_{hyd}G^{\ominus}$ values, representing the lattice Gibbs energy of the salt MX and the Gibbs energy of hydration of the ions M$^+$ and X$^-$, are both large and the difference which corresponds to $\Delta_{sol}G^{\ominus}$ is close to zero.

Whether $\Delta_{sol}G^{\ominus}$ is +ve or −ve determines whether the salt is insoluble or soluble in aqueous solutions under standard conditions. The lattice Gibbs energy of a salt is generally greater than 400 kJ mol^{-1}. The lattice Gibbs energy depends on the charges on the cation and anion and the internuclear distance between them ($\propto z^+z^-/(r^+ + r^-)$). Therefore, a solvent which is able to dissolve the salt must interact strongly with both the cation and anion through dipole interactions, hydrogen bonding, and coordinate bond formation. It is also required to have a high dielectric constant which reduces the force of attraction between the ions. The hydration Gibbs energies of the individual ions, $\Delta_{hyd}G^{\ominus}(M^+)$ and $\Delta_{hyd}G^{\ominus}(X^-)$, are inversely proportional to the radii of the individual ions, r^+ and r^-, respectively and their charges.

Table 2.61 Values of $\Delta_{sol}G^{\ominus}$/kJ mol^{-1} for some Group 1 metal salts

	F$^-$	OH$^-$	Cl$^-$	Br$^-$	I$^-$	NO$_3^-$
Li$^+$	14	−8	−41	−57	−78	−15
Na$^+$	3	−42	−9	−17	−31	−7
K$^+$	−26	−65	−5	−6	−12	0
Rb$^+$	−38	−75	−8	−7	−8	−3
Cs$^+$	−59	−84	−9	−2	0	0

The topic is excellently discussed in more detail in D. A. Johnson, *Some Aspects of Thermodynamic Inorganic Chemistry*, 2nd Ed., Cambridge University Press, 1982.

Table 2.61 gives some values of $\Delta_{sol}G^{\ominus}$ kJ mol^{-1} for a range of alkali metal salts in water. The values of $\Delta_{sol}G^{\ominus}$ are determined primarily by the relative sizes of the cations and anions, and salts which have a large mismatch between cation and anion radii are particularly soluble, i.e. have a negative value of $\Delta_{sol}G^{\ominus}$. Therefore, lithium salts of Cl$^-$, Br$^-$, I$^-$, and NO$_3^-$ are very soluble. For smaller ions such as F$^-$ and OH$^-$ it is the Rb$^+$ and Cs$^+$ salts which show greatest solubilities.

The large negative $\Delta_{sol}G^{\ominus}$ values are concentrated in the top right and bottom left sections of Table 2.61. The presence of one large ion ensures that the lattice free energy of the salt is relatively small since the lattice energy depends on $1/(r^+ + r^-)$. Although the large ion also has a smaller hydration free energy ($\Delta_{hyd}G^{\ominus} \propto 1/r$) a major contribution to $\Delta_{sol}G^{\ominus}$ can be made by the smaller ion, leading to the value of $-\Delta_{sol}G^{\ominus}$ being greater than that of $\Delta_{latt}G^{\ominus}$. The solvation free energies of lithium salts with large anions are particularly favourable and therefore such salts remain soluble even in solvents with lower dielectric constants than water, i.e. polar organic solvents such as alcohols and ketones.

It also follows from Table 2.61 that salts with size-matched cations and anions are generally the least soluble. This must not be taken to mean that such salts pack in an efficient fashion which leads to a large lattice Gibbs energy. The lattice Gibbs energies remain dominated by the term $1/(r^+ + r^-)$ and therefore the combination of large cation and large anion cannot lead to a large lattice energy. Such salts are insoluble because the hydration Gibbs energies of the individual ions do not outweigh the lattice Gibbs energy.

Maximum insolubility occurs when the radius of the cation is approximately 80 pm smaller than that of the anion.

Some corresponding data for alkaline earth metal salts are presented in Table 2.62. The spread in values in Table 2.62 is much larger than that presented in Table 2.61 for alkali metal salts, because the lattice Gibbs energies for the MX_2 salts and the Gibbs hydration energies for the individual M^{2+} ions are much larger as a result of their higher charges and their smaller relative radii. It is noteworthy that the small anions F^- and OH^- lead to more insoluble salts for the lighter Group 2 metals. In contrast the Cl^-, Br^-, and I^- salts are more soluble for Be^{2+} and Mg^{2+}.

Table 2.62 Values of $\Delta_{sol}G^{\ominus}/kJ\ mol^{-1}$ for some Group 2 metal salts

	F^-	OH^-	Cl^-	Br^-	I^-	NO_3^-
Be^{2+}	42	123	−193	−253	−294	
Mg^{2+}	58	64	−125	−159	−200	−88
Ca^{2+}	56	30	−68	−98	−128	−33
Sr^{2+}	48	−2	−41	−70	−109	−2
Ba^{2+}	38	−16	−13	−32	−66	13

The following solubility generalizations are helpful:

$CH_3CO_2^-$, NO_3^-	Soluble for most cations
SO_4^{2-}	Soluble for most cations, except Ba^{2+}, Sr^{2+}, and Pb^{2+}
PO_4^{3-}, CO_3^{3-}, SO_3^{2-}	Alkali metal salts soluble, remainder insoluble
S^{2-}	Groups 1 and 2 soluble, remainder insoluble
OH^-	Alkali metals, Sr^{2+}, Ba^{2+} soluble; the majority of others insoluble. Insolubility increases with oxidation state.

These solubility differences form the basis of the qualitative analyses procedures which are used to test for the presence of metal ions and inorganic anions in unknown mixtures of inorganic salts.

The solubilities of the ionic compounds in non-aqueous solvents tend to increase as the size of anion increases. For example, the solubilities of the salts NaX in EtOH and $(CH_3)_2CO$ increase in the order: $F < Cl < Br < I$.

Colours of compounds

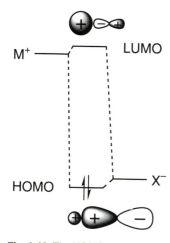

Fig. 2.40 The HOMO and LUMO of the ionic compound MX

In a compound which is predominantly ionic the large electronegativity difference between the constituent atoms ensures that there is a large energy difference between the highest energy occupied molecular orbitals localized on the anion (HOMO) and the lowest energy (unoccupied) orbitals localized on the cation (LUMO). Fig. 2.40 shows the relative energies of the HOMO and LUMO and their respective compositions in terms of the s orbital of M and the p orbital of X. The minimal covalent character is consistent with the large electronegativity difference between M and X in such a case. Consequently the energy required to promote an electron from the HOMO to the LUMO is large and only radiation in the ultraviolet region is sufficiently energetic to promote the excitation.

The resultant band observed in the electronic spectrum is described as a charge transfer band because the electron involved in the promotion is being transferred from an orbital which is localized on the anion to one that is localized on the cation. As the electronegativity difference between the atoms decreases then the energy separation between HOMO and LUMO becomes smaller, as shown in Fig. 2.41, and the charge transfer band moves towards the visible part of the electromagnetic spectrum. If the band tails into the visible region then the compound appears yellow and if the band moves further into the visible region the compound appears red. If the charge transfer band spreads right across the visible region the compound will appear black.

Therefore, it is not surprising that the majority of Group 1 and Group 2 salts are colourless and only those which have anions with transitions in their own right in the visible region appear coloured. For example, although all simple sodium salts are colourless, Na_2PtCl_6 is bright yellow because the anion $PtCl_6^{2-}$ has independent transitions which give rise to the observed colour. The effect of the electronegativity difference can be appreciated from the colours of the compounds given in Tables 2.63 and 2.64.

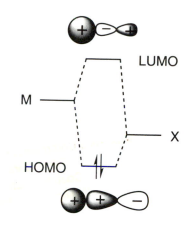

Fig. 2.41 The HOMO and LUMO of the compound MX when the electronegativity difference between M and X is small

Table 2.63 The colours of some halides

	Fluoride	Chloride	Bromide	Iodide
Ag^I	White	White	Pale yellow	Deep yellow
Hg^{II}	White	White	Pale yellow	Red
Pb^{II}	White	White	White	Yellow

Similar trends are observed for molecular and coordination compounds and some specific examples are given in Table 2.65.

Table 2.65 The colours of some molecular and coordination compounds

CCl_4	$SiCl_4$	$GeCl_4$	$SnCl_4$	$PbCl_4$
Colourless	Colourless	Colourless	Colourless	Yellow
SnF_4	$SnCl_4$	$SnBr_4$	SnI_4	
Colourless	Colourless	Yellow	Red-orange	
$AsCl_3$	$SbCl_3$	$BiCl_3$		
Colourless	Colourless	Colourless		
$AsBr_3$	$SbBr_3$	$BiBr_3$		
Colourless	Colourless	Orange-yellow		
AsI_3	SbI_3	BiI_3		
Orange-red	Yellow	Brownish violet		
WO_4^{2-}	WS_4^{2-}	WSe_4^{2-}		
Colourless	Yellow	Orange		
WO_4^{2-}	WO_2Cl_2	$WOCl_4$		
Colourless	Yellow	Red-orange		

Table 2.64 The colours of some oxides

	Oxide	Sulfide
Zn^{II}	White	White
Cd^{II}	Brown	Brilliant yellow
Hg^{II}	Red	Black

Differences in colour can have significant implications on the chemical reactivities of compounds. Most obviously, compounds which have strong absorption bands in the visible region are likely to undergo reduction at the metal centre when exposed to intense visible light. Reduction rather than oxidation is the preferred mode of reaction because the charge transfer band results in the promotion of an electron into an orbital which is localized on the metal. The photosensitivity of silver halide salts and the associated formation of reduced silver particles provides an example of such a photolytic process which is utilized in the photographic process. In general, metal halides, oxides, and sulfides, which are highly coloured or are black are more susceptible to reduction than salts which are white. Clearly the presence of low lying empty levels on the metal ion which are able to function as acceptors in the charge transfer process are also able to accept electrons from an external source in a reduction process.

The elements which are molecular (e.g. the halogens which form homonuclear diatomic molecules) have a range of occupied and unoccupied molecular orbitals and the energy differences between HOMO and LUMO is governed mainly by the efficiency of overlap between the atomic orbitals which are contributing to the molecular orbitals. Fig. 2.42 shows the p orbital contributions to the HOMO and LUMO of homonuclear diatomic molecules. Down a periodic group the overlap between these orbitals generally decreases and consequently the transitions associated with the promotion of an electron from HOMO to LUMO moves into the visible region. The halogens provide the most straightforward illustration of this trend; their colours are given in Table 2.66.

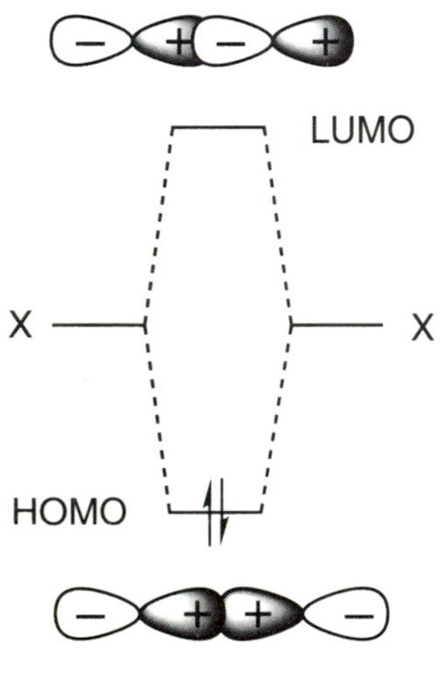

Fig. 2.42 The HOMO and LUMO of homonuclear diatomic molecules

In the halogens the actual transition responsible for the colour involves promotion of an electron from the antibonding π^* to σ^*, but the energy gap between these also decreases as the overlap decreases.

Table 2.66 Colours of the halogens

	F_2	Cl_2	Br_2	I_2
Colour of gas/vapour	Pale yellow	Green-yellow	Red-brown	Violet

The Group 16 elements (chalcogens) show a similar trend, shown in Table 2.67, although it must be remembered that in this series an identical molecular entity is not retained.

Table 2.67 Colours of the Group 16 elements

O_2 (gas)	S_8	Se_8	Te
Colourless, but blue as a liquid or solid	Yellow	Red	Black

These colour changes also have implications regarding the reactivities of the elements. Specifically a diatomic element, X_2, which absorbs strongly in the visible region is much more likely to produce radicals $X\cdot$, than a colourless gas and the process can be encouraged by shining light on the sample. The presence of such radicals can provide the initiation and propagation steps for radical reactions which in general enhance the reactivity of the element.

Similar considerations apply to the element–element bonds in more complex molecules.

$$R_3E\text{--}ER_3 \rightleftharpoons 2R_3E^\bullet$$
$$\text{Radical}$$

Light initiation of such reactions becomes easier along the series:

$$C < Si < Ge < Sn < Pb.$$

Hardness of compounds

The hardness of inorganic compounds has very important implications regarding their applications as coatings and components as drilling bits and polishing pastes. In general, the hardness of a compound is measured on a relative scale and a compound which is able to scratch a second is judged to be harder. The resulting data are put on an approximate numerical scale —Moh's scale of hardness—which range from 1 (talc)–10 (diamond). It is illustrated for some common inorganic salts in Table 2.68.

Table 2.68 Moh's scale hardness for some compounds and their M–X distances

	LiF	CaF$_2$	CdF$_2$	SrF$_2$	PbF$_2$
Hardness	3.3	6	4	3.5	3.2
M–X(pm)	202	236	234	250	257
	BeO	MgO	CaO	SrO	
Hardness	7.9	6.5	4.5	3.5	
M–X/pm	171	210	240	257	
	ZnO	ZnS	CdS	HgS	
Hardness	4.0	3.8	3.3	2.3	
M–X/pm	214	258	281	294	

The hardness, in general, reflects the lattice energies of the compounds and the inability of the lattice to deform under compression. Therefore, the hardest ionic compounds are those with small non-polarizable cations and anions and where the electronegativity difference is large. Fluorides are usually harder than oxides for the same element and oxides harder than sulfides. The hardness decreases for a series of cations down a column of the Periodic Table.

Amongst the elements themselves, the hardest solids are those which have very rigid infinite lattices and the element–element bonds are strong. Some typical examples are given in the margin with their hardnesses on Moh's scale.

At the other end of the scale, if the compound has a neutral layer structure with only van der Waals forces operating between the layers, the compound is soft. For example; graphite, 0.5; talc, 1; MoS$_2$, 1.3. The softness and the ability of the layers to move relative to each other allows these compounds to act as good lubricants.

Diamond, 10   Quartz, 7

Boron nitride, ~10   WC, 9

Silicon, 7   SiC, 9.3

Boron, 9.5

Ceramic materials

Ceramic materials are traditionally associated with rigid pottery and china formed from pliable clays at high temperatures. The modern scientific usage defines ceramics as inorganic compounds, primarily oxides, but also carbides, nitrides, silicides, and borides, which are formed in high temperature processes. These materials are of interest not only as high temperature containment materials, but because they have additional interesting and useful electrical and magnetic properties. For example, iron and chromium oxides are the active component in a range of magnetic recording media. Ceramics may also be ferro- and pyro-electric and recently a new class of 'high

For a more comprehensive introduction to ceramics and their properties see *Material Science and Engineering, An Introduction* by W. D. Callister, Jr., John Wiley and Sons, New York, 1997.

temperature' superconductors has been discovered which can operate at liquid nitrogen temperatures. Examples of some ceramics are given in Table 2.69, together with their melting point data.

Table 2.69 The melting points (°C) of some ceramic materials

Al_2O_3	2054	MgO	2826	ZrO_2	2710
BN	~2975	Si_3N_4	1900		
SiC	2830				

The great strength associated with the short bonds present in these compounds also makes them lose strength once a crack appears, i.e. they are very brittle.

The carbides and nitrides of the transition metals have very high melting points and such compounds are also very resistant to chemical attack. Examples of some transition metal carbides and nitrides are given in Table 2.70.

Table 2.70 Melting points of some transition metal carbides and nitrides

	m.p. /°C
TiC	3140
ZrC	3530
HfC	3890
TiN	2950
ZrN	2980
HfN	3300

2.6 Trends in chemical properties of compounds

The chemical properties of inorganic compounds can rarely be interpreted in terms of variations in a single atomic parameter. Many inorganic reactions are controlled by thermodynamic effects and therefore depend on the relative stabilities of the reactants and products and therefore simple interpretations based on a parameter associated with either the reactant or the product are rarely successful in rationalizing the observed trend. Nonetheless the following section attempts to bring together some relationships for simple inorganic reactions.

Acid–base properties

Brønsted acidities

This analysis ignores entropic effects which are significant when small highly charged ions are hydrated.

The acid strengths of hydrides (EH_x) in aqueous solution are affected by the following factors:

1. the strength of the E–H bond,
2. the electronegativity of E, which influences the polarity of the E–H bond, and
3. the energy of solvation of $[EH_{(x-1)}]^-$ since small anions have more favourable solvation enthalpies.

The data in Table 2.71 emphasize the important trends in pK_a values for EH_x. Down the column of the Periodic Table, the dominant trend is that the EH_x compounds become stronger acids. The strength of the E–H bond is therefore the predominant influence. Across a row the electronegativity differences predominate.

The proton affinities of the Group 15 compounds, given in Table 2.72, show how substituents can influence the basicities of phosphines. Such substituent effects are discussed in more detail on page 117.

The pK_a values of oxo acids seem to be determined primarily by the number of oxo groups (E=O) in the parent acid, as shown by the data given in Table 2.73. The additional oxo groups in a series such as HClO, HClO$_2$, HClO$_3$, and HClO$_4$ withdraw electron density from the central atom (see the structure in the margin) and encourage dissociation of the H$^+$ ion in solution.

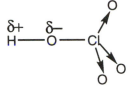

Table 2.71 pK_a values for EH$_x$

CH$_4$	NH$_3$	OH$_2$	FH
~58	39	14	3
SiH$_4$	PH$_3$	SH$_2$	ClH
~35	27	7	−7
GeH$_4$	AsH$_3$	SeH$_2$	BrH
25	~19	4	−9
SnH$_4$	SbH$_3$	TeH$_2$	IH
~20	~15	3	−10

In addition, the anion formed as a result of deprotonation may be stabilized by delocalization as indicated below.

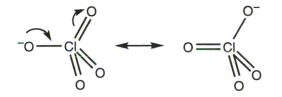

> The proton affintiy of a compound is defined as the energy released when the compound accepts a proton

These possibilities for delocalization increase with the number of oxo groups. An OH group is not able to provide the same opportunity for delocalization.

Down a column of the Periodic Table the pK_a values of oxo-acids increase, but the variations are much smaller than those noted above for EH$_x$ because in each case it is the O–H bond which is dissociating.

Where the structure change involves a change in the number of E=O bonds the pK_a values change dramatically, e.g. the difference between those of Te(OH)$_6$ and H$_2$SO$_4$ or H$_2$SeO$_4$.

For the metals on the left-hand side of the Periodic Table the increased electropositive character down a column makes the hydroxides of these metals more basic. Some data are given in Table 2.74.

Table 2.72 Proton affinities (gas phase) of some Group 15 compounds

Compound	Proton affinity /kJ mol^{-1}
PF$_3$	670
PH$_3$	804
CH$_3$PH$_2$	867
(CH$_3$)$_2$PH	908
(CH$_3$)$_3$P	959

Table 2.73 pK_a values for some oxo-acids

H$_2$SO$_3$	H$_2$SeO$_3$	H$_2$TeO$_3$
1.81	2.46	2.48
H$_2$SO$_4$	H$_2$SeO$_4$	Te(OH)$_6$
—3	—3	7.7
HClO	HBrO	HIO
7.4	8.7	11.0

Table 2.74 Values of pK_b(1) and pK_b(2) for some Group 2 hydroxides

Hydroxide	pK_b(1)	pK_b(2)
Be	10.3	
Ca	2.43	1.40
Sr	0.82	
Ba		0.64

It is interesting to compare the values of base dissociation constants of the Group 2 metal hydroxides with that of Zn(OH)$_2$, for which the value of pK_b(1) is 3.02. The increased effective nuclear charge and associated transition metal contraction makes Group 12 metal hydroxides less basic.

The chemical differences between analogous compounds for Group *N* and (*N* + 10) are discussed in some detail in Chapter 3 (page 196).

For the main group elements the oxides become less acidic down a column of the Periodic Table. For example, the +3 oxides of phosphorus and arsenic are acidic, that of antimony is amphoteric, and that of bismuth is basic. The change in anion from O^{2-}, S^{2-} to Se^{2-} makes the corresponding salts less basic, i.e. the basicity order is $Na_2O > Na_2S > Na_2Se$.

Lewis acid and base strengths

For Lewis acids such as BX_3 and AlX_3 (X = F, Cl, Br, I, or Me) the stability constants of the compounds which are formed with Lewis bases have been measured and the trends shown in Table 2.75 have been established.

Table 2.75 Orders of base strengths for some Lewis bases

Acid	Order of base strength
BBr_3	$NH_3 > PH_3 > AsH_3 \gg SbH_3$
$AlBr_3$	$Me_3N > Me_3P > Me_3As > Me_2O > Me_2S > Me_2Se > Me_2Te$
$AlMe_3$	$Me_3N > H_3N > Me_3P$
$AlMe_3$	$Me_3PO > Me_2SO > Me_2O > Me_2S$

The stability decreases as the donor atom is replaced by heavier atoms of the same group. Group 15 donors generally form stronger complexes than do Group 16 donors. The stabilities of complexes with Group 15 MX_5 Lewis acids increase with the size of the central atom. For Group 13, the anomalous behaviour of BMe_3 upsets a similar trend for the MR_3 Lewis acids.

Bond length and angle changes accompany the formation of a coordination compound compared to the values in the separate Lewis acids and bases. An example is shown in Fig. 2.43.

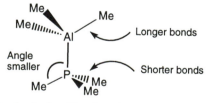

Fig. 2.43 A diagram showing the bond length and angle changes which occur when a coordination compound is formed

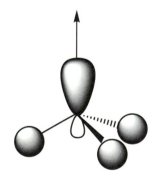

Fig. 2.44 A diagram illustrating the non-bonding molecular orbital. The orbital is slightly antibonding between the central and peripheral atoms and therefore electron donation from such an orbital strengthens the bonds slightly.

The trends associated with Lewis acids and bases are discussed in detail on page 117.

The changes can be interpreted in terms of the nodal properties of the non-bonding orbital of the Lewis base shown in Fig. 2.44. If the base is kept constant the relative Lewis acid strengths observed are given in Table 2.76.

Table 2.76 Orders of acid strengths for some Lewis acids

Base	Order of acid strength
NMe_3	$BI_3 \gg BCl_3 > BF_3$
NMe_3	$SbF_5 > AsF_5 > PF_5$
NMe_3	$BMe_3 < AlMe_3 > GaMe_3 > InMe_3 > TlMe_3$
pyridine	$AlCl_3 > AlBr_3 > AlI_3 > BBr_3 > BCl_3 > BF_3$

Stabilities of complexes

Metal cations also function very effectively as Lewis acids and are able to coordinate several Lewis bases simultaneously. Table 2.77 compares the stability constants of the Group 2 ions with simple anionic ligands. In these complexes the stabilities of the complexes is not large because the anionic ligands are merely replacing water molecules in the parent hydrated ions.

Table 2.77 Stability constants ($\log K_{LM}$) at 25°C for 1:1 complexes

	$CH_3CO_2^-$	$CH_2(CO_2^-)_2$	PO_4^{3-}	SO_4^{2-}
Mg^{2+}	1.25	1.95	1.60	2.20
Ca^{2+}	1.24	1.84	1.33	2.31
Sr^{2+}	1.19	1.25	1.0	2.3
Ba^{2+}	1.15	1.34		2.3

For $CH_3CO_2^-$, $CH_2(CO_2^-)_2$, and PO_4^{3-}, the stability constants generally decrease with increasing size of the central ion, in accordance with the expectations of an electrostatic model. However, the data for SO_4^{2-} emphasises the necessity of not drawing generalizations too widely since, for these complexes, the stability constant increases from Mg^{2+} to Ca^{2+} and then remains constant.

For polydentate ligands the chelate effect generally operates and the stability constants increase. Some representative data for oxalato, glycinato, and tartrato complexes (general structural formulae for these complexes are shown in Fig. 2.45) are given in Table 2.78.

Table 2.78 Stability constants ($\log K_{LM}$) at 25°C for 1:1 complexes

	Oxalato	Glycinato	Tartrato
Mg^{2+}	3.43	3.44	1.36
Ca^{2+}	3.00	1.43	1.80
Sr^{2+}	2.54	0.91	1.65
Ba^{2+}	2.31	0.77	1.62

The oxalato and glycinato complexes form complexes in a bidentate fashion and thereby generate five-membered chelate rings. For these ligands the stability constants fall as the ions become larger along the lines expected for an electrostatic model. The decrease is particularly marked for the glycinato complexes. For the tartrato complexes where the ligands have the ability to either wrap around the metal or indeed bridge between metal ions the stability ordering is retained for $Ca^{2+} > Sr^{2+} > Ba^{2+}$, but the Mg^{2+} shows an unexpectedly low stability. Although this behaviour creates problems in terms of providing simple periodic trends it has the very important consequence that the ligands may exhibit selective coordination towards specific metal ions. This has important implications in two areas; metal extraction and biological coordination and the transportation of metal ions. The selectivity associated with complex formation can arise from the following factors working either singly or in concert:

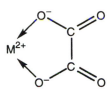

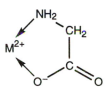

Oxalato complex

Glycinato complex

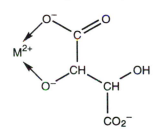

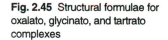

Tartrato complex

Fig. 2.45 Structural formulae for oxalato, glycinato, and tartrato complexes

1. The sizes and number of chelate rings formed between the ligand and the metal.
2. The ability of the ligand to wrap around the metal ion and the energy required to change the conformation of the ligand from that preferred in the uncoordinated state into the coordinated geometry.
3. The size of the cavity created in the ligand in its pre-coordination geometry relative to the size of the metal ion.
4. The number of donor atoms on the ligand and their donating abilities relative to the preferred coordination number of the metal.
5. The geometric preferences of the ligand and the metal ion.

As the chelate ring size increases from four to six members the angle subtended at the metal also increases. Specifically four-membered rings create angles at the metal of about 40°, five-membered rings 70–80° and six-membered rings 95–110°. As a metal ion becomes larger the preferred coordination number increases and the average angle between ligands becomes smaller. Consequently, a decrease in the sizes of the chelate rings leads to a selectivity towards the larger ion. Examples of chelate angles for four-, five-, and six-membered rings are shown in Fig. 2.46.

Table 2.79 Stability constants ($\log K_{LM}$) at 25°C for 1:1 complexes

	ionic radius/pm	Oxalato	Malonato	EDTA	TMDTA
Be^{2+}	31	4.96	6.18	9.2	10.7
Mg^{2+}	72	3.43	2.85	8.7	
Ca^{2+}	110	3.0	2.35	11.0	7.26
Sr^{2+}	117			8.63	5.28

Table 2.79 provides comparative data for oxalato and malonato complexes of the Group 2 metals. The chosen ligands are closely related, but the latter forms a larger chelating ring and therefore forms more stable complexes with the smaller ion. Table 2.79 also provides data for the hexadentate EDTA and TMDTA ligands which are virtually identical (see Fig. 2.47), but the latter forms a larger chelating ring with the nitrogen donor atoms. The TMDTA forms the more stable complex with the smaller Be^{2+} ion, which generally prefers four coordination, because it is capable of forming a larger chelating ring. In contrast EDTA which forms the smaller chelating rings forms more stable complexes with the larger Ca^{2+} and Sr^{2+} ions.

The importance of chelate ring size may also be illustrated by the stability order of the following complexes derived from dicarboxylic acids.

Oxalato	~	malonato	>	succinato	>	glutarato
$^-O_2C\text{-}CO_2^-$		$^-O_2CCH_2CO_2^-$		$^-O_2C(CH_2)_2CO_2^-$		$^-O_2C(CH_2)_3CO_2^-$
5-ring		6-ring		7-ring		8-ring

As the ligands actually chelate to the metal ion, the ring becomes progressively less stable as the ring size is increased beyond five. It is possible to form stable complexes with much larger rings containing 7–12 atoms if the appropriate geometric constraints are introduced into the rings.

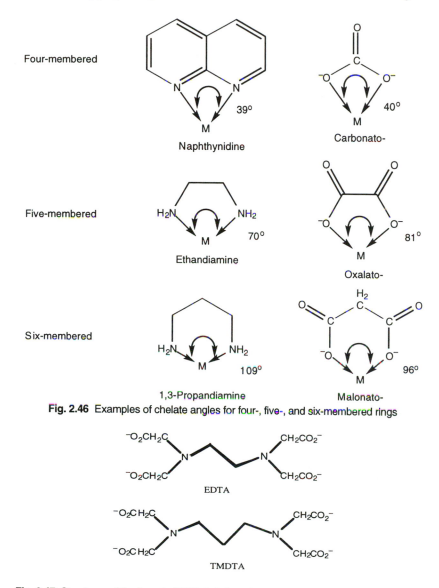

Fig. 2.46 Examples of chelate angles for four-, five-, and six-membered rings

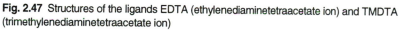

Fig. 2.47 Structures of the ligands EDTA (ethylenediaminetetraacetate ion) and TMDTA (trimethylenediaminetetraacetate ion)

The two ligands shown in Fig. 2.48 have similar skeletons, but the second example has an additional pair of oxygen donor atoms. These additional donor sites can be utilized most effectively for ions which prefer higher coordination numbers and therefore it functions in a more selective fashion for larger ions. Specifically, the complexes formed between these ligands and either copper or zinc have almost identical stability constants, but

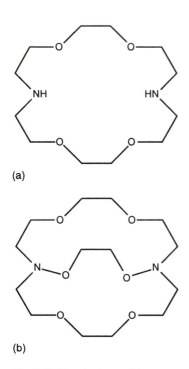

(a)

(b)

Fig. 2.48 The structures of the diaza-19-crown (a) and cryptand-222-diamine (b) ligands

For a more detailed discussion see R. D. Hancock and A. E. Martell, *Adv. Inorg. Chem.*, 1995, **42**, 89.

the ligand with additional oxygen atoms forms significantly more stable complexes with Ba^{2+}, Sr^{2+}, and Pb^{2+}.

Size selection is widely utilized in biology and ligands which mimic the ability of biomolecules to coordinate alkali metal ions include crown ethers and cryptands. These ligands offer a central cavity which accommodates a specific ion most efficiently. For example, the cryptand-222-diamine shows a selectivity towards the potassium ion which is underlined by the plot of equilibrium constant data shown in Fig. 2.49.

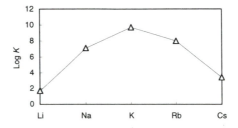

Fig. 2.49 A graph of the logarithm of the stability constants for the cryptand-222-diamine complexes of the Group 1 cations

The porphyrins and phthalocyanines, the unsubstituted basic structures of which are shown in Fig. 2.50, have even more rigid backbones and the central cavity is not only restricted in size, but also enforces a square planar geometry about the metal ion. Such ligands therefore form particularly stable complexes with metal ions which have the appropriate radii and prefer either square planar or tetragonally distorted octahedral geometries. The square planar preference is so large that ions which generally prefer tetrahedral coordination, e.g. Be^{2+} and Zn^{2+}, can be induced to take up planar geometries.

The selectivity based on geometric preferences may be illustrated by the data for trien and tren complexes given in Table 2.80. The structures of tren and trien are shown in Fig. 2.51.

Table 2.80 Stability constants ($\log K_1$) at 25°C for tren and trien complexes

	tren	trien
HL^+ (pK)	10.3	9.9
Ni^{2+}	14.8	14.0
Cu^{2+}	19.1	20.4
Zn^{2+}	14.7	12.1

Although tren is more basic than trien, Cu^{2+} forms more stable complexes with trien because the preference shown by Cu^{2+} to form planar and tetragonal octahedral complexes may be accommodated by trien, which can take up a planar geometry. The Zn^{2+} ion which forms tetrahedral complexes almost exclusively reinforces the basicity trend associated with the ligands.

For the later transition metals and the p block metals the complexes are generally more stable and the covalent contribution is greater. In these situations the stability constants generally increase down the column. Some typical data for mercaptoethanol and imidazole complexes are summarized in Table 2.81.

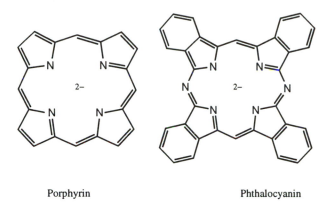

Porphyrin   Phthalocyanin

Fig. 2.50 The structures of the porphyrin and phthalocyanin −2 anions

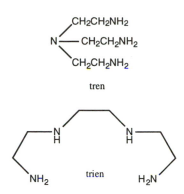

tren

trien

Fig. 2.51 The structures of the ligand molecules tren and trien

Table 2.81 Stability constants ($\log K_{LM}$) at 25°C for 1:1 complexes

	$^-SCH_2CH_2OH$	imidazole
Zn^{2+}	5.7	2.56
Cd^{2+}	6.1	2.80
Hg^{2+}	25.0	9.20

The later transition metals and the post-transition metals show the following ligand preferences:

$$PR_3 > NR_3 > AsR_3$$
$$SR_2 > OR_2 > SeR_2$$
$$I^- > Br^- > Cl^- > F^-$$

These metal ions are described as soft Lewis acids and show a preference for soft ligands. This generalization is discussed further in Chapter 5.

Kinetic factors

The observed chemical properties of a group of elements depends not only on the thermodynamic effects discussed above but also the relative rates of the reactions. Therefore, a comparative account of the chemistry of the elements requires some knowledge of the kinetics of the important general classes of reaction. It is difficult in this field to make generalizations, because in many cases the kinetic data have not been measured. Also the rate of a specific reaction, which may occur in several steps, depends on the rate of the slowest step which may be different for a group of common reactions. The mechanism of the reaction may also change down a group. However, it is useful to compare the rates for some common reaction types.

Rates of exchange in hydrated complexes $[M(OH_2)_6]^{x+}$

The rates of exchange of water molecules in such complexes may be represented by the following equation:

$$[M(OH_2)_6]^{x+} + O^*H_2 \rightleftharpoons [M(O^*H_2)(OH_2)_5]^{x+} + OH_2$$

Although this is a very simple reaction, it is representative of the more general class of ligand substitution reactions for metal complexes. For the non-transition elements the rates depend primarily on the charge/size ratio of the cation.

The data in Table 2.82 are illustrative of the general trends. In summary the rates of the exchange reactions increase going down a group and decrease with increasing charge.

Table 2.82 Rate constants, k, for exchange of OH_2 at 25°C

	k/s^{-1}		k/s^{-1}		k/s^{-1}		k/s^{-1}
$[Na(OH_2)_6]^+$	8×10^9	$[Mg(OH_2)_6]^{2+}$	1×10^5	$[Al(OH_2)_6]^{3+}$	1.8	$[Zn(OH_2)_6]^{2+}$	2×10^7
		$[Ca(OH_2)_6]^{2+}$	2×10^8	$[Ga(OH_2)_6]^{3+}$	1×10^3		
		$[Sr(OH_2)_6]^{2+}$	4×10^8	$[In(OH_2)_6]^{3+}$	2×10^5		

Rates of substitution in molecular complexes

The rates of substitution of molecular compounds of the post-transition elements also vary down the column. In general the rates of nucleophilic substitution increase from top to bottom. The reasons for this are two-fold. Firstly, the larger ions are able to expand their coordination numbers and therefore achieve an associative transition state more readily. Secondly, the central atom is becoming more electropositive and therefore attack by a nucleophile becomes more favourable. Table 2.83 summarizes the first-order rate constants for the reaction,

$$R_3MCl + {}^{36}Cl^- \rightleftharpoons R_3M^{36}Cl + Cl^-$$

in acetone/dioxan at 25°C (LiCl, 10^{-3} M), where R = C_6H_{14}.

Table 2.83 Rate constants for chloride ion exchange

	$k/10^3 \, s^{-1}$
R_3SiCl	0.007
R_3GeCl	40
R_3SnCl	4000

The comparative data for carbon are not given in Table 2.83 because R_3CCl undergoes nucleophilic substitution by a dissociative mechanism. Nonetheless, qualitative considerations suggest that in general carbon is more inert to nucleophilic substitution that the heavier elements in the group. The size of the central atom has an important influence on the mechanism of the reaction. Particularly noteworthy is the greater lability of the compounds as the central metal atom increases in size (making the expansion of the coordination number easier) and as the electronegativity of the metal decreases. Although the specific examples above indicate that the germanium compound reacts more quickly than the silicon compound, there are other examples where the rates are reversed. Qualitatively, it appears that carbon compounds are generally more inert than the corresponding compounds of Si and the heavier Group 14 elements. For example, CCl_4 does not hydrolyse readily, whereas the corresponding Si, Ge, and Sn compounds are increasingly susceptible to hydrolysis.

2.7 Modification of properties using steric and electronic effects

The previous sections has emphasised the relationships between chemical properties of compounds and the fundamental atomic properties of the central atom. One of the great strengths of modern chemistry is that the properties of a series of related compound MR_n may be significantly altered by varying the steric and electronic properties of the group R. Also the oxidation state of the central atom influences the properties of the compound.

Steric effects

Increasing the steric bulk of the group R can result in the following modifications in chemical properties.

For a general discussion of steric effects in inorganic and organometallic chemistry see D. White and N. J. Coville, *Adv. Organomet. Chem.*, 1994, **95**, 36.

Increased inertness of the compound

If the initial compound is sensitive to electrophilic or nucleophilic reagents then increasing the size of R makes the approach of the reagent towards the metal centre more difficult and the compound becomes more kinetically inert.

For example, $(Me_3Si)_3C-Zn-C(SiMe_3)_3$ may be steam-distilled whereas compounds such as $ZnMe_2$ and $ZnMe_2$ are extremely air sensitive and inflame in air. $(Me_3Si)_3CSiCl_3$ is stable towards water whereas $MeSiCl_3$ instantly hydrolyses.

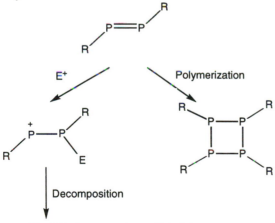

Fig. 2.52 Some reactions of RP=PR compounds

Kinetic trapping of multiply bonded compounds

Multiply bonded compounds of the heavier p block elements are thermodynamically less stable than the corresponding compounds of the second row elements and also more susceptible to attack by electrophiles and oligomerization processes which lead to rings and chains as shown in Fig. 2.52.

Such processes can be blocked if R is a large group and this strategy has led to the isolation and structural characterization of the multiply bonded compounds whose structures are shown in Fig. 2.53.

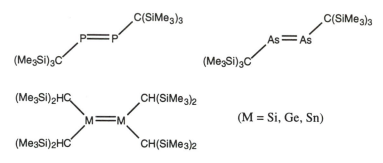

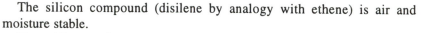

Fig. 2.53 Examples of multiply bonded compounds

The silicon compound (disilene by analogy with ethene) is air and moisture stable.

The trifluormethyl substituted phenyl ring illustrated in Fig. 2.54 is particularly effective at stabilizing compounds with low coordination numbers.

Extremely large groups are required when multiple bonds to atoms which have no substituents are to be kinetically protected as shown by the examples in Fig. 2.55.

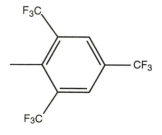

Fig. 2.54 The 2,4,6-tri(trifluoromethyl)phenyl group

The application of these steric principles has recently resulted in the first example of a Ga≡Ga bond, G. H. Robinson *et al.*, *J. Amer. Chem. Soc.*, 1997, **119**, 5471.

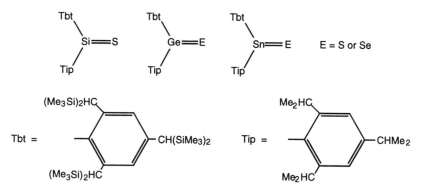

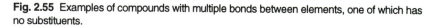

Fig. 2.55 Examples of compounds with multiple bonds between elements, one of which has no substituents.

Stabilization of low coordination numbers and radicals
Large groups limit the extent of polymerization and therefore can lead to the stabilization of unusual coordination numbers and valences. For example:

$$[(Me_3Si)_3C-Li-C(SiMe_3)_3]^-$$
Linear two-coordinate lithium

This may be contrasted with the Li_4Me_4 which has a tetrahedron of lithium atoms with face bridging methyl groups, as shown in Fig. 2.56.

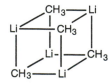

Fig. 2.56 The structure of $Li_4(CH_3)_4$

The lower-valent Group 14 compounds may similarly be stabilized as the examples given in Fig. 2.57 show.

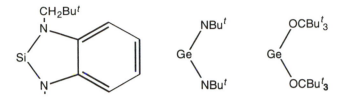

Fig. 2.57 Some lower-valent compounds of silicon and germanium

Sn[N(SiMe₃)₂]₃
Ge[N(SiMe₃)₂]₃
P[N(SiMe₃)₂]₂
As[N(SiMe₃)₂]₂

In addition kinetically inert radicals of these elements may be isolated using this strategy. Examples are given in the margin.

Stabilization of alternative geometries

It is energetically favourable to place sterically demanding ligands as far apart as possible and therefore where this conflicts with the preferred geometry of the central atom the shape of the compound may be altered by increasing the steric bulk of the ligand. For example, $Sn(\eta\text{-}C_5H_5)_2$ is angular whereas $Sn(\eta\text{-}C_5Ph_5)_2$ is linear.

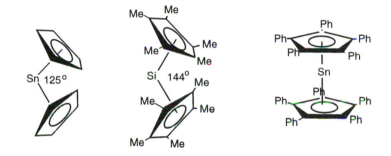

For a discussion of bulky cyclopentadienyl ligands see H. Schuman and C. Jaruach, *Adv. Organomet. Chem.*, 1991, **33**, 291.

Fig. 2.58 The structures of $Sn(\eta\text{-}C_5H_5)_2$, the angular form of $Si(\eta\text{-}C_5Me_5)_2$, and $Sn(\eta\text{-}C_5Ph_5)_5$

The equilibrium is so finely balanced for $Si(\eta\text{-}C_5Me_5)_2$ that both linear and angular forms may be isolated. The structures of $Sn(\eta\text{-}C_5H_5)_2$, the angular form of $Si(\eta\text{-}C_5Me_5)_2$, and $Sn(\eta\text{-}C_5Ph_5)_5$ are shown in Fig. 2.58.

Table 2.84 Geometries of some tertiary amines and ethers

Molecule	Geometry	Bond angle	Molecule	Geometry	Bond angle
Me₃N	Pyramidal	108°	Me₂O	Angular	111°
(H₃Si)₃N	Planar	120°	(H₃Si)₂O	Angular	148°
(H₃Ge)₃N	Planar	120°	(H₃Ge)₂O	Angular	140°
			(H₃Sn)₂O	Angular	140°

The electronic or steric origins of the observed geometries in the molecules given in Table 2.84 is less clear cut.

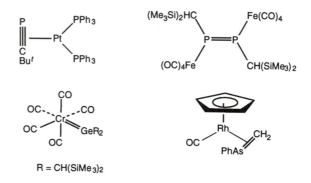

d_π–p_π bonding

Clearly the planar and linear geometries are preferred for the larger group, but also because of π-bonding between the filled p orbitals on N and O and either the d orbitals on Si, Ge, Sn or the antibonding σ orbitals between Si, Ge, and Sn and H.

Similarly, $N(Pr^i)_3$ and $N(SiMe_3)_3$ are planar rather than pyramidal—the geometry usually associated with trivalent nitrogen. Geometric modification has an interesting influence on the basicities of the compound. For example, $N(Pr^i)_3$ is much more basic than NMe_3 because the lone pair now occupies a p orbital rather than an sp^3 hybrid.

Stabilization of unstable compounds by complex formation

The occurrence of low valent and multiply bonded compounds of the type illustrated above has led to an exploitation of their coordination chemistry with transition metals. Some illustrative examples of metals stabilizing relatively otherwise unstable main group multiply bonded molecules are shown in Fig. 2.59.

Fig. 2.59 Examples of metal-stabilized main group multiply bonded molecules

Although, steric effects retard nucleophilic and electrophilic reactions at the central atom they can accelerate ring forming reactions by placing the atoms to be joined in the same region of space and with the appropriate conformation, i.e. thereby making the entropic contribution to the activation energy more favourable. For example, there is a well documented *gem*-dimethyl effect which places the substituents in the right conformation for ring formation. The cyclization rates given in the margin for the reactions shown in Fig. 2.60 provide a specific example.

	Cyclization rate
R = R′ = H	300 hours at 124°C
R = H, R′ = Me	81 hours at 80°C
R = R′ = Me	20 minutes at 20°C

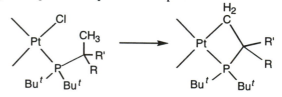

Fig. 2.60 Cyclization reactions for some platinum complexes

Electronic Effects

Acid-base properties

The properties of compounds may be influenced significantly by the donor or acceptor properties of the substituents bonded to the central atom. Table 2.85 summarizes the pK_a values for some organic acids in non-aqueous solvents. The dissociation of H^+ is promoted by electron withdrawing CN and NO_2 groups which stabilize the carbanion formed in the process.

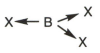

Fig. 2.61 σ-electron withdrawal from the boron atom to the more electronegative ligands; F > Cl > Br > I because of electronegativity differences

Table 2.85 pK_a values of organic acids in non-aqueous solvents

	pK_a		pK_a
$(NC)_3CH$	−5	Ph_3CH	30
$(O_2N)_3CH$	0	$PhCH_3$	35
$(O_2N)CH_3$	10	CH_4	~44

The boron halides BX_3 form complexes with NMe_3 and the stability order is $BBr_3 < BCl_3 < BF_3$. Mesomeric π-bonding effects are the dominant influence in much the same manner as that described above. The back donation from filled p orbitals on the halogen to the empty p orbital on boron is most effective for fluorine and least effective for bromine because of the better match in orbital size bonds to more effective overlap.

If the acceptor is a proton the basicities of phosphines and amines decrease in the orders:

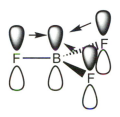

Fig. 2.62 π-back donation from F atoms to the central boron atom; F > Cl > Br > I because of better overlap

$$PMe_3 > PH_3 > PF_3$$
$$NMe_3 > NH_3 > NF_3$$

The Lewis bases, PR_3, become better donors as the electron donating properties of the R group increase as long as the Lewis acid is a simple acceptor and the steric effects are not important.

The rates of oxidative addition reactions such as the following:

$$PR_3 + XY \rightleftharpoons PR_3XY$$

$$XY = Cl_2, Br_2, I_2, RX$$

are increased by electron donating groups, R, and, for example, phenyl derivatives are less reactive than the corresponding alkyl compounds.

However, when the Lewis acid is a transition metal in a low oxidation state it is necessary not only to take into account the donor ability of the phosphine, but also its efficiency at accepting electron density from filled d orbitals on the metal, and the steric demands of the ligand. The orders of these three factors for some phosphine ligands are as follows.

σ-bonding ability

$$PBu'_3 > P(OR)_3 > PR_3 \approx PPh_3 > PH_3 > PF_3 > P(OPh)_3$$

π-accepting ability

$$PF_3 > P(OPh)_3 > PH_3 > P(OR)_3 > PPh_3 \approx PR_3 > PBu^t_3$$

Steric demand

$$PBu^t_3 > PPh_3 > P(OPh)_3 > PMe_3 > P(OR)_3 > PF_3 > PH_3$$

The observed equilibrium constant also reflects the steric demands of the ligand which are estimated using Tolman's cone angle concept (see Chapter 5).

For Lewis acids steric effects are also important as indicated by the enthalpies for complex formation of the pyridine and substituted pyridine complexes of boron trimethyl given in Fig. 2.63.

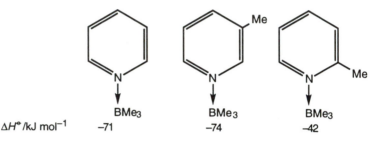

| ΔH°/kJ mol^{-1} | -71 | -74 | -42 |

Fig. 2.63 Enthalpies of complex formation for pyridine/substituted pyridine complexes of boron trimethyl

When π-bonding is less important then the substituent electronegativity is the most important factor in deciding the relative strengths of the Lewis acids. For example,

$$SiF_4 > SiCl_4 > SiBr_4 > SiI_4$$

If the halides are replaced by methyl or aryl substituents then the Lewis acidity is effectively removed. For example, in tin chemistry, $[SnCl_6]^{2-}$ may be formed from $SnCl_4$ and Cl^-, but the related $[SnMe_6]^{2-}$ anion is unknown. The Lewis acidities also have a direct influence on the rates of reaction of such compounds with moisture. Thus $SnCl_4$ is readily hydrolysed and fumes in air, whereas $SnMe_4$ is unreactive towards water under normal conditions. The mechanism of hydrolysis presumably involves initial attack by the water molecules on the tin atom. More generally the Lewis acidities of the molecules R_nMX_{4-n} follow the orders:

$$Si \approx Ge < Sn < Pb$$

$$I < Br < Cl < F$$

$$R_3MX < R_2MX_2 < RMX_3$$

The Lewis acidity of the metal centre can also have a profound influence on the structure adopted by the compound in the solid state. For example, Me_3SnBr, Me_3SnCl, and Me_3SnF have monomeric tetrahedral structures in the gas phase, but their relative Lewis acid strengths: $Me_3SnF > Me_3SnCl > Me_3SnBr$ influence their relative tendencies to aggregate in the solid state. This aggregation arises from the donation of lone pairs on the halides on other molecules to form bridges. The compound Me_3SnBr remains essentially molecular, Me_3SnCl forms weak Sn–Cl–Sn bridges and Me_3SnF forms an infinite linear polymer based on Sn–F–Sn bridges and trigonal bipyramidal coordination environments as shown in Fig. 2.64. These structural differences have a profound influence on the m.ps and b.ps of this series of compounds as indicated by the data given in Table 2.86.

Table 2.86 Melting and boiling points of Me_3SnX (X = Cl, Br or I) and their structures

	m.p.(°C)	b.p.(°C)	Structure
Me_3SnF	375 decomp.	–	One-dimensional chains
Me_3SnCl	42	154	Molecular
Me_3SnBr	27	163.5	Molecular
Me_2SnF_2	400 decomp.	–	Two-dimensional sheet
$MeSnF_3$	321–7		Three- dimensional infinite structure

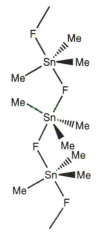

Fig. 2.64 The structure of polymeric $[Me_3SnF]$

As the number of halide ligands is increased then not only is the Lewis acidity increased but also the availability of donor sites on the halogen. For example, Me_2SnF_2 forms a two-dimensional structure (shown in Fig. 2.65), and $MeSnF_3$ has a three-dimensional infinite structure.

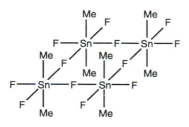

Fig. 2.65 A portion of the infinite 2-dimensional structure of $[Me_2SnF_2]$

The ability to form bridges and their stabilities depends on the bridging groups, for example, for the simple dissociation process

$$[R_2AlX]_2 \rightleftharpoons 2R_2AlX$$

which is shown for R = X in Fig. 2.62. The dimer stability follows the order:

$$X = R_2N > RO > Cl > Br > PhC{\equiv}C > Ph > Me > Et > Pr^i > Bu^t$$

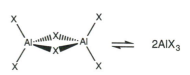

Fig. 2.66 The dissociation equilibrium of Al_2X_6

The donor abilities for the groups and the reorganization energies involved in transforming the trigonal coordination environment in the monomer into the tetrahedral environment in the dimer are both important factors.

Electronegative substituents can often stabilize organometallic compounds by strengthening the metal-carbon bond, e.g. $Sn(CF_3)_4$ is more stable than $SnMe_4$. This influence has been used to greatest effect to form the first examples of organoxenon compounds. The compounds $[Xe(C_6F_5)(NCCH_3)]^+$ and $Xe(CF_3)_2$ are known, although the corresponding C_6H_5 and CH_3 derivatives are too unstable to be isolated.

Radical reactions

The formation of radicals depends primarily on the strength of the bond being ruptured and this can be influenced by the steric and electronic properties of the substituents. For example, in organolead compounds PbR_4, the thermal stabilities which affect the relative abilities of the compounds to form organo-radicals follows the order:

$$PhPh_4 > PbMe_4 > PbEt_4 > PbPr^i_4$$

Repulsive effects between sterically demanding groups can encourage the dissociation of homolytic metal-metal bonds as in the following example.

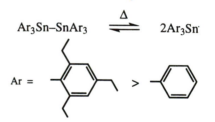

Effect of oxidation state

The chemical and physical properties of compounds are influenced by the oxidation state of the central atom. Table 2.87 contrasts the properties of some chlorides of the Group 14 metals.

The following differences are apparent.

1. The higher oxidation state halides are molecular in nature and retain the tetrahedral structure in the liquid and solid phases. Therefore, they are low melting point liquids.

2. In this oxidation state the Ge, Sn, and Pb atoms are sufficiently large to expand their coordination numbers and therefore they are all reasonably strong Lewis acids. This property also makes them susceptible to nucleophilic attack by water and therefore they are readily hydrolysed and fume in air.

3. The compounds in the lower oxidation state have both a lone pair and an empty p orbital and are therefore amphoteric as Lewis acids/bases.

4. The ions are sufficiently Lewis acidic that in the solid state the compounds adopt polymeric structures, hence their higher melting points.

Table 2.87 Some properties of the chlorides of Ge, Sn, and Pb

Compound	Melting point (°C)	Properties
$GeCl_4$	–50	Colourless liquid, fumes in air, good Lewis acid. Molecular.
$GeCl_2$	Decomposes to $Ge + GeCl_4$	Colourless solid, hydrolyses slowly in H_2O. Amphoteric.
$SnCl_4$	–33	Fuming liquid, very readily hydrolysed, soluble in organic solvents. Molecular. Very strong Lewis acid.
$SnCl_2$	246	White compound, soluble in water to form a molecular hydrate, $SnCl_2(OH_2)_2$. Soluble in polar organic solvents. Amphoteric. In solid state forms $SnCl_3$ based chains.
$PbCl_4$	–15	Yellow oil which fumes in air. Molecular. Soluble in organic solvents. Good Lewis acid.
$PbCl_2$	501	Infinite structure with 9-coordinate Pb. Colourless solid. Soluble in H_2O and polar organic solvents.

5. The lower oxidation state compounds either only hydrolyse slowly, e.g. Ge, or form hydrates, e.g. Sn and do not fume in air, because they are not sufficiently acidic to release HCl.

Similar property trends are observed for other compounds of the Group 13-15 elements. For example, the oxides become more strongly acidic as the oxidation state is increased. Also, the higher oxidation state compounds are generally more strongly oxidizing (or less strongly reducing) than the corresponding lower oxidation state compounds. For example, Pb^{IV} compounds are more strongly oxidizing than Pb^{II}, and Sn^{IV} compounds are less strongly reducing than Sn^{II} compounds.

For the later groups of the Periodic Table the higher oxidation compounds become more stable as the group is descended and the generalizations developed above have to be modified somewhat. Table 2.88 contains a summary of some of the properties of the Group 16 and 17 fluorides and oxides.

Although the highest oxidation state fluorides have a significant residual positive charge on the central atom the compound SF_6 is inert to hydrolysis and IF_7 is relatively inert. In these compounds the high coordination numbers are sufficiently close to the point of saturation that nucleophilic attack by water is kinetically hindered.

These compounds are also less effective fluorinating agents than the lower fluorides, e.g. IF_5 is a stronger fluorinating agent than IF_7. When the coordination number is reduced, but the oxidation state maintained then the compounds become much more susceptible to hydrolysis, e.g. SO_3 vs. SF_6. Also the compounds become more reducing as the oxidation state is reduced, e.g. SO_2 vs. SO_3.

The standard heats of formation for the iodine fluorides provide a thermodynamic basis for the qualitative observation that the fluorides of iodine are more stable as the oxidation state increases.

Table 2.88 Some properties of iodine and sulfur fluorides and oxides

Compound	Melting point ($^\circ$C)	$\Delta_f H^\ominus$ /kJ mol^{-1}	Properties
IF_7	5	−962	Molecular vapour which slowly hydrolyses. Strong fluorinating agent
IF_5	9	−847	Liquid, molecular, very strong fluorinating agent. Reacts violently with water
IF_3		−500	Yellow solid, decomposes above -30°C. Reacts very violently with water
IF		−95	Dark solid, disproportionates above −40°C. Decomposes with water
SF_6	−50		Colourless, non-flammable gas. Stable to hydrolysis
SF_4	−121		Readily hydrolysed. Reacts very vigorously with H_2O
SF_2			Colourless very unstable gas which readily disproportionates to SF_4 and S
SO_3	17		Colourless solid/liquid which reacts explosively with water to form sulfuric acid. Oxidizing
SO_2	−76		Colourless pungent gas. Mild reducing agent, amphoteric Lewis acid/base. Soluble in H_2O
SO			Short lived, unstable diradical, only stable in matrices or in plasmas

Important parameters

1. Atomic sizes of atoms and ions

2. Ionization energies

3. Electron capture enthalpies

4. Polarizabilities and polarizing powers of ions

5. Electronegativities

6. Bond enthalpies of single and multiple bonds

7. Relative contributions of s and p orbitals to the chemical bonds

This chapter demonstrates that the Mendeleev-based Periodic Table, when combined with the quantum mechanical description of the atom, provides a useful framework for understanding modern inorganic chemistry. However, the Periodic Table is not a Rosetta stone which provides a simple translation of all the properties of all the elements, but resembles a family tree. It sets out the basic family relationships, which in a chemical context means the common electronic configurations of columns of atoms. An appreciation of the relationships between the atomic properties and the chemical and physical properties of the elements and their compounds requires further work. Specifically, it is important to understand how the atomic and molecular parameters summarized in the margin vary for columns of atoms. The chemical and physical properties of a group of compounds depends on a subtle interplay of these parameters. By focusing on specific properties the approach used in this chapter represents an attempt to outline how a modern chemist uses the available data on the atomic properties to interpret trends and account for exceptions. The procedures are neither rigorous nor quantitative but their sensible application gives the chemist a pragmatic description of the relationship between the trends and the atomic parameters which currently seem most relevant for their interpretation. The day may come when all the properties of interest could be calculated using accurate quantum mechanical computer programmes, but it is still a little way off.

In recent years chemists have been able to manipulate, as never before, the properties of a class of compounds by varying the electronic and steric properties of the substituents attached to the central atoms. This chemical fine-tuning has added a new dimension to the Periodic Table and has allowed chemists to modify the properties of a series of compounds in a much more predictable manner.

2.8 Tables of properties and summaries of trends of the elements of Groups 1, 2, and 13 to 17

The previous sections have attempted to illustrate how the vertical trends associated with the elements and their compounds may be interpreted using the fundamental properties of atoms. It must be remembered, however, that chemists do not set out to discover trends—their primary goal is to discover new compounds and to study the physical and chemical properties of the compounds once they have been isolated in a pure state. It is only possible to discuss trends associated with the relative stabilities, boiling points, geometries, etc., of compounds after the analogous compounds of a periodic group of elements have been isolated. Also, the application of a compound either for a specific commercial process or as a reagent in an organic or inorganic synthetic procedure needs a very specific knowledge of the particular properties of that compound. Therefore, a familiarity with the detailed properties of compounds is an inherent feature of inorganic chemistry. The following sections therefore provide detailed information on the properties of the important compounds of the s and p block elements.

The information has been organized in a tabular form in order to assist the reader in making comparisons between elements of the same Periodic Group. Each table is accompanied by a short commentary which summarizes the important general trends for that group of elements. It is hoped that this combination of practical information and broad trends will assist in the assimilation of the essential factual basis of the subject. Finally, the industrial processes leading to the 20 most widely used inorganic chemicals and their applications are summarized in order to emphasize the importance of inorganic chemicals in the outside world.

Summary of Group 1 trends

1. All the metals are malleable and they become softer down the column. Lithium is sufficiently hard that it is necessary to cut it with a knife, but rubidium and caesium have the consistency of putty.

2. The reactivities of the metals towards O_2 and water increases down the column. Lithium reacts slowly with water, sodium vigorously, potassium causes the evolved hydrogen to ignite, and rubidium and caesium react explosively. When Li is burned in oxygen (1 atmosphere) it forms Li_2O primarily, sodium forms the peroxide, Na_2O_2, and K, Rb, and Cs form superoxides, MO_2.

3. Lithium does not replace the proton in $PhC{\equiv}CH$ whereas the remaining elements do.

4. They all form a wide range of salts which exhibit typical ionic properties—they have high melting points and are water soluble, giving conducting solutions. The halides are ionic and are not hydrolysed. The oxides and hydroxides are basic and the hydrides are ionic, basic, and strong reducing agents.

5. The thermal stabilities of the carbonates, nitrates, sulfates, peroxides, and superoxides increase down the column.

6. The reverse order of reactivity is observed for the reactions of the alkali metals with N_2 and carbon. Lithium is unique in reacting with N_2 to form the purple nitride, Li_3N, and only Li and sodium react with carbon to form the carbides M_2C_2 (M = Li or Na). Both of these reactions are thermodynamically favourable because of the high lattice enthalpies of the nitride and carbide anions associated with small cations.

7. The high charge/size ratio for lithium leads to other anomalies, e.g. LiH is more thermally stable than the other hydrides of the Group; the carbonate is much less stable. Also, lithium salts are much more soluble in organic solvents. Lithium and sodium form a number of hydrated solid salts, potassium forms a few and Rb and Cs none.

8. The solubilities of the hydroxides in water increase down the column:

$$LiOH < NaOH < KOH < RbOH < CsOH$$

9. Alkali metals do not readily form complexes with ligands such as NH_3 and CN^-, but polydentate ligands such as crown ethers and cryptands with oxygen and nitrogen donor atoms do form stable complexes. The bonding in the complexes is primarily electrostatic and the relative sizes of the cation and the cavity is important. For example, for complexes with the 18-crown-6 ether the order of stability is:

$$Li^+ < Na^+ < K^+ > Rb^+ > Cs^+$$

Smaller ring sizes lead to a preference for coordination to Na^+ and ultimately Li^+ (12-crown-4 and 15-crown-5) and larger ring sizes for Rb^+ and Cs^+ (24-crown-8).

10. Lithium alkyls are widely used as reagents in organic chemistry because they are readily synthesized, they are soluble in organic solvents and they provide a slightly more reactive source of carbanions than Grignard reagents. These properties arise because the alkyl lithium compounds form oligomers in the form of rings or polyhedra which have the hydrocarbon residues dominating the surface. In contrast, sodium and potassium alkyls are more ionic and generally adopt infinite 3-D structures in the solid state. They are not soluble in hydrocarbons and are extremely air and moisture sensitive.

Anomalous properties associated with lithium

The high enthalpy of atomization of lithium metal, the high hydration enthalpy of Li^+ and its small size make it stand out from the other members of the series in the following ways.

1. Lithium reacts very slowly with water, yet it is the only alkali metal to form a nitride with N_2.

2. Lithium salts are generally less soluble in water, but more soluble in organic solvents.

Group 1

	Lithium	Sodium
Physical properties	White metal, least dense known; soft, but hardest of the group.	Soft, fusible, volatile metal; denser than lithium, but softer.
Flame test	Carmine red	Yellow
Action of air		
(a) cold	Tarnishes	Tarnishes quickly owing to oxidation; white coating of hydroxide formed in moist air. Stored in oil.
(b) when heated	Burns at 200^oC, to form Li_2O, with dazzling flame.	Burns readily to form Na_2O and Na_2O_2.
Action on water	Slow reaction yielding LiOH and hydrogen.	Vigorous reaction, does not ignite if in small quantity .
Oxide	Li_2O—reacts slowly with water to LiOH; m.p. 1570^oC, anti-fluorite structure.	Na_2O—white deliquescent solid, combines vigorously with water, m.p. 1132^oC, anti-fluorite structure. Component of glasses and fluxes.
Hydroxide	LiOH—white crystalline solid, m.p. 471^oC, not deliquescent and not very soluble in water, hygroscopic, absorbs CO_2 from air. Sparingly sol. in EtOH.	NaOH—strongly alkaline and caustic substance, very soluble and deliquescent. Stable to heat. Absorbs CO_2 from air, very sol. in EtOH. Very important industrial alkali. Thalium iodide structure type, m.p. 323^oC.
	All strongly alkaline and caustic substances, v. sol. and deliquescent.	
Nitride	Li_3N—formed by direct combination of elements at 800^oC. Purple hexagonal crystals. Very water sensitive, liberating NH_3.	Nitride not formed.
Chloride	LiCl—deliquescent (good desiccant) and very soluble in water; forms hydrates. Sol. EtOH and acetone, m.p. 610^oC, NaCl structure.	NaCl—anhydrous above 0.15^oC, stable, and not deliquescent, m.p. 801^oC.
Sulfate(VI)	$Li_2SO_4.H_2O$—fairly soluble in water, insol. EtOH, m.p. 859^oC.	Na_2SO_4—hygroscopic, very soluble in water, used as drying agent as solid, m.p. 884^oC.
Nitrate(V)	$LiNO_3$—decomposes to oxide when heated.	$NaNO_3$—very soluble in water. Sol. EtOH. Decomposes on heating into $NaNO_2 + O_2$.
Carbonate	Li_2CO_3—white solid, anhydrous and only sparingly soluble in water. Decomposes on strong heating to Li_2O and CO_2.	Na_2CO_3—forms hydrates, chiefly $Na_2CO_3.10H_2O$ efflorescent. Very soluble in water and on heating, m.p. 850^oC. Decomp. to Na_2O. Sp. sol. EtOH. Used in paper, soap, glass industries.

Potassium	Rubidium	Caesium
Soft, fusible, volatile metals like lithium, increasing in density and softness, but decreasing in melting point.		

Lilac	Red-violet	Like Rb

Tarnish quickly owing to oxidation; white coating of hydroxide formed in moist air. Stored under oil.

Burn with increased readiness to peroxides and superoxides.

Very vigorous, the hydrogen igniting at once.	Explosive reaction.	
K_2O—like Na_2O. Anti-fluorite structure. m.p. 740°C, hygroscopic, v. sol. water, sol. EtOH.	Rb_2O—like Na_2O. Anti-fluorite structure, pale yellow. m.p. 400°C (decomp.).	Cs_2O—anti-fluorite structure, orange, m.p. 490°C.
KOH—β-TlI structure, deliquescent, v. sol. water, sol. EtOH, reacts readily with CO_2.	RbOH—deliquescent, v. sol. water, m.p. 301°C, only absorbs CO_2 at high pressures.	CsOH—deliquescent pale yellow, v. sol. water, m.p. 315°C, only absorbs CO_2 at high pressures.

All strongly alkaline and caustic substances, very soluble and deliquescent. Stable to heat.

Nitrides not formed from N_2 and metal.

KCl—deliquescent, v. sol. water, sp. sol. EtOH. m.p. 773°C, NaCl structure.	NaCl structure, m.p. 718°C v. sol. water, sp. sol. EtOH. Forms polyhalides in solid state.	CsCl, structure m.p. 646°C v. sol. water, sol. EtOH. Forms polyhalides in solid state.
K_2SO_4—like Na_2SO_4, but forms no hydrates and is not very soluble in water.	Rb_2SO_4—m.p. 1060°C v. sol. water.	Cs_2SO_4—m.p. 1005°C hygroscopic, v. sol. water, insol. EtOH.
KNO_3—saltpetre, very soluble in hot water, but much less so in cold. On heating gives $K_2O + NO + O_2$.	$RbNO_3$—like KNO_3. v. sol. water, sp. sol. acetone. m.p. 310°C.	$CsNO_3$—like KNO_3. v. sol. water, sol. acetone. m.p. 414°C.
K_2CO_3—hygroscopic and very soluble in water, insol. EtOH. Decomp. 901°C to give oxide.	Rb_2CO_3 m.p. 837°C. Like K_2CO_3—very soluble in water and not decomposed by heat.	Cs_2CO_3—deliquescent, sol. in water, decomp. 610°C with CO_2 loss. Like K_2CO_3—very soluble in water and not decomposed by heat.

3.   Its salts are much less thermally stable because, for lithium, the lattice energy of the oxide which is formed is much greater than those of the original salts.

$$Li_2CO_3 \xrightarrow{700^{\circ}C} Li_2O + CO_2 \quad (Na, K,... \text{ reaction} > 800^{\circ}C)$$

$$2LiOH \rightarrow Li_2O + H_2O \quad (Na, K,... \text{ no reaction})$$

4.   The greater electronegativity of lithium leads to the formation of less polar organometallic compounds with oligomeric structures, rather than the 3-D infinite structures for the heavier elements of the Group.
5.   The high charge/size ratio of Li^+ leads to more stable complexes with simple ligands containing oxygen and nitrogen donors.

Summary of Group 2 trends

1.   Beryllium is the least reactive and does not react with water even at red heat and does not react with N_2. Magnesium only reacts at a reasonable rate with steam, calcium and strontium readily tarnish in moist air, and barium tarnishes readily.
2.   The ionic character of the compounds increases down the Group. Beryllium forms highly covalent compounds generally with sp^3 tetrahedral geometries, e.g. $[BeF_4]^{2-}$, $[BeCl_2]_\infty$ (infinite linear polymer). Magnesium forms more polar compounds with 6-coordination. Calcium, strontium, and barium form increasingly ionic compounds with higher coordination numbers (8 is particularly common).
3.   The organometallic compounds of beryllium are covalent and rather unreactive. Magnesium forms two important series of organometallic compounds, $RMgX$ and R_2Mg, which provide convenient sources of carbanions for organic syntheses. The organometallic compounds of calcium, strontium, and barium are generally more reactive and are insoluble in organic solvents.
4.   The oxides become progressively more basic down the Group. BeO and $Be(OH)_2$ are amphoteric and react with acids and strong bases such as NaOH. MgO is basic and $Mg(OH)_2$ weakly basic and do not dissolve in NaOH solution. The oxides of calcium, strontium, and barium are basic and their hydroxides are strongly basic. The solubilities of the hydroxides in water follow the order:

$$Ba(OH)_2 > Sr(OH)_2 > Ca(OH)_2 > Mg(OH)_2$$

5. BeX_2 (X = F, Cl, Br or I) are covalent polymers which are readily hydrolysed and are Lewis acids forming adducts BeX_2L_2 (L = Lewis base). Magnesium, calcium, strontium, and barium halides are essentially ionic and are soluble in water.

6. BeH$_2$ is a covalent polymer, magnesium hydride is partially ionic and the hydrides of calcium, strontium, and barium are very ionic and hydridic in their properties.

7. Mg^{2+} and Ca^{2+} show the greatest tendency to form complexes especially with ligands which have oxygen donor atoms. For small highly charged anions the order of stability is generally:

$$Mg^{2+} > Ca^{2+} > Sr^{2+} > Ba^{2+}$$

but for the anions, NO$_3^-$, SO$_4^{2-}$, and IO$_4^-$ the stability order is:

$$Ba^{2+} > Sr^{2+} > Ca^{2+} > Mg^{2+}$$

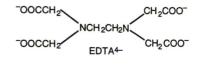

The most important complexes of these metals are with EDTA^{4-}. The order of stability for this and related polydentate ligands is:

$$Mg^{2+} < Ca^{2+} > Sr^{2+} > Ba^{2+}$$

The calcium complex, [Ca(edta)]$^{2-}$, is particularly important because it is water soluble and allows edta to solubilize calcium carbonate. Polyphosphates, e.g. P$_2$O$_7^{4-}$ and P$_3$O$_{10}^{5-}$ are able to function similarly to solubilize hard water deposits of CaCO$_3$.

The crown polyethers and cryptate ligands also form stable complexes with Ca^{2+} and Mg^{2+}.

8. Both magnesium and calcium have important biological roles. The former occurs in chlorophyll in which it is coordinated to a tetrapyrrole molecule and the latter has an important structural role cross-linking polymer chains, e.g. in collagen.

Because of their important biological roles Ca^{2+} and Mg^{2+} have to be transported through cell membranes and ligands which wrap round the metal ions and retain a hydrophobic surface are able to achieve this, e.g. the *l*-aspartate complex of Mg^{2+}.

9. The thermal stabilities of the nitrates, carbonates, and peroxides increase down the column.

10. The solubilities of the sulfates, nitrates, and chlorides increase down the group.

11. The solubilities of the halides in alcohols increase down the group.

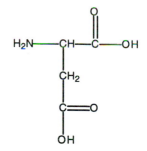

l-aspartic acid

Anomalous nature of beryllium

In common with lithium, and for the same underlying causes, beryllium exhibits some anomalous properties.

1. The high enthalpy of atomization of beryllium causes it to be harder, higher melting, more dense, and less reactive than the heavier members of the group.

Group 2

	Beryllium	Magnesium
Physical properties	Soft, silvery white lustrous metal. Brittle at room temperature, ductile at red heat. Highly toxic.	White, strong, malleable, light.
Flame test	No colour	No colour
Action with air		
(a) Cold	Unreactive	Not affected by dry air, but is corroded by moist air.
(b) When heated	Ignites only as fine powder to give Be_3N_2 and BeO.	Burns, brilliant white flame to MgO.
Reaction with water	Only reacts above red heat. Dissolves readily in non-oxidizing acids and alkalis. Passivated by oxidizing acids.	Slow in the cold, much more quickly when boiling, forming $Mg(OH)_2$. Heated metal burns in steam to MgO.
Oxide	BeO—very refractory m.p. $2570^{\circ}C$ and inert. Amphoteric, dissolves in conc. H_2SO_4 and fused KOH. Sparingly soluble in water.	MgO—white infusible powder. Basic, not reduced easily.
	Enthalpy of hydration becomes more negative $\longrightarrow$	
Peroxide	Non-existent	
Hydroxide	$Be(OH)_2$—formed on boiling $BeCl_2$ + OH^-. Insoluble in water. Dissolves in alkalis to give $Be(OH)_4^{2-}$ and carboxylic acids to give $[Be_4O(O_2CR)_6]$.	$Mg(OH)_2$—white powder, slightly soluble in water and alkaline. Soluble in aqueous NH_4Cl.
Chloride	$BeCl_2$—colourless and deliquescent. Lewis acid.	$MgCl_2.6H_2O$—deliquescent and very soluble in water.
Carbonate	Not stable as $BeCO_3$. Can only be obtained as hydroxycarbonate.	$MgCO_3$—white insoluble powder. Soluble in aqueous CO_2 solution and in ammonium salts. Decomposes on heating to MgO and CO_2.
Sulfate(VI)	$BeSO_4$—colourless crystals. Decomp. to BeO > $635^{\circ}C$. Sol. in water.	$MgSO_4.7H_2O$—Epsom salt, soluble in water.
Nitrate(V)	$Be(NO_3)_2$—deliquescent white amorphous solid soluble in water.	$Mg(NO_3)_2.6H_2O$—very soluble in water. On heating gives the oxide, NO_2 and O_2.

Calcium	Strontium	Barium
White, fairly hard, very light	White, light (like calcium), malleable silvery metal	Fairly soft white metal
Red	Carmine-red	Apple green
Tarnishes in moist air	Becomes yellow bronze in air due to oxide coating	Tarnishes readily in air. In powder form ignites spontaneously
Burns, reddish light, to CaO	Burns to SrO	Burns readily to BaO
Readily evolves hydrogen and forms $Ca(OH)_2$ with cold water	Slightly more vigorous than calcium	More vigorous than strontium
CaO—white amorphous with very high m.p. Reacts with most non-metals on heating. With water reacts vigorously to form $Ca(OH)_2$.	SrO—resembles CaO. Reacts vigorously with water.	BaO—lower m.p. than CaO. Basic and reactive. Combines vigorously with water and CO_2. Forms BaO_2 when heated in air.
	Enthalpy of hydration becomes more negative $\longrightarrow$	
CaO_2—prepared from H_2O_2 and $Ca(OH)_2$. Octahydrate exists. Like BaO_2 but not as stable	SrO_2—preparation and properties similar to CaO_2	BaO_2—prepared by heating BaO in air. Evolves O_2 at red-heat. With acids in the cold forms H_2O_2.
$Ca(OH)_2$—white amorphous powder slightly soluble in water to give "lime-water". Has the properties of an alkali.	$Sr(OH)_2$—resembles $Ca(OH)_2$, but is more soluble in water.	$Ba(OH)_2$—forms a crystalline hydrate. Fairly soluble in water to give a strongly alkaline solution.
$CaCl_2.6H_2O$—deliquescent and extremely soluble in water. Anhydrous salt very useful for drying liquids and gases.	$SrCl_2.6H_2O$—very soluble in water, but much less deliquescent than the calcium salt. Anhydrous salt less hygroscopic than $CaCl_2$.	$BaCl_2.2H_2O$—less soluble in water than $SrCl_2$. Neither hygroscopic nor deliquescent.
$CaCO_3$—exists in crystalline or anhydrous forms. Insoluble in water, but dissolves in CO_2 solution. Decomposed to oxide at high temperatures.	$SrCO_3$—even less soluble than $CaCO_3$. Does not decompose until $1200^\circ C$. Dissolves in an aqueous solution of CO_2.	$BaCO_3$—very insoluble in water. Requires a slightly higher temperature than does $SrCO_3$ for decomposition. Dissolves in an aqueous solution of CO_2.
$CaSO_4$—white powder slightly soluble in water. Stable to heat. Forms two hydrates; $CaSO_4.0.5H_2O$ —Plaster of Paris, $CaSO_4.2H_2O$ —gypsum.	$SrSO_4$—less soluble in water than $CaSO_4$ and does not form a hydrate.	$BaSO_4$—very insoluble in water. Fairly soluble in conc. H_2SO_4. Like Ca and Sr sulfates, is reduced to sulfide by red-hot carbon.
$Ca(NO_3)_2.4H_2O$—deliquescent and very soluble in water. Decomposed at red-heat into CaO, NO_2, and O_2.	$Sr(NO_3)_2.4H_2O$—slightly deliquescent and soluble in water. Decomposes on heating like $Ca(NO_3)_2$.	$Ba(NO_3)_2$—does not form a hydrate – not very soluble in water and not deliquescent.

2. The high size/charge ratio of Be^{2+} leads to compounds which are more covalent and complexes which are more stable than those of the remaining Group 2 cations. Many of its compounds have anomalously low melting points, enthalpies of formation, and are more soluble in organic solvents. The compounds are stronger Lewis acids. The halides are hygroscopic and fume when exposed to moist air.

3. Unlike magnesium and the heavier metals of the group beryllium oxide and hydroxide are amphoteric.

4. Beryllium salts are much less thermally stable because the high lattice energy of the oxide lowers the Gibbs energy change for the decomposition reaction. Similarly it does not form a peroxide or superoxide. With ethyne, beryllium forms the carbide, Be_2C, rather than the ethynide presumably because the lattice energy of the carbide is very favourable.

Summary of Group 13 trends

1. The chemistry of boron is quite different from that of the heavier Group 13 elements. It differs from aluminium in the following ways.

(a) Its oxide and hydroxide are acidic whereas those of aluminium are amphoteric.

(b) Boron is a semi-conductor which has various polymorphs based on icosahedral boron cages, whereas aluminium is a metal with a close packed structure. Boron is very inert and only attacked by hot concentrated oxidizing acids.

(c) No simple salts of B^{3+} are known, whereas those of Al^{3+} are numerous and well documented.

(d) Boron forms a wide range of hydrides which have cage structures. $(AlH_3)_4$ has a polymeric structure which resembles that of AlF_3.

(e) The stereochemistries of many boron compounds are based on trigonal sp^2 and tetrahedral sp^3 geometries. In the latter the octet rule is obeyed. Aluminium forms many compounds with tetrahedral, trigonal bipyramidal, and octahedral geometries.

(f) Multiple $p_\pi-p_\pi$ bonding in boron–nitrogen, boron–oxygen, and boron–fluorine compounds is much more significant than that for the corresponding aluminium compounds. $\{BN\}_x$ for example adopts a graphite structure.

2. Aluminium, gallium, indium, and thallium all form a range of compounds in the +3 oxidation state and compounds in the +1 oxidation state become progressively more stable down Group 13.

3. The oxides of aluminium and gallium are amphoteric and indium and thallium oxides are more basic.

4. The octahedral aqua-ions $[M(OH_2)_6]^{3+}$ are acidic and the pK_a values for the equilibria:

$$[M(OH_2)_6]^{3+} \rightleftarrows [M(OH_2)_5(OH)]^{2+} + H^+$$

are Al, ~5; Ga, 3; In, ~4; Tl, 1; showing that the Al ion is least acidic and the Tl ion the most acidic.

5. The MX_3 compounds are Lewis acids and the Lewis acid strengths decrease in the order: Al > Ga > In.

6. The stability of the hydrides decreases down the Group and there are no stable Tl–H compounds. Extraordinary precautions are required to exclude air and moisture in order to isolate Ga_2H_6.

7. Aluminium is resistant to corrosion because of an impermeable oxide layer, but is soluble in non-oxidizing mineral acids. Gallium, In, and Tl dissolve readily in acids, but thallium dissolves only slowly in H_2SO_4 and HCl.

8. AlN, GaN, and InN have wurtzite structures analogous to cubic BN, but show no analogue of the graphite-type structure of BN.

Summary of Group 14 trends

1. *Stabilization of (+2) oxidation state relative to (+4) of the elements down the group.*

    The halides, oxides, and sulfides of the M^{2+} ions become more stable on descending the group. For example $SiCl_4$, $SiBr_4$, and SiI_4 are all stable. $PbCl_4$ decomposes at 105°C and PbI_4 does not exist. Similarly the ease of oxidation of the M^{2+} halides increases down the column.

    $PbCl_2$ may only be converted to $PbCl_4$ by heating in a stream of chlorine.

    Similarly PbO_2 is an oxidizing agent, whereas SnO_2, GeO_2, SiO_2 are not.

    The stabilities of the M^{II} and M^{IV} organometallic derivatives of the elements behave differently. $PbEt_4$ which can be readily stored is more stable than $PbEt_2$, which is not isolable as a solid.

    The compounds in the lower oxidation state are in general more ionic, less likely to form molecular structures, the halides are less readily hydrolysed and the oxides are less acidic.

2. *Hydrides and alkyls become less stable down the column.*

    The M–H and M–C mean bond enthalpies decrease down the column and consequently the hydrides and alkyls become thermodynamically less stable and kinetically more reactive. Carbon, of course, forms a very wide range of hydrides, silicon forms primarily SiH_4 and Si_2H_6 which are spontaneously inflammable. The higher silanes decompose readily to Si_2H_6. The H–H bond polarities are opposite to C–H.

    The silanes are strong reducing agents. The germanes GeH_4, Ge_2H_6, and Ge_3H_8 are less flammable than SiH_4 and are resistant to hydrolysis. SnH_4 decomposes at 0°C to Sn and PbH_4 is extremely unstable.

    The organometallic derivatives of silicon and germanium are very similar. They are more reactive than the carbon analogues because the M–C bonds are more polar and the central atom can expand its coordination number more easily.

    The rates of hydrolysis are in the order:

$$Pb \gg Sn \ggg Ge > Si$$

Group 13

	Boron	Aluminium
Elements	Amorphous black powder, metallic appearance when crystalline. Crystalline forms very hard and refractory. Very unreactive, reacts with NaOH at 500°C, dissolves in H_2SO_4/HNO_3. Amorphous more reactive. m.p. 2190°C, b.p. 3660°C. At high temperatures reacts with most metals.	White and lustrous, dulled by air. Light, but has great tensile strength. Malleable, ductile, and has good electrical and thermal conductivities. m.p. 660°C, b.p. 2494°C. Face centred cubic structure. Once thin film of oxide coating has formed not attacked by air or water. Very reactive as fine powder, burns in air. Reacts with HCl and Cl_2 if oxide layer is removed. HNO_3 and H_2SO_4 do not react readily. Metal dissolves in alkalis. Powerful reducing agent with other metal oxides (thermite reaction), used for welding iron. Reacts with S, P, and N_2 at high temperatures, with carbon at 1600°C.
Hydrides	Extensive series of volatile boron hydrides with the formulae B_nH_{n+4}, B_nH_{n+6}, and $B_nH_n^{2-}$ are known. See Chapter 4 for their structures. B_2H_6 (m.p. -165.5°C, b.p. -92.5°C) burns with a green flame very exothermically. Bridge bonds broken to form adducts, BH_3L stability : L = $PF_3 > CO > Et_2O > Et_2S > C_5H_5N > Me_3N > H^-$. $NaBH_4$ widely used as reducing agent.	No volatile aluminium hydrides are known. $[AlH_3]_n$ colourless solid with infinite structure similar to AlF_3 based on octahedral Al. Forms molecular adducts Me_3NAlH_3 and $(Me_3N)_2AlH_3$ (trigonal bipyramid). $LiAlH_4$ and $NaAlH_4$ widely used as reducing agents in non-aqueous solvents. AlH_6^{3-} also known in Li_3AlH_6 and Na_3AlH_6.
Oxides	B_2O_3 (m.p. 450°C) glassy material, acidic oxide which reacts with water to give $B(OH)_3$, which is a weak monobasic acid, $pK_a \sim 9$. The borate anion $B(OH)_4^-$ condenses to form a wide range of oligomeric borate anion salts, many of which are based on $B_3O_3^{3-}$ hexagonal rings, e.g. $B_3O_6^{3-}$, $B_4O_9^{6-}$, $B_5O_{10}^{5-}$, $B_6O_{10}^{2-}$. Borates and H_2O_2 give peroxoborates with bridging peroxo-groups, $[(HO)_2B(O_2)_2B(OH)_2]^{2-}$.	Alumina and corundum Al_2O_3 have infinite structures with Al^{3+} in octahedral sites. γ-alumina has a defect spinel structure. Aluminium hydroxide $Al(OH)_3$ (Gibbsite) has a layer structure. IThe oxide is amphoteric dissolving in alkalis to form aluminates. The $Al(OH_2)_6^{3+}$ ion is acidic ($pK_a \sim 5$) and hydrolysis of solutions leads to oligomeric species, $Al_6(OH)_{15}^{3+}$, $Al_8(OH)_{22}^{2+}$, $Al_3(OH)_{11}^{2-}$. Forms double salts, alums, $M^IM^{III}(SO_4)_2 \cdot 12H_2O$, ($M^I$ = Na, K, Rb, Cs, M^{III} = Al), corresponding compounds of Ga^{III} and In^{III} are known. Al_2O known only in the gas phase.
Halides	BF_3, colourless non-flammable gas, m.p. -128°C, b.p. -100°C, soluble in water and organic solvents, only hydrolyses slowly. Lewis acid, forms complexes with O, N, S, and P donors, catalyst for Friedel Crafts reactions. B_2F_4 (m.p. -56°C, b.p. -34°C) and B_2Cl_4 have B–B bonds and sp^2 hybridized boron atoms. Pyrolysis gives clusters of the type B_nX_n (X = Cl or Br). Solvolysis and hydrolysis give $B_2(OR)_4$ and $B_2(OH)_4$. BCl_3, colourless gas, readily hydrolysed, m.p. -107°C, b.p. 13°C. Lewis acid. BBr_3, colourless liquid, readily hydrolysed, m.p. -46°C, b.p. 91°C. Soluble in non-protic inorganic and organic solvents. Catalyst in polymerization reactions. Lewis acid.	AlF_3, colourless crystals, subl. 1270°C, sparingly soluble in water. Insoluble in organic solvents. Infinite LaF_3 structure. $AlCl_3$, colourless hexagonal plates, moisture sensitive, m.p. 192°C, subl. 180°C. Infinite $CrCl_3$ structure (octahedral Al). Very moisture sensitive. Lewis acid and Friedel Crafts catalyst. $AlBr_3$, colourless crystals, m.p. 97°C, b.p. 250°C. Dimeric molecular structure in solid and liquid. Lewis acid. Friedel Crafts catalyst. Forms weak adducts with aromatic hydrocarbons and olefins. Reacts violently with water.

Gallium	Indium	Thallium
Soft, blue-white metal. Stable in air and not attacked by water. Largest liquid range of all metals, m.p. 30°C, b.p. 2403°C. Easily oxidized by O_2 at 1000°C. Reacts only slowly with conc. mineral acids. Oxidizing acids reacts with metal readily. Reacts readily with the majority of non-metals.	Soft silvery white, malleable and ductile metal, m.p. 156°C, b.p. 2070°C. Stable in dry air, but is oxidized by moist air. Soluble in most acids. Reacts with non-metals and especially with halogens. Does not dissolve in alkalis.	Soft, greyish-white ductile metal. Tarnishes readily in air. Reacts with steam or moist air to form TlOH. Dissolves readily in dil. H_2SO_4 and HNO_3, not very soluble in aqueous HCl. Does not dissolve in alkalis. m.p. 304°C, b.p. 1457°C. Very toxic.
Ga_2H_6 (m.p. –21.4°C, b.p. 139°C, decomp. > 130°C) has been isolated and structurally characterized by electron diffraction in the gas phase. It has a diborane structure. Aggregates in the solid. Forms adducts $GaH_3(NMe_3)$ and $GaH_3(NMe_3)_2$. Very reactive towards air and moisture.	$[InH_3]_n$ not well characterized.	$[TlH_3]_n$ not well characterized.
Ga_2O_3 high temperature α and low temperature γ-forms which are similar to the aluminium oxides. γ-Ga_2O_3 has tetrahedrally and octahedrally coordinated Ga^{3+}. The hydrated GaO.OH and $Ga(OH)_3$ are similar to the aluminium compounds. $pK_a [Ga(OH_2)_6]^{3+}$ ~ 3. Ga_2O also known.	In_2O_3 has only a single yellow modification. $In(OH)_3$ $pK_a[In(OH_2)_6]^{3+}$ ~ 4	Tl_2O_3 (m.p. 716°C) brown-black. Heating to 100°C gives Tl_2O.
GaF_3, colourless sublimable crystals, m.p. > 1000°C. Infinite ReO_3 type structure with octahedral Ga. Sparingly soluble in water. Lewis acid, forms octahedral complexes, e.g. GaF_6^{3-}, $GaF_3(NH_3)_3$. $GaCl_3$, colourless moisture sensitive crystals, m.p. 76°C. Dimeric in solid, liquid, and gas phases. Soluble in hydrocarbons and ethers. Lewis acid, e.g. $GaCl_3(NH_3)_3$ known. $GaBr_3$, colourless moisture sensitive crystals, m.p. 123°C. Dimeric, Lewis acid.	InF_3, colourless sublimable crystals, has infinite ReO_3 structure in solid (octahedral In). Moisture sensitive. Lewis acid forming $InF_3(NH_3)_3$ and $InF_3(OH_2)_3$. $InCl_3$, yellow deliquescent crystals, m.p. 586°C, subl. 500°C. Dimeric in vapour and liquid. Infinite in solid state, octahedral In atoms. Lewis acid, e.g. $InCl_3(DMSO)_3$, $InCl_3(OH_2)_3$, $InCl_3(thf)_2$, and $InCl_3(PPh_3)_2$. $InBr_3$, colourless solid which turns yellow on heating, m.p. 420°C, subl. 371°C. Infinite solid state structure based on octahedral In atoms. Dimeric in gas and liquid phases. Lewis acid which forms many complexes.	TlF_3, colourless solid, m.p. 550°C (decomp.), infinite structure based on BiF_3. Hydrolyzes readily to $Tl(OH)_3$. $TlCl_3$, colourless crystals, m.p. 155°C, infinite structure ($CrCl_3$ structure, octahedral). Very soluble in water. Oxidizing agent for olefins. Soluble in EtOH, Et_2O. Lewis acid forming complexes, e.g. $TlCl_3(py)_2$, $TlCl_3(DMSO)_2$. $TlBr_3$, very soluble, strong oxidizing agent, e.g. of olefins to glycols. TlF, colourless solid, distorted NaCl structure, m.p. 322°C, b.p. 826°C. TlCl, colourless solid, CsCl structure, photosensitive, m.p. 430°C, b.p. 720°C. Sparingly soluble in water. Tl_2Cl_4, $Tl[TlCl_4]$, colourless needles, soluble in EtOH, HCl. TlBr, Light sensitive yellow solid, CsCl structure, m.p. 460°C, b.p. 815°C. Sparingly soluble in water. Tl_2Br_4, $Tl[TlBr_4]$, off-white crystals, insoluble in water and organic solvents.

Group 14

	Carbon	Silicon
Element	Non-metallic appearance depends on allotrope-diamond; colourless crystals; graphite; metallic lustre, C_{60}, C_{70} black powders. Electrical conductivity depends also on allotropic form-diamond; insulator; graphite; conductor. Graphite is the thermodynamically more stable allotrope at room temperature and pressure.	Non-metallic but has a metallic lustre as a crystalline solid, which has diamond structure—there is no structural analogue to graphite or C_{60}. m.p. $1410^{\circ}C$, b.p. $2355^{\circ}C$. Very important assemiconductor in electronics industry.
Reactions with O_2, H_2O, acids, and alkalis	Only attacked by oxidizing agents, burns in air to give CO_2 (or CO if oxygen deficiency). Burns in fluorine.	Attacked by oxidizing acids and concentrated alkalis. Burns in fluorine. Reacts with Cl_2 at $300^{\circ}C$, S at $600^{\circ}C$, P at $1000^{\circ}C$.
Hydrides	Whole range of saturated C_nH_{2n+2} and unsaturated hydrides with double and triple C≡C bonds. Lower members are flammable gases higher members flammable oils, and waxes. Oxidized to $CO_2 + H_2O$ on burning CH_4.	More limited range of hydrides Si_nH_{2n+2} and multiply bonded derivatives only isolated when the multiple bonds are protected by large groups. (SiH_4 b.p. $-112^{\circ}C$). Spontaneously inflammable.
Oxides	CO neutral very good ligand for stabilizing metals in low oxidation states, very toxic; CO_2 acidic oxide forms limited range of complexes with transition metals. Both CO and CO_2 insert into metal–hydrogen and metal–carbon bonds—CO_2 also into metal–oxygen and nitrogen bonds. C_3O_2 gas (foul smelling).	SiO_2 crystalline solid (quartz, cristobalite, tridymite), m.p. $1716^{\circ}C$, relatively unreactive towards acids, Cl_2 and metals dissolves in HF, strong alkalis and is attached by F_2. Basic material in glass, refractory, and ceramics industries. Forms the basis of a wide range of silicates based on SiO_4^{4-} tetrahedra linked in rings, chains. The alumino-silicates (zeolites) have cavities of molecular dimensions and consequently are catalytically important and are also used as ion exchange materials. SiO is a high temperature phase which disproportionates at room temperature, m.p. $>1702^{\circ}C$. Insol. in H_2O, sol. in HF/HNO_3.

Germanium	Tin	Lead
Structure based on diamond, brittle grey white, lustrous. m.p. 945°C, b.p. 2850°C. Semiconductor important in transistor technology.	α-form (grey tin) has cubic diamond structure transforms at 14°C to β-form (white tin) distorted close packed structure. Normal allotrope at room temperature. Soft, pliable, silvery white metal. m.p. 232°C, b.p. 2625°C alloys with Cu(bronze), Sb(pewter).	A blue-grey dense metal, m.p. 325°C, b.p. 1150°C. It is soft and very malleable and ductile.
No reaction with H_2O, dil. acids or alkalis. Dissolves slowly in hot conc. H_2SO_4. (Reacts explosively with HNO_3). Reacts with air at 400°C. Ignites in F_2, Cl_2, Br_2.	Reacts with conc. HCl to give $SnCl_2$, oxide film limits reactions with O_2. Reacts with steam to give H_2 and alkalis and acids similarly. Reacts with F_2, Cl_2, Br_2, I_2, S and Se.	In air quickly oxidizes to form thin oxide film, pyrophoric when finely divided. It reacts with Cl_2 to give $PbCl_2$ and S to give PbS. It reacts with water only when oxygen is also present. Not attacked by non-oxidizing acids, but will dissolve slowly in sulfuric and acetic acid in the presence of air. It dissolves in warm conc. H_2SO_4. Soluble in alkaline solutions. Reacts with Cl_2 and F_2 at room temperature.
GeH_4 colourless gas (b.p. -89°C), dec. 230°C; flammable, insol. in H_2O, reacts with aq. NaOCl, HCl. Sol. in hot HCl and aqueous NH_3. Ge_2H_6(b.p. -29°C). Synthesis GeO_2 + $LiAlH_4$. Less flammable than silanes, but easily oxidized to GeO_2. Resistant to hydrolysis and aqueous basic solutions.	SnH_4 colourless gas (b.p. -53°C). Synthesis $SnCl_4$ + $LiAlH_4$, thermally very unstable decomp. to Sn at 0°C. Stable in dilute aqueous acid and base.	Hydride very unstable and poorly characterized.
GeO (GeO_2+ Ge) yellow powder easily oxidized. Disprop. at 700°C. GeO_2—hexagonal form quartz structure; tetragonal form 6-coord. Ge. Tetragonal stable r.t. insol. cold H_2O, sol. in hot H_2O. m.p. 1086°C. Hexagonal form sol. in H_2O, acids, alkalis m.p. 1115°C. with conc. HCl gives $GeCl_4$. Oxo-anions much less common than silicates in part because Ge favours octahedral coordination. This geometry also found in $Ge(OH)_6^{2-}$.	SnO_2 rutile structure octahedral Sn, colourless insol. in H_2O m.p. 1650°C. Amphoteric SnO blue black crystals. Pb structure sq. pyramidal with apical lone pair; oxidized to SnO_2 in air at 220°C, also red modification. Oxo-anions limited to compounds such as $K_2SnO_3.3H_2O$ containing $Sn(OH)_6^{2-}$.	PbO(yellow solid) m.p. 897°C, insol. in H_2O in CH_3CO_2H, layer structure based on square pyramids of Pb with lone pair apical. Amphoteric. $$PbO + 2HNO_3 \rightarrow Pb(NO_3)_2 + H_2O$$ $$PbO + 2NaOH \rightarrow Na_2PbO_2 + H_2O$$ PbO_2 synthesized by oxidation of $Pb(OH)_2$ with OCl^-, dark brown rutile structure (m.p. 290°C decomp.), oxidizing agent, liberates O_2 on heating. Insol. in H_2O. Pb_3O_4 (red), Pb_2O_3 (yellowish red) mixed valence oxides (Pb^{II}, Pb^{IV}) $Pb(OH)_2$ alkaline dissolves in base.

Group 14 (cont.)

	Carbon	Silicon
Halides	CF_4 (b.p. –128°C) very stable.	SiF_4 (b. p –86°C) readily hydrolysed reacts with F^- to form $[SiF_6]^{2-}$.
	CCl_4 (b.p. 76°C) liquid widely used as solvent not readily hydrolysed.	$SiCl_4$ (b.p. –58°C) readily hydrolysed by H_2O.
	CBr_4 (b.p. 190°C) decomposes near boiling point, yellow solid at room temperature.	$SiBr_4$ (b.p. 154°C) colourless fuming liquid, hydrolysed by H_2O.
	CI_4 (m.p. 171°C) red solid decomposes readily under light or by heating.	SiI_4 (m.p. 121°C) colourless solid hydrolysed by H_2O.
		SiF_2 isolated by matrix isolation techniques (SiF_4 + Si). Polymerizes at room temperature.
Sulfides	Carbon disulfide (CS_2) flammable liquid which slowly polymerizes under pressure to form $(CS)_n$ black solid polymer.	SiS_2 white crystals chain structure based on SiS_4 tetrahedra.
		SiS Pale yellow powder which is easily hydrolysed (FeS+SiO_2 synthesis) burns in air.
Carbides (C and Si only)	Wide range of carbides known: salt-like with electropositive metals which hydrolyze to give hydrocarbons Be_2C, Al_4C_3 have C^{4-} ions, CaC_2, $Al_2(C_2)_3$ have C_2^{2-}. Covalent carbides with elements with similar electronegativities Interstitial carbides with transition metals with very high melting points—unreactive.	SiC–ZnS structure with many modifications. Refractory, very hard (9.5 Moh scale). Abrasive material. Colourless when pure but usually grey metallic sheen with appropriate impurities important extrinsic semi-conductor. Insol. in H_2O, sol. in fused KOH.

Organotin compounds more readily expand their coordination geometries and more readily form cationic species. Organolead compounds decompose readily at 100-200°C by free radical processes.

3.   *Catenation*
The element–element mean bond enthalpies decrease in the order

$$C–C > Si–Si > Ge–Ge > Sn–Sn > Pb–Pb$$

and therefore the range of ring and polyhedral molecules diminishes down the group. Carbon not only forms an extensive range of chain and ring compounds, but also polyhedral molecules such as prismane, C_6H_6, and cubane, C_8H_8. Analogous compounds are known for Si, Ge, and Sn if the hydrogens are replaced by bulky organic substituents.

Germanium	Tin	Lead
GeF_4 gas stable to $1000^\circ C$, fumes in air, sublimes $-37^\circ C$.	SnF_4 subl. $704^\circ C$, polymeric octahedral coordination.	$PbCl_2$ colourless sol in hot H_2O, m.p. $501^\circ C$. 9 coord. Pb (tricapped trigonal prism).
$GeCl_4$ colourless fuming liq. (dec. in H_2O). b.p. $84^\circ C$.	SnF_2 8-membered Sn_4F_4 rings with trigonal bipyramidal Sn.	$PbBr_2$ colourless sol. in H_2O (m.p. $373^\circ C$.
$GeBr_4$ grey white crystals, m.p. $26^\circ C$, decomp H_2O.	$SnCl_4$ (b.p. $114^\circ C$).	
$GeBr_2$ Monomer in gas phase, $GeBr_4$ + Ge, disprop. at $150^\circ C$, hydrolyses to $Ge(OH)_2$ m.p. $143^\circ C$. (Bent molecule in gas phase).	$SnCl_2$ formed from Sn + HCl. (m.p. $246^\circ C$), polymeric nine coordinate in solid based on $SnCl_3$ chains, colourless crystals, reducing agent.	PbI_2 sol. in hot H_2O yellow solid (m.p. $402^\circ C$) CdI_2 structure octahedral .
GeI_4 red brown cubes, decomp. in H_2O, m.p. $144^\circ C$.	$SnBr_2$ bright yellow crystals, hydrolysed by H_2O, m.p. $216^\circ C$, sol in H_2O and organic polar solvents.	$PbCl_4$—yellow oil, fumes in air, hydrolysed by H_2O, sol. in organic solvents, m.p. $-15^\circ C$. Decomp, exothermically to $PbCl_2$ + Cl_2.
GeF_2 polymeric, hygroscopic crystals (m.p. $111^\circ C$).	SnI_2—red needles m.p. $320^\circ C$ layer structure Sn.	
GeI_2—CdI_2 structure oct. Ge, orange plates, sol in H_2O, decomp. on heating, sol in HI, subl. $240^\circ C$ under vacuum.		
GeS_2 white powder structure based on Ge_3S_3 rings. Decomp. in H_2O on heating to give GeO_2 sol. in alkali with decomp.	SnS_2. CdI_2 layer structure octahedral Sn, air-stable gold yellow crystals insol. in H_2O , sol. in alkali and aqua regia m.p. $600^\circ C$ (dec.).	PbS_2—red-brown powder readily decomposes in PbS. CdI_2 structure —octahedral Pb.
GeS layer structure similar to isoelectronic P, yellow-red. In sol. hot H_2O, sol in HCl alkali. m.p. $530^\circ C$.	SnS distorted NaCl structure, dark blue-grey crystals. Insol. H_2O, m.p. $882^\circ C$, semi conductor. Sol in HI and conc. acid, subl. $240^\circ C$ under vacuum.	PbS(galena), black powder, metallic crystals, m.p. $1114^\circ C$, insoluble in dilute acids but sol. in conc. HCl giving $PbCl_2$. NaCl structure semiconductor.

However, few examples exist for Pb, which form compounds containing the anionic Zintl polyhedral anion Pb_9^{4-} analogous to Sn_5^{2-}.

4. *Multiply bonded compounds*
   The ability of the elements to form multiple bonds diminishes in the series.

$$C\text{–}C > Si\text{–}Si > Ge\text{–}Ge > Sn\text{–}Sn > Pb\text{–}Pb$$

because the p_π–p_π overlaps become less favourable. This has the following manifestations:

(a) For the elements below C the allotropes which would structurally resemble graphite which has a delocalized two dimensional π-system are not observed.

(b) There are no simple analogues of ethene (C_2H_4) and ethyne (C_2H_2) and compounds of Si, Ge, and Sn with multiple bonds may only be isolated when there are bulky organic substituents on the group 14 atoms. Furthermore, the Ge and Sn compounds do not have planar geometries.

(c)   The analogues of CO_2 and CS_2 have polymeric structures rather than triatomic molecular geometries.

(d)   The heavier elements do not form analogues of carbides with C_2^{2-} and C_3^{2-} multiply bonded ions.

5.   *The elements become progressively more metallic down the column.*
Carbon especially in its diamond and polyhedral forms is a typical non-metal, silicon is a semiconductor and tin and lead are typical metals. Although tin has one modification (grey tin) which is isostructural with Ge, Si, and diamond. Lead only occurs in close packed structural forms.

6.   *The oxides become more basic down the column.*
CO_2 and SiO_2 are acidic oxides, SnO_2 is amphoteric and GeO_2 is mainly acidic with slight amphoteric character. The Si–O mean bond enthalpies are particularly large and this leads to a wide range of silicates. In general for the elements below carbon the M–O bonds are sufficiently strong that the oxides are susceptible to hydrolysis.

7.   *The typical coordination numbers increase down the group.*
For carbon the tetrahedral geometry predominates unless multiple bonds are formed. For the heavier elements the tetrahedral geometry is also widespread but the larger size of the central atoms leads to the formation of compounds with higher coordination numbers, e.g.

SiF_5^-	trigonal bipyramidal
SiF_6^{2-}	octahedral
$SnCl_5^-$	trigonal bipyramidal
$SnPh_2(NO_3)_2(OPPh_3)$	pentagonal bipyramidal—7-coordinate
$Sn(NO_3)_4$	dodecahedral 8-coordinate

The increased facility to achieve the higher coordination numbers is also reflected in the transition from molecular to polymeric, e.g. CF_4, SiF_4, and GeF_4 are molecular, whereas SnF_4 and PbF_4 have infinite lattices based on octahedral metal centres.

Of course all of these compounds with coordination numbers greater than four have electron configurations which exceed the octet rule. The MX_4 compounds (M = Si, Ge, Sn, Ph) are therefore Lewis acids, although they do not have empty p valence orbitals.

The higher coordination numbers are also reflected in the geometries of the oxo-anions:

CO_3^{2-}	SiO_4^{4-}	$Ge(OH)_6^{2-}$	$Sn(OH)_6^{2-}$	$Pb(OH)_6^{2-}$
Trigonal planar	Tetrahedral	Octahedral		

The chlorides of Ge, Sn, and Pb react with aqueous HCl to form the $[MCl_6]^{2-}$ anions, whereas $SiCl_4$ hydrolyses and CCl_4 is unreactive. However, SiF_4 does form $[SiF_6]^{2-}$ with HF.

8.   The heavier elements form a wider range of complexes and more cationic complexes. Si, Ge, Sn, and Pb all form oxalato-complexes $[M(ox)_3]^{2-}$ and cationic complexes $[M(acac)_3]^+$.

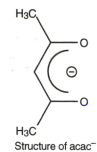

Structure of acac⁻

9.   *Ease of reduction of halides.*
      Although C–Cl and Ge–Cl bonds in four valent compounds are reduced to the corresponding hydrides by Zn and HCl, Si–Cl and Sn–Cl bonds are not.

10. SiH_4 hydrolyses in the presence of trace amounts of base more readily than CH_4, GeH_4, and SnH_4.

Summary of Group 15 trends

1.   The hydrides become less thermally stable and more reactive down the column. The M–H and M–CH$_3$ mean bond enthalpies decrease in strength down the group and consequently the hydrides and alkyls become less stable.

2.   The majority of nitrogen compounds are covalent. The main exceptions being those based on the nitride ion.

3.   *Multiple bond formation*
      Nitrogen forms very strong multiple p_π–p_π bonds to itself and neighbouring elements belonging to the same row, e.g.

$$C\equiv N^-\quad N\equiv N\quad N\equiv O^+$$

      Compounds of P, As, and Bi with multiple bonds may be obtained if large groups are introduced into the molecules, e.g. P_2R_2 and As_2R_2. Similar multiply bonded compounds of Sb are not known.

4.   *Coordination numbers*
      The coordination numbers increase down the group. For nitrogen 3 and 4 coordination predominate. Phosphorus and arsenic in addition form octahedral complexes and higher coordination numbers are observed for Sb and Bi.

5.   *Increased metallic character*
      The elements become increasingly metallic down the column in their chemical and physical properties. Down the group the oxo cations, e.g. SbO^+ and BiO^+ become more prevalent.

Group 15 (Nitrogen and Phosphorus)

	Nitrogen
Element	Very stable diatomic gas (m.p. $-210^{\circ}C$, b.p. $-196^{\circ}C$) with a high dissociation energy (946 kJ mol^{-1}) and kinetically inert because of lack of dipole moment, unavailability of lone pairs, and large HOMO–LUMO gap. Reacts with metallic lithium at room temperature to give Li_3N and its π-acceptor abilities are exploited by low oxidation state transition metal complexes to form N_2 complexes which are more reactive. Indeed the molybdenum complex $Mo(NPr^iPh)_3$ reacts with N_2 to split the N–N bond and give the nitrido-complex $MoN(N^iPrPh)_3$.
	Only reacts with O_2 and H_2 at high temperatures and in the presence of catalyst to give NO and NH_3 respectively.
	Does not react with sulfur.
	Does not react with acids such as HNO_3.
Hydrides	NH_3 (ammonia) colourless and pungent smelling gas (m.p. $-78^{\circ}C$, b.p. $-34^{\circ}C$) very soluble in H_2O giving weak base $K_b= 1.8x10^{-5}$. Hydrogen bonds strongly, liquid ammonia is commonly used as a solvent of high dielectric constant for reductions with alkali metals, where unlike water it does not instantly react with them but forms coloured solutions containing the metal cations and solvated electrons. It burns in air to give N_2 and in the presence of a catalyst to give NO. Forms a wide range of complexes with Lewis acids and transition metal ions. Forms many ammonium salts which are generally water soluble and volatile because of their dissociation into NH_3 and the mineral acid.
	N_2H_4 (hydrazine) (m.p. $2^{\circ}C$, b.p. $114^{\circ}C$) polar colourless oily liquid, which is very soluble in H_2O, where it is less basic than NH_3 ($K_b = 9x10^{-7}$). Endothermic enthalpy of formation and decomposes violently to N_2 in the presence of a transition metal catalyst. Powerful and useful general reducing agent particularly as the hydrate $N_2H_4.H_2O$. It burns in air very exothermically and therefore has been used as a rocket fuel.
Halides	NF_3 (m.p. $-209^{\circ}C$, b.p. $-129^{\circ}C$) colourless stable unreactive gas. Low dipole moment makes it a poor ligand towards Lewis acids. More reactive at higher temperatures and is used as fluorinating agent for metal halides, e.g. $AlCl_3$, reacts vigorously with organics. NF_4^+ salts are powerful oxidizing agents. Slightly soluble in H_2O where it is not readily hydrolysed.
	N_2F_4 (m.p. $-40^{\circ}C$, b.p. $70^{\circ}C$) more reactive than NF_3 and more widely used as fluorinating agent. Slowly hydrolysed by H_2O. Explosive and inflammable.
	NCl_3 (m.p. $-40^{\circ}C$, b.p. $70^{\circ}C$) dense liquid which readily hydrolyses to NH_3 and HOCl. Soluble in $CHCl_3$ and CCl_4. Explosive.
	NBr_3 very unstable even at low temperatures, deep red explosive solid. Stable in dilute aqueous solutions, soluble in $CHCl_3$.
	NI_3 Only exists as NH_3 adduct ($NI_3.NH_3$) explosive to touch and insoluble in water.

Phosphorus

Occurs as several allotropes, white phosphorus containing P_4 molecules, red phosphorus cross-linked chains, black phosphorus two dimensional sheets. In all of these the phosphorus atoms are three coordinate pyramidal.

White phosphorus m.p. 44°C—soluble in organic solvents, inflames in air.

Red phosphorus m.p. 600°C—insoluble in organic solvents, stable in air.

Black phosphorus m.p. 610°C—insoluble in organic solvents, stable in air.

Phosphorus burns to give P_4O_6 and P_4O_{10}, reacts with S to give P_4S_3 and P_4S_{10} and with halogens to give the halides. They do not react with non-oxidizing acids, but react with HNO_3 to give H_3PO_4. With concentrated ammonia they form PH_3 and HPO_2^-.

With electropositive metals they react to give phosphides (P^{3-}) and with transition metals to form a wide range of compounds with differing stoichiometries.

PH_3 (phosphine or phosphane), (m.p. -133°C, b.p. -88°C), less thermally stable than NH_3, very soluble in H_2O and soluble in organic solvents, made from ionic phosphides and acid. Not spontaneously inflammable when pure, but inflames with traces of P_2H_4. Very toxic and has garlic odour. Hydrogen bonding much weaker than ammonia and therefore lower b.p. also has lower dielectric constant and therefore not used as solvent. Weaker base than ammonia , but forms some phosphonium salts.

P_2H_4 (m.p. -99°C, b.p. 67°C) decomposes even at -30°C to give insoluble polymers. Toxic and spontaneously inflammable.

PF_3 (m.p. -152°C, b.p. -102°C) very stable colourless gas, not readily hydrolysed, but is attacked by OH^-. Very poisonous. Acts as a good π-acceptor ligand in low oxidation state transition metal compounds.

PCl_3 (m.p. -74°C, b.p. 76°C) colourless liquid, which is readily and vigorously hydrolysed by H_2O to H_3PO_4. Poisonous, corrosive, and forms transition metal complexes as a ligand and acts as base with BX_3. Widely used as reagent in inorganic and organic chemistry. Reacts with O_2 to give $POCl_3$.

PI_3 (m.p. -61°C), red crystalline solid, sol. in CS_2, decomposes in moist air. Forms complexes with BX_3 and is used as a deoxygenating reagent.

The trihalides, except PF_3, may be obtained directly from the elements (in excess) and the halogens. The pentahalides are obtained when excess halogen is present.

PF_5 (m.p. -75°C, b.p. -83°C), thermally stable colourless gas which fumes in moist air. Poisonous. Very strong Lewis acid towards hard bases such as amines and ethers. Complexes hydrolyze readily.

PCl_5 (m.p. 167°C, sublimes at about same temperature), colourless solid which exists as $PCl_4^+PCl_6^-$ in solid state. Readily hydrolysed to H_3PO_4. Used as chlorinating agent.

PBr_5 (b.p. 106°C, decomposes) reddish yellow solid which exists in solid state as $PBr_4^+Br^-$.

PI_5 (m.p. 40°C) brown black crystals which exist as $PI_4^+I^-$ in solid state. Soluble in organic solvents. Very readily hydrolysed.

Group 15 (Nitrogen and Phosphorus, cont.)

	Nitrogen
Oxides	N_2O, nitrous oxide, (m.p. $-91^{\circ}C$, b.p. $-89^{\circ}C$) colourless gas, neutral unreactive, supports combustion at higher temperatures. Soluble in organic solvents, sparingly sol. in H_2O. Used as anaesthetic—laughing gas. Thermodynamically unstable with respect to elements, but only decomposes if catalyst present. Generally unreactive to alkali metals and halogens. Forms a limited number of transition metal complexes behaving as a simple two electron donor ligand through nitrogen.

NO, nitrogen monoxide or nitric oxide, (m.p. $-164^{\circ}C$, b.p. $-152^{\circ}C$) paramagnetic colourless gas at room temperature, dimeric diamagnetic in solid state. Reacts very rapidly with O_2 to give brown fumes of NO_2, nevertheless has an important biological role as a neurotransmitter and controlling blood pressure. Oxidation and reduction lead respectively to NO^+ and $N_2O_2^{2-}$ the latter is a strong oxidizing agent. Nitric oxide forms a very wide range of nitrosyl complexes the majority of which have linear M–N–O geometries analogous to that found in carbonyl complexes, the minority have angular (M–N–O approximately 120°) geometries.

NO_2, nitrogen dioxide, (m.p. $-11.2^{\circ}C$), brown paramagnetic gas at room temperature in H_2O it dissolves to give HNO_2 and HNO_3, disproportionates in acid and alkali. In equilibrium with diamagnetic, colourless N_2O_4:

$$2NO_2 \rightleftharpoons N_2O_4$$

In solid state the dimer occurs exclusively, and in the liquid and gas both species are present and dissociation is only complete at about $140^{\circ}C$. Reasonably strong oxidizing agent. Liquid N_2O_4 (b.p. $21.2^{\circ}C$) is used as a solvent particularly for the synthesis of metal nitrates in an anhydrous form.

N_2O_3 blue liquid which readily dissociates into NO and NO_2. In water it dissolves to give HNO_2. Soluble in organic solvents.

N_2O_5 colourless solid which exists as $NO_2^+NO_3^-$. Dissolves in H_2O to give HNO_3, but also soluble in $CHCl_3$ and CCl_4.

Acids	HNO_2, nitrous acid, not known in a pure form, but aqueous solutions may be obtained by acidifying nitrite salts. Weak acid $pK_a = 3.13$. These are unstable to heat give NO_3^- and NO. Many nitrite salts are known and often they are water soluble. The nitrite anion commonly acts as a ligand towards transition metal ions.

HNO_3, nitric acid, (m.p. $-41^{\circ}C$, b.p. $83^{\circ}C$) stronger acid $pK_a = -1.20$, aqueous solutions used as strong oxidizing agents and with HCl (*aqua regia* for dissolving gold and platinum). Many nitrate salts known, the majority of which are soluble in H_2O, NO_3^- can act as a mono- or bidentate ligand towards many metals.

Nitrides /phosphides	Nitrides exist for the majority of metals.

Ionic Li_3N, Ca_3N_2, which hydrolyze readily to give NH_3 and $M(OH)_x$.

Covalent with infinite 2-dimensional, e.g. BN analogue of graphite, and infinite 3-dimensional, e.g. diamond form of BN. Also, AlN, GaN, Si_3N_4 very stable and strong ceramic material.

Transition metal nitrides are high melting point conducting ceramics.

Phosphorus

P_4O_6 (m.p. 23°C, b.p. 175°C), colourless crystals containing isolated molecules. Acidic oxide soluble in alkalis and organic solvents. Reducing agent.

P_4O_{10} (m.p. 420°C, 360°C sublimes), colourless solid, very hygroscopic and forms very concentrated syrupy solutions with H_2O. Very widely used dehydrating agent in inorganic and organic chemistry. Will also extract elements of H_2O from acids to give the parent anhydride.

H_3PO_2 (hypophosphorous acid) (m.p. 27°C, decomposes 130°C) is a strong monobasic acid ($pK_a = 2.0$). which is also a reducing agent.

H_3PO_3 (phosphorous acid) (m.p. 74°C, decomp. 180°C) dibasic strong acid ($pK_a(1) = 1.8$, $pK_a(2) = 6.6$) hygroscopic and very soluble in H_2O. Obtained from hydrolysis of PCl_3 and P_4O_6. Reducing agent.

H_3PO_4 (orthophosphoric acid) (m.p. 42°C), colourless crystals very soluble in H_2O and also ethanol. Very syrupy when concentrated solutions, not strongly oxidizing except at very high temperatures. Tribasic acid—$pK_a(1)$ 2.15, $pK_a(2) = 7.1$ and $pK_a(3) = 12.4$. Phosphate salts and phosphate esters are of great importance in biology and as minerals.

Phosphides are known for most elements. Ionic for groups 1 and 2. Wide range of stoichiometries for phosphides of the transition metals, e.g. Fe_3P, Fe_2P, FeP, FeP_2, and FeP_4. Covalent phosphides with molecular structures, e.g. GaP zinc blende structure, important as semiconducting materials.

Group 15 (Arsenic, Antimony and Bismuth)

	Arsenic	Antimony	Bismuth
Elements	Yellow less dense allotrope, very reactive, soluble in CS_2. Metallic allotrope grey, dull metallic lustre, brittle, sublimes easily (m.p. $814^{\circ}C$ under 36 atmospheres pressure). Puckered sheets of arsenic atoms. Semi-conductor.	White crystalline, lustrous metal, m.p. $631^{\circ}C$, b.p. $1380^{\circ}C$. Most important and stable allotrope is isostructural with metallic arsenic. For a metal the resistivity is relatively high.	Silvery lustrous metal, m.p. $271^{\circ}C$, b.p. $1450^{\circ}C$. Brittle and very crystalline. Isostructural with metallic arsenic. High resistivity for a metal.
Stability in air	Air stable, burns on heating to As_4O_6.	Air stable, but on heating reacts to give Sb_4O_6.	Tarnishes slightly in air, burns in oxygen forming Bi_2O_3.
Reaction with Cl_2	With Cl_2 gives $AsCl_3$, with F_2 can obtain AsF_5.	With Cl_2 gives $SbCl_3$.	With chlorine gives $BiCl_3$.
Reaction with S	Reacts with S to give As_4S_3.	Reacts with S to give Sb_2S_3.	Reacts with S to give Sb_2S_3.
Reactions with acids	With conc. HNO_3 gives H_3AsO_4.	No reaction with dilute acids, with hot conc. HNO_3 gives Sb_2O_5, reacts with HF to give fluorides and with H_2SO_4 to give $Sb_2(SO_4)_3$.	No reaction with dilute acids, but with oxidizing acids gives $Bi(NO_3)_3$ and $Bi_2(SO_4)_3$.
Hydrides	AsH_3 thermally unstable (m.p. $-116^{\circ}C$, b.p. $-62^{\circ}C$)—very toxic gas; flammable, readily oxidized at As_4O_6.	SbH_3, colourless gas (m.p. $90^{\circ}C$, b.p. $-18^{\circ}C$), very soluble in CS_2, organic solvents, very readily decomposes at room temperature. Very toxic. Burns in air to give Sb_4O_6. Forms mirror of antimony metal on heating.	BiH_3 very unstable and decomposes above $-45^{\circ}C$. (m.p. $-67^{\circ}C$, b.p. $-67^{\circ}C$ extrapolated). Toxic.
Oxides	As_2O_3 (m.p. $312^{\circ}C$) puckered layers of pyramidal AsO_3 units, in gas exists as As_4O_6. As_4O_6 (m.p. $278^{\circ}C$) converted to As_2O_3 on heating, has structure analogous to P_4O_6. Reacts with H_2O 'As(OH)$_3$' poorly characterized, and with bases to form $AsO(OH)_2^-$, $As(OH)O_2^{2-}$, and AsO_3^{3-}.	Sb_2O_3 colourless (m.p. $665^{\circ}C$) double infinite chains of SbO_4 tetrahedral corner sharing. Sb_4O_6 colourless low temp. form of Sb_2O_3. Insoluble in water or dilute acids. Soluble in conc. acids. Also in strong bases to give AsO_3^{3-} salts. Amphoteric.	Bi_2O_3 yellow solid insol. in H_2O. Soluble in acids. Layer structure based on 8-coord. Bi atoms. Unstable, dissolves in acids to give Bi^{III} salts.
	As_2O_5 has AsO_6 octahedra and AsO_4 tetrahedra forming an infinite lattice. Deliquescent colourless solid, v. soluble in H_2O. Cannot be formed in $As+O_2$, readily loses O_2 on heating to give As_2O_3. Forms H_3AsO_4 when dissolved in H_2O.	Sb_2O_5 formed from Sb_2O_3 $+O_2$ at high temperatures and pressures. SbO_6 octahedra in solid state, connected by corners and edges. Insol. in H_2O, sl. soluble in KOH, no reaction with HCl. Toxic yellow solid.	Bi_2O_5 poorly defined light brown-yellow solid, decomp. $350^{\circ}C$ to Bi_2O_3.

Group 15 (Arsenic, Antimony, and Bismuth, cont.)

	Arsenic	Antimony	Bismuth
Acids	H_3AsO_4 unlike H_3PO_4 oxidizing acid. Forms fewer condensed arsenates than phosphates $pK_a(1)$ 2.3, $pK_a(2)$ 6.8, $pK_a(3)$ 11. $As(OH)_3$, alkali metals salts very soluble	SbO_4^{3-} unknown, exists as $Sb(OH)_6^-$ in basic solutions of $Sb(V)$. No well defined oxo-acid. $Sb_2O_3.xH_2O$ dissolves in bases to form $MSbO_2$ salts.	Bi^{III} has no acidic properties and forms no stable oxo anions. $Bi(OH)_3$ white-pale yellow solid. Non-acidic, slightly soluble in H_2O.
+3 Fluorides	AsF_3(m.p. -6^oC, b.p. 63^oC, b.p. 63^oC) molecular in gas phase, in solid state has extended structure, Soluble in organic solvents, weak fluorinating agent. Does not form stable transition metal complexes.	SbF_3(m.p. 192^oC, b.p. 376^oC). Molecular in gas phase, distorted octahedral geometry in solid. Colourless deliquescent crystals, sol. in H_2O, insoluble in organic solvents. Lewis acid, does not form stable transition metal complexes.	BiF_3 (m.p. 725^oC, 900^oC) solid, 8-coordinate infinite structure, insoluble in organic solvents.
+3 Chlorides	$AsCl_3$(m.p. -16^oC, b.p. 103^oC). Colourless crystals, fume in air. Toxic.	$SbCl_3$(m.p. -73^oC, b.p. 223^oC). Colourless microscopic crystals, which fume in air, Lewis acid, poisonous, highly corrosive. Forms complexes with aromatics. Forms insoluble $SbOCl$ in water.	$BiCl_3$(m.p. 233^oC, b.p. 447^oC). Colourless deliquescent solid, 9-coordinate infinite lattices. Lewis acid, forms complexes with aromatics and NH_3. Forms insoluble $BiOCl$ in water.
+5 Fluorides	AsF_5 (m.p. -80^oC, b.p. -53^oC), colourless gas, hydrolysed by H_2O. Strong Lewis acid.	SbF_5 (m.p. 70^oC, b.p. 150^oC), tetramer based on octahedral SbF_6 units in liquid and gas phases. Strong Lewis acid, fluorinating agent. Reacts vigorously with water. Corrosive and toxic.	BiF_5(m.p. 154^oC) linear octahedral chains of BiF_6 units in solid. Reacts explosively with H_2O giving ozone and OF_2. Powerful fluorinating agent. Weaker Lewis acid that SbF_5. Very toxic.
+5 Chlorides	$AsCl_5$ only stable below -50^oC.	$SbCl_5$ (m.p. 3^oC, b.p. 140^oC) decomposes above 140^oC. Colourless liquid, fumes in air. Dimeric in solid, edge shared octahedra Sb_2Cl_{10}. Soluble in organic solvents. Poisonous and corrosive.	

6. *Oxides become more basic down the group*
    Phosphorus and arsenic oxides are acidic, antimony oxide is amphoteric, and that of bismuth is basic.
7. The halides become more ionic and increasingly adopt infinite structures in preference to molecular ones.
8. Catenation occurs in the order: N < P > As> Sb > Bi.
    Phosphorus forms a wide range of ring and cage compounds because of the favourable value of D(P–P).

9. *Negative oxidation states:*
   For nitrogen the N^{3-} ion is well established in ionic nitrides of the electropositive elements. The anionic derivatives of the heavier elements frequently retain element–element bonds, e.g. Sb_4^{2-}, Bi_4^{2-}, Sb_7^{3-}, and As_{11}^{3-}.

10. The donor/acceptor behaviour of R_3M is discussed in detail on p. 113. Briefly:
    R_3M donor ability: $N < P > As > Sb > Bi$
    Steric effects $N > P > As > Sb > Bi$ and increase with the bulk of the substituents:
    $$R_3P > PR_2H > PRH_2 > PH_3$$
    π-acidity: $R_3P > R_3As > R_3Sb > R_3Bi$
    Lewis acidity of +5 fluorides:
    $$PF_5 > AsF_5 > SbF_5$$

11. Hydrolysis of halides. In the +5 oxidation state PF_5 not readily hydrolysed, AsF_5 hydrolysed, SbF_5 vigorously reacts with water. BiF_5 reacts explosively with water. In the +3 oxidation state NF_3 unreactive, PF_3 reacts only with OH^- not OH_2, AsF_3, SbF_3 soluble in water, BiF_3 insoluble in water, soluble in inorganic acids.

12. Stabilization of +3 oxidation state relative to +5.
    The trend is less well defined than that for Group 14 and in fact an alternation in stabilities is observed. This may be illustrated by the following halide stabilities.

	N	P	As	Sb	Bi
EF_5	×	√	√	√	√
ECl_5	×	√	u	√	×
EBr_5	×	√	×	√	×
EI_5	×	√	×	√	×
EF_3	√	√	√	√	√
ECl_3	u	√	√	√	√
EBr_3	u	√	√	√	√
EI_3	u	√	√	√	√

√: known and stable ×: unknown u: known but unstable

The +5 oxidation state halides are unknown for N, well defined for P, only stable for As as the fluoride, well defined for the fluoride and chloride of antimony, and only known for the fluoride for Bi.

For both oxidation states the stability order is: $F > Cl > Br > I$ as anticipated by the ordering of the mean bond enthalpies, i.e. fluorine forms the strongest bonds and iodine the weakest.

The oxides show a similar trend. Bi^V and N^V oxides and oxoacids are strongly oxidizing whereas P^V is very stable and As^V and Sb^V are mildly oxidizing.

Summary of Group 16 trends

1. The elements become progressively more metallic down the column. Polonium has chemical and physical properties characteristic of a metal and tellurium lies on the borderline.

2. Chemically the metallic character of the heavier elements is reflected in their increased tendency to form cationic species, the ionic character and basicities of their oxides, and their increased tendency to form complexes.

3. The non-metal character of the earlier members of the Group is the molecular nature of the stable elemental allotropes, the ability to form anions, e.g. O^{2-} and S^{2-}, which result from completing the octet. Compounds resulting from these anions, e.g. Na_2E, CaE (E = O, S, Se...) become progressively more covalent down the Group as a result of the decreasing electronegativity of the chalcogens.

4. The thermal stabilities of the hydrides EH_2 decrease down the column primarily because of the decreasing EH mean bond enthalpies.

5. The abilities of these hydrides to form hydrogen bonds decreases rapidly down the Group after oxygen and this has a dramatic effect on the boiling and melting points of the hydrides.

6. The formation of molecular compounds with strong multiple bonds is particularly important for oxygen which forms strong p_π–p_π bonds, e.g. CO, d_π–p_π bonds with transition metals, e.g. OsO_4 and main group atoms, e.g. R_3PO and even p_π–f_π bonds in UO_2^{2+}. Multiple bonding is less significant for the heavier elements, although sulfur and selenium provide some examples of multiply bonded compounds with p_π–p_π and d_π–p_π bonding, e.g. selenoketones.

7. The increasing size of the atoms leads to compounds with progressively larger maximum coordination numbers. Oxygen usually has a coordination number of 2 or 3 with a few examples of 4 coordination, sulfur exhibits a maximum coordination number of 6 and higher coordination numbers of 8 have been observed for Te.

    These maximum coordination numbers have an impact on the case of hydrolysis of the halides, e.g. the rate of hydrolysis is:

    $$TeF_6 > SeF_6 > SF_6$$

    Also, the octahedral anionic complexes $[MX_6]^{2-}$ (X = halide) are more commonly observed for Se, Te, and Po.

8. Oxygen has a strong preference for the –2 formal oxidation state, whereas the heavier elements exhibit oxidation states of 2, 4, and 6. The alternation effect, which is discussed earlier in this chapter, is observed for this Group.

9. The tendency towards catenation reaches a maximum at sulfur which forms a wide range of ring and chain compounds. e.g. S_n, XS_nX, $O_3S(S_n)SO_3^{2-}$, and S_8^{2+}.

Group 16

	Oxygen	Sulfur
Element	Dioxygen (O_2) colourless gas which forms 21% by volume of the Earth's atmosphere, m.p. $-218^{\circ}C$, b.p. $-183^{\circ}C$. Condenses to give blue liquid and solid. Paramagnetic. Colourless, odourless, and tasteless. Essential for life and combustion processes. Reacts with almost all elements except noble gases. Slightly soluble in water, more soluble in organic solvents. Very soluble in fluorocarbons making them useful as temporary blood substitutes.	Yellow solid, soluble in C_6H_6 and CS_2 insoluble in H_2O. Forms many allotropes based on S_n rings also metastable forms including plastic and fibrous sulfur with S_n chains. Most stable allotropes based on S_8 rings which crystallize in orthorhombic (m.p. $113^{\circ}C$) and monoclinic ($119^{\circ}C$) crystalline forms. Mined in elemental form using Frasch processes, but also found in many sulfide minerals. Burns in air with blue flame to form SO_2, dissolves in oleum to give blue solutions containing S_4^{2+} and S_8^{2+} cations.
	Allotropic form, O_3 ozone, m.p. $-193^{\circ}C$, b.p. $-112^{\circ}C$; colourless gas with characteristic odour. Unlike O_2 it is diamagnetic. As liquid it is blue and violet black when solid. Slightly soluble in H_2O. Molecular geometry is angular (O–O–O $= 117^{\circ}$; O–O 128 pm) Formed by passing silent electric discharge through O_2. Powerful oxidant, which is used for sterilizing water. Formed in upper atmosphere by U.V. radiation of O_2 and plays an important role there as a U.V. filter. It is destroyed by reactions with Cl atoms and NO, the former resulting from chlorofluorohydrocarbons used as refrigerants and aerosol propellants and the latter from high flying supersonic aircraft.	
Halides	F_2O, colourless gas, m.p. $-224^{\circ}C$, b.p. $-145^{\circ}C$, powerful oxidizing agent. Highly toxic, corrosive and explosive.	SF_2, colourless gas, very unstable towards disproportionation to $S + SF_4$.
	F_2O_2, thermally unstable brown gas decomp. to give F_2 and O_2, m.p. $-164^{\circ}C$, b.p. $-57^{\circ}C$. Strong oxidizing and fluorinating agent.	S_2F_2, $F_2S{=}S$, colourless gas stable $> 200^{\circ}C$, m.p. $-165^{\circ}C$, b.p. $-11^{\circ}C$. Hydrolyzed by water.
	Cl_2O, Cl_2O_6, Cl_2O_7, Br_2O, BrO_2, I_2O, I_2O_4, I_2O_5, and I_4O_9 are discussed in the Group 17 Table.	SF_4, colourless gas, fluorinating agent. Readily hydrolysed. Soluble in aromatic solvents. m.p. $-121^{\circ}C$, b.p. $-38^{\circ}C$. Toxic and corrosive.
		SF_6, colourless odourless gas. Very stable and inert. Insoluble in water and not hydrolysed. Used as a gaseous dielectric, m.p. $-51^{\circ}C$, subl. $-65^{\circ}C$.
		S_2F_{10}, colourless liquid. Stable to hydrolysis, m.p. $-53^{\circ}C$, b.p. $27^{\circ}C$. Very toxic.
		SCl_2, red liquid, m.p. $-122^{\circ}C$, b.p. $60^{\circ}C$ (decomp.), decomposes on heating to give S_2Cl_2. Used as a chlorinating agent.
		SCl_4, $[SCl_3^+]Cl^-$, unstable white solid, decomp.at $-30^{\circ}C$. Readily hydrolysed.

Selenium	Tellurium	Polonium
Has six allotropic forms, α, β, γ, grey 'metallic', red amorphous, and black; α, β, and γ have Se_8 rings, most stable is the grey metallic which has helical chains, the black form has complex structure with large rings. The grey form has m.p. 217°C, b.p. 685°C. Semiconductor and photoconductor leading to its use in photocopiers. Sparingly soluble in CS_2, dissolves in oleum to give Se_4^{2+}. Burns in air with a purple flame.	Occurs in a 'metallic' form with helical chains of Te atoms, semi-conductor, brittle and poor conductor of heat. Grey powder, crystallizes with metallic appearance. Insoluble in organic solvents, m.p. 450°C. Burns readily in air with a blue flame.	Typical metal with a simple cubic structure which is unique. ^{210}Po occurs naturally in uranium minerals and is intensely radioactive ($t_{1/2}$ = 138 days) pure α emitter which damages solutions and solids and requires special safety precautions. Formed in gram quantities by nuclear reactions. Silvery white metal, m.p. 254°C.
SeF_2, only identified in matrix isolation studies. SeF_4, colourless fuming liquid, soluble in organic solvents, m.p. −10°C, b.p. 102°C. Hydrolyses violently. Fluorinating agent. Good non-aqueous solvent. Toxic. SeF_6, thermally stable colourless gas and solid at low temperature, m.p. −47°C, sublimes ~ −35°C. Does not hydrolyze. Poisonous. Gaseous electrical insulator. $SeCl_2$, unknown as solid, but present in $SeCl_4$ vapour. $SeCl_4$, yellow crystals, m.p. 305°C, subl. 196°C. Toxic. Hydrolyses in moist air. Soluble in organic solvents. Ionic solid.	TeF_4, colourless crystals, red vapour, m.p. 130°C, b.p. 374°C. Polymeric bridged structure, can behave as Lewis acid and base. Readily hydrolysed. Toxic. TeF_6, colourless gas with terrible odour, m.p. −39°C, subl. −39°C. Hydrolyses in water to H_6TeO_6. Toxic. Lewis acid, forming TeF_8^{2-} with F^-. $TeCl_2$, black solid, red vapour, m.p. 208°C, b.p. 328°C. Hydrolyses and disproportionates in water to TeO_2 and Te. Hydrolyses in moist air.	$PoCl_2$, dark red hygroscopic solid, subl. at 190°C. Readily oxidized to Po^{IV}. $PoCl_4$, pale yellow liquid, readily hydrolysed, m.p. 300°C, b.p. 390°C.

Group 16 (cont.)

	Oxygen	Sulfur
Oxides		SO, prepared from SO_2 and S in gas discharge. Very short lived diradical.
		SO_2, toxic, colourless gas with a strong and characteristic smell. m.p. $-75^{\circ}C$, b.p. $-10^{\circ}C$, soluble in H_2O. Formed by burning sulfur and widely found in atmosphere as a result of burning sulfur-containing fossil fuels. Catalytically oxidized to SO_3 which when dissolved in water leads to droplets of H_2SO_4 (acid rain).
		SO_3, occurs as solid with three crystalline modifications, m.p. $17^{\circ}C$ (α), $33^{\circ}C$ (β), and $62^{\circ}C$ (γ) forms, because of alternative oligomeric and cyclic ring structures. Prepared on large industrial scale from catalytic oxidation of SO_2 and then converted into sulfuric acid. Reacts violently with H_2O and therefore fumes in air and is very hygroscopic. Very corrosive. Sulfonating agent. Reasonably strong oxidizing agent, e.g. $S \rightarrow SO_2$, $PCl_3 \rightarrow POCl_3$. Acts as Lewis acid to py, Me_3N.
Acids		H_2SO_3, does not exist as isolable entity, formed in small amounts when SO_2 dissolves in H_2O (major species $SO_2.7H_2O$). Reasonably acidic $pK_a(1)$ 1.9, $pK_a(2)$ 7.2, reducing agent, oxidized by air to sulfuric acid. Forms many sulfite salts, e.g. $NaHSO_3$ (generally with larger cations), Na_2SO_3, etc.
		H_2SO_4, viscous colourless liquid, m.p. $10^{\circ}C$, b.p. $290^{\circ}C$. Very strong acid $pK_a(1)$ -3, $pK_a(2)$ 2.0 ($25^{\circ}C$). Not a strong oxidizing agent, but is very corrosive. Forms wide range of salts, e.g. $NaHSO_4$, Na_2SO_4 etc. Produced commercially on millions of tonnes scale by the Contact Process.
Hydrides	OH_2, water	SH_2, hydrogen sulfide
	Colourless, odourless, and tasteless liquid, which is an essential medium for life. 76% of water on the earth is found in oceans and rivers, 20% in sedimentary rocks and only 1% as ice, m.p. $0^{\circ}C$, b.p. $100^{\circ}C$, pK_a 14.0 ($25^{\circ}C$).	Foul smelling (characteristic 'bad egg' smell) and very poisonous gas, more toxic than HCN, m.p. $-83^{\circ}C$, b.p. $-59^{\circ}C$. Synthesized from sulfides and mineral acids, soluble in H_2O and EtOH. Readily oxidized. $pK_a(1)$, 6.9, $pK_a(2)$, 14.2 ($20^{\circ}C$). S^{2-} only exists in very alkaline solutions.
	O_2H_2, hydrogen peroxide	H_2S_n ($n = 2-6$) also known.
	Colourless liquid m.p. $-0.4^{\circ}C$, b.p. $150^{\circ}C$. More acidic than water. Strong oxidizing agent in acidic and basic aqueous solutions. Decomposes catalytically in presence of metal ions generating O_2.	

Selenium	Tellurium	Polonium
SeO_2, colourless crystals which are very soluble in H_2O, m.p. 34°C, subl. 317°C. Very toxic and corrosive. Has infinite chains in solid state. Oxidizing agent especially in organic chemistry for oxidation of alkenes and alkynes. Formed by burning the element in air or treating it with concentrated HNO_3.	TeO_2, colourless crystals, α form m.p. 733°C rutile-like infinite structure with pyramidal TeO_5 fragments; β-form yellow layer structure based on edge sharing TeO_4 units. Soluble in NaOCl, HCl but almost insoluble in H_2O (amphoteric).	PoO_2 exists as fcc (fluorite) and tetragonal forms and formed from Po + O_2 at 250°C. Sublimes at 885°C.
SeO_3, colourless, hygroscopic solid, m.p. 165°C (decomp). Tetrameric as solid. Thermodynamically unstable with respect to SeO_2 and O_2. Stronger oxidizing agent than SO_3.	TeO_3 Yellow orange solid synthesized by dehydration of $Te(OH)_6$ at 300°C. Decomp. 400°C to TeO_2 and O_2. Acidic oxide. Strong oxidizing agent. Insoluble in H_2O, but forms selenates on heating with solid bases.	PoO_3 not known.
H_2SeO_3, colourless crystalline solid, layers of SeO_3 limited by hydrogen bonds, m.p. 70°C, soluble in H_2O but decomposes above 70°C. Oxidizing agent. Reasonably acidic $pK_a(1)$, 2.6, $pK_a(2)$ 8.3 (25°C). Oxidized by halogens to H_2SeO_4 can also be reduced to Se by $NaBH_4$, SO_2, or H_2S. Forms salts e.g. $NaHSeO_3$ and Na_2SeO_3.	H_2TeO_3, colourless crystals slightly soluble in H_2O. Acidic $pK_a(1)$ 2.7, $pK_a(2)$ 7.7 (250°C). May be oxidized to H_2TeO_4 and reduced to Te readily. Forms salts, e.g. Na_2TeO_3.	$Po(OH)_4$, pale yellow solid which is soluble in NaOH. Basic.
H_2SeO_4, colourless deliquescent crystals, m.p. 58–62°C, very soluble in H_2O, strong acid, $pK_a(1)$ 1.7 (25°C) made from SeO_2 and H_2O_2. Also stronger oxidizing agent than H_2SO_4. Liberates O_2 on heating above 200°C.	H_2TeO_4, colourless hygroscopic crystals, m.p. 136°C hydrated form $H_2TeO_4(H_2O)_2$ is actually $Te(OH)_6$. Weak dibasic acid $pK_a(1) \sim 7$. forms salts e.g. $NaTeO(OH)_5$ and $K_2TeO_2(OH)_4$, also Ag_6TeO_6. Strong oxidizing agent but kinetically slow.	
SeH_2 hydrogen selenide	TeH_2, hydrogen telluride	PoH_2, very poorly characterized.
Very toxic evil smelling gas, m.p. −64°C, b.p. −42°C which is soluble in H_2O. $pK_a(1)$ 3.9; $pK_a(2)$ 11.0 at 25°C. Readily oxidized by O_2 (flammable) depositing colloidal red selenium.	Highly poisonous and evil smelling gas, oxidized by O_2 to elemental tellurium, m.p. −49°C, b.p. −2°C. Thermally unstable above 0°C, and decomposes in H_2O. $pK_a(1)$ 2.6, ~11.	

Group 17

	Fluorine, F_2	Chlorine, Cl_2
Colour of vapour	Pale yellow gas	Green-yellow gas
Melting point / $^{\circ}C$	–233	–103
Boiling point / $^{\circ}C$	–188	–34.6
Smell (safety hazard)	Very irritating, but less so down the group: $F_2 > Cl_2 > Br_2 > I_2$	
Solubility in water g / 100 mL	reacts with water $\rightarrow H^+ + F^- + HOF$	0.82, slight hydrolysis
Density (l or s)	1.1 (liquid at 85K)	1.51 (liquid at 293K)
Bleaching	Destroys fabric	Excellent
Support of combustion	Extremely vigorous	Excellent
Action on H_2	Explodes at $-253^{\circ}C \rightarrow HF$	Explodes in sunlight $\rightarrow HCl$
Action on metals	Most metals burn	Some metals burn
Action on non-metals	All except N_2 and O_2 react	All except N_2, O_2, and C react
Action on oxidizing agents	Nil	Nil
Oxidizing action	Extremely vigorous	Vigorous
Oxides	F_2O colourless gas, yellow liquid, m.p. $-224^{\circ}C$, b.p. $-145^{\circ}C$. Powerful oxidizing agent. Highly toxic , corrosive, and explosive. F_2O_2, made from electric discharge of $O_2 + F_2$, thermally unstable brown gas, m.p. $-164^{\circ}C$, b.p. $-57^{\circ}C$. Decomposes to $O_2 + F_2$ at $-100^{\circ}C$.	Cl_2O, made from $Cl_2 + HgO$, powerful oxidizing agent. Reddish brown liquid, m.p. $-121^{\circ}C$, b.p. $2^{\circ}C$. Soluble in organic solvents, reacts with H_2O to give HOCl. ClO_2, orange green gas, m.p. $59^{\circ}C$, b.p. $11^{\circ}C$. Powerful oxidizing agent, unstable in light. Paramagnetic, insoluble in water. Used widely for bleaching wood pulp and water purification. Cl_2O_6, ($ClO_2^+ClO_4^-$), red liquid, m.p. $4^{\circ}C$, b.p. $203^{\circ}C$. Dissociates to ClO_3 as a gas, very powerful oxidant of organic compounds. Cl_2O_7, anhydride of $HClO_4$ (synthesis: $P_2O_5 + HClO_4$). Colourless liquid, m.p. $-92^{\circ}C$, b.p. $82^{\circ}C$. Most stable of chlorine oxides, but still explosive! Reacts explosively with organic compounds.

Bromine, Br$_2$	Iodine, I$_2$
Red-brown liquid	Violet solid
–7.2	113.7
59	184.5
Safety hazard becomes less $\longrightarrow$	
3.5	0.015
3.12 (liquid at 293K)	4.93 (solid)
Medium	Nil
Medium	Slight
Combines on Pt spiral $\rightarrow$ HBr	Partial and reversible $\rightarrow$ HI
All metals corrode	Most metals corrode
Only P, As, S, Se, Te, F, Cl, and I react	Only P, As, F, Cl, and Br react
Nil	HNO$_3$ gives I$_2$O$_5$
Medium	Slight
Br$_2$O, yellow solid, m.p. –18°C (decomp.), soluble in CCl$_4$, synthesized from Br$_2$ + HgO, decomposes above –60°C.	I$_2$O, made from I$_2$ + O$_3$ in Ac$_2$O. Reactive intermediate.
BrO$_2$,dark yellow solid, decomp. at 0°C. Explodes on heating, unstable above –40°C.	I$_2$O$_4$, (IO$^+$IO$_3^-$) yellow crystals decomp. at 130°C to I$_2$O$_5$ + I$_2$.
	I$_2$O$_5$, colourless hygroscopic crystals, insoluble in organic solvents. Formed by dehydration of HIO$_3$ at 240°C. Stable up to 300°C. Oxidizes H$_2$S, HCl, CO under mild conditions.
	I$_4$O$_9$, I(IO$_3$)$_3$, yellow crystals, hygroscopic, decomposes at 75°C to I$_2$O$_5$ + I$_2$.

Group 17 (cont.)

	Fluorine	Chlorine
Oxo-acids and salts	HOF, colourless gas, m.p. $-117^{\circ}C$ obtained from $F_2 + H_2O$ at $-40^{\circ}C$ spontaneously decomposes at room temperature to $HF + O_2$.	HOCl, formed from $Cl_2O + H_2O$. Powerful bleach, soluble in water, $pK_a = 7.3$. Disproportionates in water solution to $Cl^- + ClO_3^-$ at $75^{\circ}C$.
		$HClO_3$, fairly stable in aqueous solution at room temperature and as salts, $pK_a \sim 2.7$. Strong oxidizing agent.
		$HClO_4$, synthesized from $KClO_4 + H_2SO_4$, m.p. $-112^{\circ}C$, b.p. $130^{\circ}C$, $pK_a \sim -3$. Very soluble in water, constant boiling mixture at $203^{\circ}C$, 72%. Powerful oxidant, explosive with organic compounds.
		Stability increases down the group and with percentage oxygen.
Fluorides		ClF, colourless solid, pale yellow liquid, m.p. $-156^{\circ}C$, $-100^{\circ}C$. Reacts violently with water and organic compounds. Fluorinating agent.
		ClF_3, colourless gas, m.p. $-76^{\circ}C$, b.p. $12^{\circ}C$. Very reactive, explosive with water, violently with organic compounds.
		ClF_5, colourless gas, very vigorous fluorinating agent, m.p. $-103^{\circ}C$, b.p. $-13^{\circ}C$.
Solubility	Soluble in water	Insoluble in water, soluble in NH_4OH.
Action on NaOH	Gives F_2O	Gives $OCl^- + ClO_3^-$

Summary of Group 17 trends

1. All the elements are diatomic and molecular and the boiling and melting points increase as a result of the increasing van der Waals interactions between the diatomic molecules for the heavier elements.

2. The elements are typical non-metals in their physical and chemical properties. They form anionic compounds based on X^- (X = Halogen) which is associated with the completed octet.

   The ionic compounds MX become progressively less ionic as the relative atomic mass of X increases, because of the decreasing electronegativity of the halogens. Iodine has the greatest tendency to form cationic species, e.g. I_2^+, I_5^+, because it has the lowest ionization energy. The cation Br_2^+ is known in $Br_2^+Sb_3F_{16}^-$ and Br_5^+ has been reported.

Bromine	Iodine
HOBr, formed from $Br_2 + H_2O + AgNO_3$ (HgO). Strong bleaching agent, disproportionates at room temperature, $pK_a = 8.8$.	HOI, unstable due to very rapid disproportionation.
HBrO_3, stable in aqueous solution and as salts, $pK_a \sim 0.7$. Strong oxidizing agent.	HIO_3, colourless crystals, most stable of HXO_3 compounds, m.p. $110^\circ C$ (decomp.) from $HIO_3 + HNO_3$. Very soluble in water, insoluble in organic solvents.
HBrO_4, formed by oxidation of BrO_3^- with XeF_2 or F_2. 6M aqueous solution stable up to $100^\circ C$. Very strong oxidant.	HIO_4, formed by dehydration of H_5IO_6 at $100^\circ C$. In solution exists as IO_4^-, IO_5^{3-}, IO_6^{5-}, and $(HO)_2I_2O_8^{4-}$.
In oxygen compounds, I_2 displaces Cl_2.	H_5IO_6, colourless crystals, m.p. $122^\circ C$, $pK_a(1) = 1.6$. Powerful oxidizing agent in aqueous solutions. Periodates are strong and rapid oxidizing agents.
	In hydrogen compounds, Cl_2 displaces I_2.
The salts are more stable than the acids	
BrF, red/brown gas, m.p. $-33^\circ C$, b.p. $20^\circ C$.	IF, dark solid, disproportionates above $-40^\circ C$. Decomposed by water.
BrF_3, powerful fluorinating agent, straw coloured liquid, m.p. $9^\circ C$, b.p. $127^\circ C$. Reacts violently with water and organic compounds.	IF_3, yellow solid, decomp. above $-30^\circ C$ to $I_2 + IF_5$, reacts explosively with water.
BrF_5, colourless liquid, m.p. $-61^\circ C$, b.p. $41^\circ C$. Very vigorous fluorinating agent. Reacts violently with water.	IF_5, yellow liquid, m.p. $9^\circ C$, b.p. $105^\circ C$. Reacts violently with water. Strong fluorinating agent.
	IF_7, vapour hydrolyses slowly, colourless gas, strong fluorinating agent.
Insoluble in water, slightly soluble in NH_4OH	Insoluble in water and NH_4OH
Gives $OBr^- + BrO_3^-$	Gives $OI^- + IO_3^-$

3. The atoms also form strong covalent bonds with other non-metals. The mean bond enthalpies for E–X bonds are particularly large for fluorine and therefore a wide range of molecular fluorides is known and fluorine is particularly effective at bringing out the highest valencies of the non-metals and highest oxidation states of the metals.

4. The oxidizing ability of the halogens decreases markedly down the Group; $F_2 > Cl_2 > Br_2 > I_2$ and only iodine is oxidized by nitric acid.

5. The stabilities of the hydrogen halides decreases down the Group, but their acid strengths increase.

6. Only H–F forms strong hydrogen bonds and this is reflected in the boiling and melting points of the hydrogen halides.

7. The halogens form many interhalogen compounds with the less electronegative halogen surrounded by the more electronegative halogens. Neutral, anionic, and cationic interhalogen compounds are known. ICl and IBr are widely used in organic synthesis and are commercially available.

   The most extensive series of compounds exists for iodine, e.g. IF_7, IF_5, ICl_4^-, ICl_2^-. Fluorine does not form any interhalogen compounds where it occupies the central position within the molecule.

8. Oxygen fluorides are extremely strong and reactive oxidants and have been explored as potential rocket fuels, the oxides become less reactive down the column and more numerous. Iodine forms a particularly wide range of oxides.

9. The perhalates, EO_4^-, are only known for Cl, Br, and I. They exhibit an alternation in their oxidizing abilities and the perbromates are particularly strong oxidizing agents.

10. In the highest oxidation state (+7) the relative oxidizing ability is:

$$Br > I > Cl$$

and results in the formation of the corresponding +5 oxoanions.

			$E^\oplus$/V
ClO_4^-	$\rightleftarrows$	ClO_3^-	1.20
BrO_4^-	$\rightleftarrows$	BrO_3^-	1.85
IO_4^-	$\rightleftarrows$	IO_3^-	1.63

In common with the examples discussed above the oxoanions are less strongly oxidizing in basic solution, at $a_{OH^-} = 1$, for ClO_4^-/ClO_3^- $E^\oplus = 0.37$ V, and even then readily disproportionate to HOCl and HOClO.

The +3 oxidation state is only significant for chlorine and the +1 oxidation state is well established for Cl, Br, and I. Solutions of alkali metal hypochlorites and hypobromites are prepared by passing Cl_2 or Br_2 respectively into cold solutions of the Group 1 hydroxide. They are strong oxidizing agents:

			$E^\oplus$/V, $a_{OH^-} = 1$
ClO^-	$\rightleftarrows$	Cl^-	0.89
BrO^-	$\rightleftarrows$	Br^-	0.76
IO^-	$\rightleftarrows$	I^-	0.48

The hypohalite ions disproportionate according to the equation:

$$3XO^- \rightleftarrows 2X^- + XO_3^-$$

the equilibrium constants are 10^{27} for ClO^-/Cl^- (the reaction is slow at room temperature), 10^{15} for BrO^-/Br^-, and 10^{20} for IO^-/I^-. HOF has been prepared from ice + F_2 but is very reactive, decomposing to HF + O_2.

The top twenty industrially produced inorganic substances (US)

Compound	Quantity produced in US 1995 (billions of kg per annum)	Mode of synthesis	Properties	Uses
H_2SO_4	43	SO_2 from burning S or FeS_2, then catalytic oxidation to SO_3(V_2O_5/ Pt catalysts) at 400°C. SO_3 dissolved in conc. H_2SO_4.	Colourless oily liquid, b.p. 340°C. Strong acid sol. in all proportions with H_2O.	Making fertilizers such as ammonium sulfate and super phosphate. Synthesis of organic sulfuric acids, rayon. Petroleum refining. Lead acid batteries.
N_2	31	Fractional distillation from liquid air.	Colourless gas, b.p. −196°C.	Used in the synthesis of ammonia, used as a coolant. Also as an unreative gas.
O_2	24	Fractional distillation from liquid air.	Colourless gas, b.p. 183°C.	Used in combination with hydrocarbon gases in welding torches. Also in open-hearth furnaces for steel production to augment O_2 in air. Used as a respiratory gas. Also in production of *synthesis gas* ($CO+H_2$).
CaO	19	Obtained from calcining $CaCO_3$.	Colourless alkaline solid. Addition of H_2O results in the liberation of much heat giving $Ca(OH)_2$.	Preparation of cement and as a general purpose alkaline particularly in agriculture. Used in paper and glass manufacture, leather tanning, and sugar refining.
NH_3	16	Haber process—N_2 + H_2 at 500–600°C under pressure over an Fe catalyst.	Colourless pungent gas, b.p. −33°C.	NH_3 and ammonium salts are used directly as fertilizers. Also used as refrigerant gas. Converted in large quantities by catalytic oxidation in HNO_3.
H_3PO_4	12	Made from mineral Ca_3PO_4 and sulfuric acid.	Colourless crystals, m.p. 42°C. Water soluble and resistant to reduction and oxidation	Used in soft drinks, dental cements, rust-proofing, and for making phosphate salts which are used in water softeners, detergents, and fertilizers.
$NaOH$	12	Electrolytically manufactured from conc. NaCl solutions (Castner–Kellner process).	Colourless solid m.p. 323°C. Very soluble in H_2O.	Industry's most important alkali, used for neutralizations, phenol synthesis. Manufacture of $NaOCl$, Na_3PO_4, Na_2S, etc. and most importantly with CO_2 to give Na_2CO_3.

Compound	Quantity produced in US 1995 (billions of kg per annum)	Mode of synthesis	Properties	Uses
Cl_2	11	Electrolytically manufactured by the electrolysis of brine (conc. NaCl solution).	Greenish yellow gas, b.p. $-34^{\circ}C$; strongly oxidizing.	Used widely as a bleaching agent in paper industry and as a disinfectant. Important for the synthesis of organo-chlorine compounds.
Na_2CO_3	10	Either from NaOH and CO_2 or NaCl + $CO_2/NH_3 \rightarrow NaHCO_3$ $\rightarrow$ heat Na_2CO_3 (Solvay Process).	Colourless solid, v. soluble in H_2O, m.p. $850^{\circ}C$.	Used widely in paper pulp in soap and detergent industries to produce sodium salts. Raw material for glass making
HNO_3	8	Catalytic oxidation of NH_3.	Colourless liquid miscible with H_2O, m.p. $-42^{\circ}C$. Strong acid and oxidizing agent.	Used widely for the synthesis of inorganic and organic nitro-compounds which are used in the production of dyes and explosives. Metal nitrates used as fertilizers.
NH_4NO_3	7	$NH_3 + HNO_3$	Colourless hygroscopic solid, v. sol. H_2O, EtOH, NH_3.	Explosives, fertilizers.
CO_2	5	Produced by recovery of gases from furnaces or fermentation processes, or from heating carbonates or treating them with acids.	Colourless odourless gas. Reduced to CO by C or H_2.	Used in manufacture of Na_2CO_3, in soft drinks, as a fire extinguisher gas. Solid CO_2 widely used as coolant (dry ice $-78^{\circ}C$). As a super critical liquid for the extraction of caffeine.
HCl	3	Produced commercially as a side product of the chlorination of organic hydrocarbons.	Colourless gas, b.p. $-85^{\circ}C$ dissolves readily in water to give a strong acid.	Used for the extraction of metals, food processing, and synthesis of organo-chlorine compounds.
$(NH_4)_2SO_4$	2	Made from NH_3 and H_2SO_4.	Colorless crystalline solid, Sol. in water.	Widely used as a fertilizer.
Carbon black	1.4	Burning hydrocarbons in a deficiency of air.	Black amorphous solid, high surface area. Adsorbs a wide range of gases.	Used as a pigment and reinforcing agent for example in rubber tyres. Electrical and lubricating application. Moderator in nuclear reactors.
KOH	1.4	Electrolysis of concentrated KCl solutions.	Colourless solid, m.p. $406^{\circ}C$.	Used in glass manufacture and soap industry.
TiO_2	1.4	K_2TiF_6 decomposed by NH_3 and ignited.	Colourless solid, m.p. $1843^{\circ}C$.	Brilliant white pigment used in paints.

Further reading

Inorganic Chemistry, D. F. Shriver, P. W. Atkins and C. H. Langford, 2nd Edn., Oxford University Press, Oxford, 1994.

Basic Inorganic Chemistry, F. A. Cotton, G. Wilkinson and P. L. Gaus, 3rd Edn., John Wiley, New York, 1995.

Descriptive Inorganic Chemistry, G. Rayner-Canham, W. H. Freeman, New York, 1995.

Chemistry of the Elements, N. N. Greenwood and A. Earnshaw, 2nd Edn., Butterworth-Heinemann, Oxford, 1997.

Advanced Inorganic Chemistry, F. A. Cotton and G. Wilkinson, 5th Edn., John Wiley, New York, 1988.

Main Group Elements, A. G. Massey, Ellis Horwood, Chichester, 1990.

Structural Inorganic Chemistry, A. F. Wells, 7th Edn., Clarendon Press, Oxford, 1984.

Dictionary of Inorganic Compounds, Exec. Ed., J. E. McIntyre, Chapman and Hall, 1994. A database of 40 000 inorganic compounds-—also available on CD-ROM.

Periodicity and the s and p Block Elements, N. C. Norman, Oxford University Press, 1997.

Inorganic Chemistry of the Main Group Elements, R. B. King, VCH-Weinheim, 1994.

Organometallics, A Concise Introduction, Ch. Elschenbroich and A. Salzer, VCH-Weinheim, 1989.

The Elements, J. Emsley, 3rd Edn., Oxford University Press, Oxford, 1998.

Periodic Table on the World Wide Web at
`http://www.shef.ac.uk/chem/chemdex/periodic-tables.html`

For a more general introduction to the World Wide Web as a chemical information tool see P. Murray Rust, H. S. Rzepa and B. J. Whitaker, *Chem. Soc. Rev.*, 1997, **26**, 1.

3 Horizontal and diagonal trends for the s and p block elements

3.1 Introduction

Although the Mendeleev classification emphasizes vertical relationships within the Periodic Table, it is also important to recognize the consequences of variations in properties across the Periodic Table. For the pre- and post-transition elements these comparisons are complicated by the systematic variations in the valencies and oxidation states of the elements across a series. There are, nonetheless, some interesting horizontal relationships. The final section of this chapter also draws attention to some important diagonal relationships.

For the transition elements, the lanthanides, and the actinides, compounds with common formulae are formed in large numbers and the horizontal comparisons are very important. These latter comparisons are discussed in Chapter 5. This chapter focuses attention on some of the more significant horizontal and diagonal relationships for the s and p block elements.

3.2 Variations in atomic properties

Ionization energies

The variation in first ionization energies for those elements which have $ns^x np^y$ ($x = 1$ or 2, $y = 0 - 6$) electronic configurations are illustrated in Figs 3.1 and 3.2.

(a)   (b)

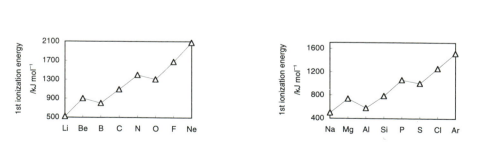

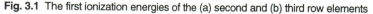

Fig. 3.1 The first ionization energies of the (a) second and (b) third row elements

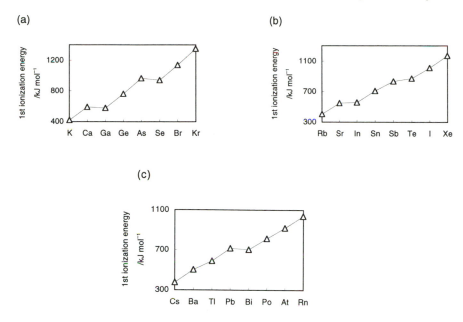

Fig. 3.2 The first ionization energies of the s and p elements in the (a) fourth, (b) fifth, and (c) sixth rows of the Periodic Table. The first ionization energy of Tl ($5s^2 5p^1$) does not show the discontinuity which is generally observed. This has been attributed to relativistic effects.

All rows of the Periodic Table show the same basic pattern—an overall increase in ionization energies across the series with superimposed 'regular' discontinuities. The overall increase in ionization energies across the series results from the increase in effective nuclear charge and the contraction in atomic radius which results from it.

The first ionization energies of atoms with $n s^2 n p^1$ and $n s^2 n p^4$ ground state electronic configurations are lower than might be estimated on the basis of a simple extrapolation. The first discontinuity is attributed to the lower binding energies of the np orbitals relative to the ns orbitals and the second from the effects of pairing electrons in the same orbital of the p subshell. The fourth electron coming into the p shell has to be placed in an orbital which is already occupied and therefore it experiences additional electron–electron repulsion. Furthermore, it has a spin which must oppose the spins of the other three electrons and therefore the repulsion is not mitigated by changes in exchange energy.

The changes in exchange energies which accompany ionization for atoms with p¹ to p⁶ configurations are given in Table 3.1. The atoms with p¹ and p⁴ electronic configurations lose no exchange energies on ionization and therefore they lie on the straight line which represents the overall increase in ionization energies in Fig. 3.3. In contrast p³ and p⁶ atoms lose 2K units of exchange energy and therefore their ionization requires more energy and they lie above the straight line in the figure. The other electronic configurations lose 1K unit of exchange energy. When these exchange energy effects are superimposed on the overall increase in orbital binding energies, the observed trend in ionization energies is reproduced as shown in Fig. 3.3.

Exchange energies are defined and discussed in Chapter 1.

Table 3.1 Changes in exchange energy for ionization processes: $M(g) \rightarrow M^+(g)$

M(g) configuration	Change in exchange energy
p¹	0
p²	K
p³	2K
p⁴	0
p⁵	K
p⁶	2K

The effect of the changes of exchange energy as the atom is ionized is shown in Fig. 3.3.

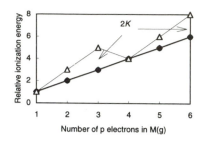

Fig. 3.3 A diagram showing the effect of exchange energy variations on an otherwise general linear increase in ionization energy as the number of p electrons changes

The spread in ionization energies across the series becomes smaller as the principal quantum number increases and the atoms become larger. Therefore, the enormous range of chemical properties across the second series (Li, Be, B, C, N, O, F, and Ne) is diminished somewhat as n increases. So, when the value of n reaches 5 the elements Rb, Sr, In, Sb, and Te are essentially metallic, although of course they show very different chemical properties. Also, the noble gas, xenon, actually forms a wide range of compounds, many of which are similar to those formed by the adjacent halogen, iodine, whereas F and Ne could not be more different in their chemical properties.

Phosphorus and sulphur are more alike than nitrogen and oxygen and some of the chemical similarities of the former pair of elements are emphasized in Table 3.2. The differences in valencies of adjacent main group elements often mask these similarities.

Table 3.2 A comparison of the properties of phosphorus and sulfur

Phosphorus	Sulfur
Both exist as easily melted solids soluble in chloroform	
Both burn readily in oxygen and in chlorine, union is exothermic	
Both give oxides soluble in water to give acids	
$P_4O_6 \longrightarrow H_3PO_3$	$SO_2 \longrightarrow H_2SO_3$
$P_4O_{10} \longrightarrow H_3PO_4$	$SO_3 \longrightarrow H_2SO_4$
The higher oxide and acid in each case are the more stable	
The lower oxide and acid in each case are reducing agents	
Both elements unite directly with metals giving:-	
phosphides	sulfides
These compounds may both act upon water to give hydrides:-	
PH_3	H_2S
Both elements are soluble in NaOH to give hydrides	
PH_3, basic, liberated	H_2S, acidic, unites with the NaOH
Both elements exhibit allotropy	
Chiefly tervalent and pentavalent	Chiefly bivalent, tetravalent, and hexavalent

Fig. 3.4 shows the variations in the cumulative ionization energies for the elements Al, Si, and P respectively. These cumulative ionization energies clearly define the valence electrons of the atoms, i.e. the ionization energy required to remove an electron from the $(n - 1)$ shell is prohibitively large. They also define the commonly observed oxidation states of the elements in their compounds as $x + y$ and $x + y - 2$, where the electron configuration of the neutral atom is $ns^x np^y$. The graphs show significant changes in gradient for the formation of the M^{x+y-2} and M^{x+y} ions.

The ionization data illustrate the importance of the subshell structure of the Periodic Table as well as the shell structure. In this instance the subshell corresponds to the formation of ions with a complete ns subshell. The chemical implications of this subshell structure is discussed in the previous chapter under the description 'inert-pair' effect.

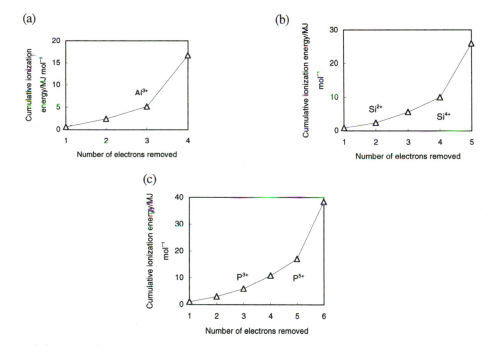

Fig. 3.4 The cumulative totals of ionization energies of (a) aluminium, (b) silicon, and (c) phosphorus plotted against the total number of electrons removed

Fig. 3.5 show the variations in the cumulative ionization energies for the sulfur and chlorine atoms. The less dramatic changes in the slope of the graph are consistent with the observation that sulfur forms compounds in oxidation states 2, 4, and 6 and chlorine forms compounds in oxidation states 1, 3, 5, and 7. Therefore, the atoms on the right hand side of the Periodic Table are able to form compounds where the central atom has an incomplete p shell.

The cumulative ionization energy trends for S and Cl do not in themselves exclude the possibility of forming compounds with unpaired electrons, i.e. sulfur in oxidation states 1, 3, and 5. However, such radicals tend to be reactive and if the S–S bond enthalpy is large enough they form dimers with sulfur-sulfur bonds. Typical examples are shown in Fig. 3.6.

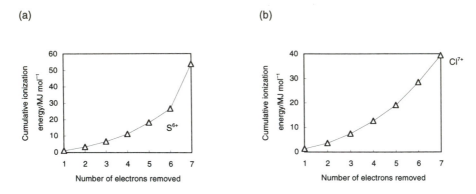

Fig. 3.5 The variation in the cumulative total of ionization energies of the (a) sulfur and (b) chlorine atoms as the number of removed electrons increases

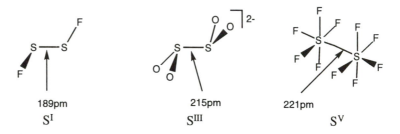

Fig. 3.6 Molecules and ions of sulfur containing element–element bonds

The formal oxidation state of the central atom is not influenced by the formation of element–element bonds. Therefore, the radical and the dimer have the same formal oxidation state. They do however, have different valencies. The valencies of the sulfur atoms in the above examples are 2, 4, and 6 respectively.

Chlorine, with its relatively weak Cl–Cl bond, forms a wider range of radicals, particularly in its oxides, e.g. ClO and ClO_2, with chlorine in the formal oxidation states 2 and 4. However, such paramagnetic molecules are relatively rare compared with the wide range of chlorine molecules which have been isolated and where the formal oxidation state is 1, 3, 5, or 7.

It is worth noting that the isolation of specific compounds, which can be readily stored in significant quantities at room temperature, depends on kinetic as well as thermodynamic factors. In general, radicals tend to be more reactive than molecules with even numbers of electrons and therefore they readily form dimers or abstract H or Cl atoms from other molecules. Both of these reactions result in diamagnetic products.

Summary
The observed oxidation states of the p block elements are dominated by the closed shell, s^0p^0 and s^2p^6, and subshell, s^2, electronic configurations of the atoms. However, the later p block elements also form compounds where the formal oxidation state

leads to ions with s^2p^2 and s^2p^4 electron configurations, e.g. SF_2 and XeF_4 (s^2p^2), and XeF_2(s^2p^4).

Molecules with unpaired electrons (radicals) tend to be rare, because they dimerize readily to form dimers or chains with element–element bonds.

The preferred oxidation states and valencies of the p block elements are discussed further in Section 3.4.

Electron attachment enthalpies

Fig. 3.7 illustrates the enthalpy changes for electron attachment of elements with $ns^x np^y$ electronic configurations. The discontinuities noted above for ionization energies are noticeable for the following processes:

$$ns^2 + e \rightarrow ns^2np^1 \text{ (Be, Mg, Ca, Sr, Ba, Ra)}$$
$$ns^2np^3 + e \rightarrow ns^2np^4 \text{ (N, P, As, Sb, Bi)}$$

and the electronic origins are similar.

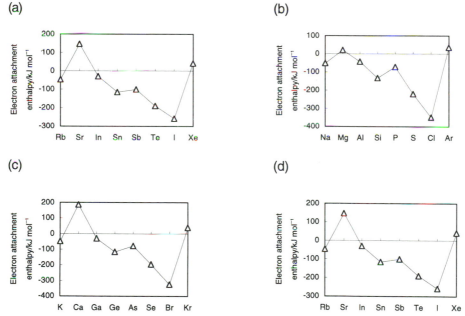

Fig. 3.7 The electron attachment enthalpy changes for the s and p elements of the (a) second, (b) third, (c) fourth, and (d) fifth rows of the Periodic Table

Particularly noteworthy features of Fig. 3.7 are the favourable electron attachment enthalpies associated with the halogen atoms which are just one electron short of a closed shell configuration. The electron attachment enthalpies also become less favourable from right to left across the Periodic Table and this accords with the observation that for the second row elements only the ions F^-, O^{2-}, N^{3-} are observed in infinite solid state structures which can be described as unambiguously ionic.

The electron attachment enthalpies for the noble gases are so unfavourable that they cannot be measured, calculated values are given in the figure.

The other noteworthy features of Fig. 3.7 are the relatively favourable electron attachment enthalpies for the Group 1 elements which result from the filling of the s subshells.

$$ns^1 \quad + \quad e \rightarrow \quad ns^2$$

These metals do indeed form compounds which contain the M^- anions (M = Na, K, and Rb). The recombination of M^- and M^+ to form metal is prevented by complexing the M^+ ion with a polydentate ligand. A specific example of such a complex, [sodium-2.2.2-crypt] sodide, is illustrated in Fig. 3.8.

Electronegativities

The electronegativity variations across the series reflect the ionization energies and electron attachment enthalpies discussed above. In particular the electronegativities increase across the series. The spread in electronegativities which is illustrated in Fig. 3.9 diminishes for successive rows of the Periodic Table.

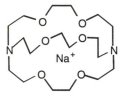

Na⁻

Fig. 3.8 The structure of [Na(2.2.2–crypt)]⁺Na⁻

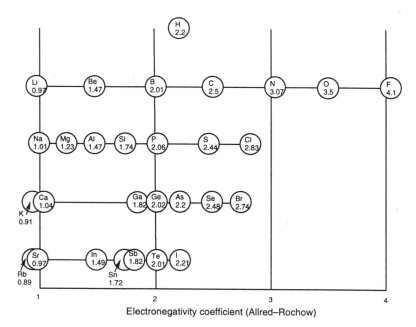

Electronegativity coefficient (Allred–Rochow)

Fig. 3.9 The Allred–Rochow electronegativity coefficients of the first five rows of s and p block elements

The other noteworthy feature of Fig. 3.9 is the marked alternation in electronegativities of the Group 13–16 elements. The chemical implications of this alternation effect have been discussed in some detail in Chapter 2.

The difference in electronegativity between the atoms in a compound determine the ionic character of the bonds in that compound. For example, organometallic compounds which contain alkyl or aryl radicals bonded to metal and metalloid atoms are widespread. The electronegativity differences indicated in Fig. 3.9 suggest that those with B, N, P, S, Cl, Ge, As, Se, Br, Te, and I will be essentially covalent because the electronegativity difference between carbon and these elements is less than 0.5. On the other hand, the organometallic compounds of Groups 1 and 2 are mainly ionic. These predictions are borne out and illustrated in the diagram of the Periodic Table in Fig. 3.10. Similar correlations may be developed for other classes of compound, e.g. hydrides, oxides, fluorides, sulfides, etc.

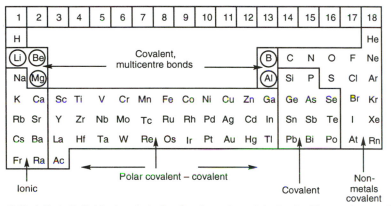

Fig. 3.10 A Periodic Table demonstrating the elements participating in different classes of organometallic compound formation

Atomic sizes

The covalent radii of the elements decrease across the series as expected from the increase in the effective nuclear charges. The variation is illustrated schematically in Fig. 3.11.

The sizes of the cations and anions derived from these elements are also illustrated in Fig. 3.11. It is noteworthy that the radii of the cations are always smaller than the covalent radii and the radii of the anions are always larger.

3.3 Variations in physical properties of the elements

Metallic character

Across any row of the s and p block elements the electropositive character of the element decreases and the electronegativity increases. Consequently the properties of the elements always undergo a transition from metallic to non-metallic. Those properties which typify metals and non-metals are summarized in Table 3.3.

The characteristics of metals and non-metals summarized in the table are by no means definitive and are not universally applicable. Many exceptions may be quoted for specific properties. This is hardly surprising since the definition of a metal is trying to encompass metals which are as different as sodium and gold. Perhaps the most generally applicable criterion for a metal is based on its electrical conductivity.

Table 3.3 The main differences between metal and non-metal elements. Comparison of metal and non-metal properties

Metals	Non-metals
Close packed structures with high coordination numbers	Low-connectivity structures which are loosely packed
All infinite structures	Molecular, layer, or infinite structures
Physical properties	
Good conductors	Poor conductors
Malleable, ductile, elastic	Brittle
Lustrous, hard, high tensile strength	Non-reflective, low tensile strength
Very large range of densities and melting points	Very large range of densities and melting points
Chemical properties	
Tend to form cations	Tend to form anions
Reducing agents	More electronegative are oxidizing agents
More electropositive react with water, liberating H_2	Never liberate H_2 from acids
Least electropositive require oxidizing acids to liberate H_2	
With other metals form alloys	With metals form anions
With oxygen more electropositive react to form oxides	Form covalent oxides
Properties of compounds	
Oxides – more electropositive form basic oxides; less electropositive form amphoteric oxides	Oxides are covalent and acidic
Chlorides – ionic, water soluble forming solvated ions	Chlorides – molecular, readily hydrolysed
Hydrides – more electropositive are ionic reducing agents; less electropositive form structures with bridging hydrogen atoms	Hydrides either non (NH_3)- or weakly (PH_3)-reducing
Ions are Lewis acids which form a wide range of complexes	Form anions which are Lewis bases and which therefore function as ligands

Atomic electrical conductivity is the conductivity of a block of solid 1 cm^2 in cross-section and of sufficient length to include 1 mole of elemental atoms.

Metals have atomic conductivities greater then 3×10^{-4} ohm^{-1} cm^{-4} and their conductivities decrease with temperature.

Metalloids or semi-metals have small but measurable conductivities which increase with temperature. The conductivities of these elements are generally very sensitive to the presence of impurities.

Non-metals are generally insulators.

Specific examples of properties which do not conform to the simple classification scheme developed above are summarized below.

1. Graphite and iodine are both lustrous although they are non-metals.
2. Graphite is a good conductor of electricity and is widely used as a battery electrode.

3. Some metals are rather brittle, e.g. Zn, As, and Sb and not very malleable.

4. Occasionally metals form anions, e.g. Na^-, Au^-, Pb_9^{4-} and non-metals form cations, e.g. S_4^{2+}, Se_8^{2+}, and I_2^+.

5. Sometimes the oxides of metals are acidic rather than basic, e.g. arsenic and antimony oxides.

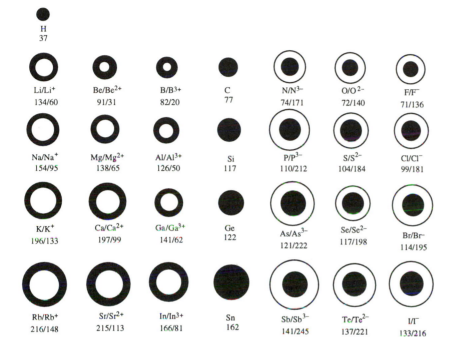

Fig. 3.11 A representation of the radii of the atoms of the s and p block elements (full circles) and their common ions (open circles)

Metallic radii and densities

From left to right across a row of the Periodic Table the sizes of the atoms generally decrease because the effective nuclear charges of the atoms increase. The metallic radii for the third series of the Periodic Table, which is the first to contain the d block elements, given in Table 3.4 confirm this trend for the pre-transition elements and the transition elements with the exception of manganese and copper which have larger radii than anticipated on the basis of a simple extrapolation. The reasons for the anomalous behaviour of manganese and copper are discussed in Chapter 5.

The trends in metallic radii for the post-transition elements are less straightforward, in part because the elements do not form related simple close-packed metallic lattices. Gallium possesses a complex structure in which each atom has one neighbour at 244 pm, two at 270 pm, two at 273 pm, and two at 279 pm. Germanium has a diamond structure, grey metallic arsenic has a layer structure in which each atom forms three strong bonds to its three nearest neighbours, and the grey metallic form of selenium has helical chains of atoms with significant interactions between adjacent chains.

Table 3.4 The metallic radii and densities of the fourth row elements

	K	Ca	Sc	Ti	V	Cr	Mn	Fe
Metallic radius/ pm	235	197	164	147	135	130	135	126
Density / kg m^{-3}	860	1540	3000	4500	6100	7200	7440	7860
	Co	Ni	Cu	Zn	Ga	Ge	As	Se
Metallic radius/ pm	125	125	128	137	141	137	139	140
Density / kg m^{-3}	8860	8900	8920	7130	5910	5320	5730	4790

The densities of this row of the Periodic Table (also summarized in Table 3.4) follow the anticipated trend from K to Cu where the decreasing sizes of the atoms and increasing relative atomic masses lead to a continuous increase in density of the elements.

The lower density of zinc compared with copper reflects the completion of the s and d metallic band structures and an expansion of the metallic radius.

Table 3.5 Comparison of the third row elements and their compounds

	Na	Mg	Al	Si
Electron configuration	$3s^1$	$3s^2$	$3s^2 3p^1$	$3s^2 3p^2$
Electronegativity	1.01	1.23	1.47	1.74
1st ionization energy/kJ mol^{-1}	495	738	577	787
Electron attachment enthalpy/kJ mol^{-1}	−21	67	−26	−135
Structure		Infinite close packed		Infinite semi-conductor (diamond)
m.p. ($^{\circ}$C)	98	651	660	1410
Bonding in compounds	Ionic	Ionic	Polar-covalent, Polarity decreasing $\longrightarrow$	
Predominant oxidation states (valencies)	1	2	3	4
Oxides	Na$_2$O ionic, basic	MgO ionic, basic	Al$_2$O$_3$ amphoteric	SiO$_2$ acidic
Hydrides	NaH ionic	MgH$_2$ ionic	AlH$_3$ polar-covalent, hydridic	SiH$_4$ covalent, neutral
Chlorides	NaCl ionic	MgCl$_2$ ionic	AlCl$_3$ polar-covalent	SiCl$_4$ covalent, readily hydrolysed

The densities of the elements from Ga to Se in which there is filling of the 4s band do not follow a simple trend, once again because they do not have analogous close packed structures.

3.4 Variations in chemical properties

Table 3.5 summarizes how the specific properties of the elements Na–Ar vary across the row and illustrates how the metal to non-metal transition is exemplified in practice.

Oxidation states and valencies

One of the triumphs of quantum theory is the way in which the filling of the quantized shells by electrons leads to atomic electronic configurations which may be used to interpret the observed oxidation states and valencies of the atoms. This is clearly evident from the data presented in Table 3.5, which shows that the preferred oxidation state (or valency) is equal to the number of valence electrons for Groups 1 and 2. For Groups 13, 14, and 15, the d subshell is effectively core-like and the group valency, N, is 10 less than the formal group numbering recommended by IUPAC. For groups 13–15 alternative $N - 2$ valencies are observed which are associated with the presence of a lone pair of electrons on the central atom. The occurrence of such compounds is described as the 'inert pair' effect, the thermodynamic origins of which were discussed in Chapter 2. For Groups 16, 17, and 18 compounds with N, $N - 2$, and $N - 4$ valencies are observed.

In older versions of the Periodic Table, when the main groups (i.e. s and p block elements) were numbered from 1 to 8, the Group valency of an element was considered to be equal to the Group number, N, for elements up to and including Group 4 (which is now Group 14) and equal to $8 - N$ for the elements of Groups 5 to 8 (which are now Groups 15 to 18). This arises from the inclusion of the transition series in the Group numbering with their full complement of ten d electrons.

P	S	Cl	Ar	
$3s^2 3p^3$	$3s^2 3p^4$	$3s^2 3p^5$	$3s^2 3p^6$	Electron configuration
2.05	2.45	2.85		Electronegativity
1060	1000	1255	1520	1st ionization energy / kJ mol^{-1}
−60	−196	−348		Electron attachment energy/kJ mol^{-1}
Molecular P_4	Molecular S_8	Molecular Cl_2		Structure
44	119	−101	−189	m.p. ($^{\circ}$C)
	Polar-covalent,	Covalent		Bonding in compounds
Polarity decreasing $\longrightarrow$				
				Predominant oxidation states (valencies)
(3 or 5)	(2, 4, or 6)	(1, 3, 5, or 7)		
P_4O_{10}	SO_3	Cl_2O_7		Oxides
Acidic	Acidic	Acidic		
PH_3	SH_2	ClH		Hydrides
Covalent PCl_5 PCl_3	Covalent, protonic SCl_2 SCl_4	Covalent, protonic Cl_2		Chlorides
Covalent, higher halides particularly easily hydrolysed				

For Groups 17 and 18 the $N - 6$ valency is also observed. As noted previously, the occurrence of these alternative valencies reflects the relative cumulative ionization energies illustrated in Figs 3.4 and 3.5. The higher oxidation states and valencies are generally observed for fluorides and oxides because these elements form the strongest bonds.

It is important to emphasize that for a particular row of elements with the electronic configurations $ns^x np^y$ ($x = 1-2$; $y = 1-8$) the ionization energies of the atoms increase greatly and the properties of the elements reflect this difference. At the beginning of the series where the atoms have low ionization energies the atoms form metals where the electrons are easily delocalized throughout the structure and the resulting metals are very reactive and strongly reducing. At the other extreme, the ionization energies of the noble gases are so large that compound formation is thermodynamically unfavourable, except for Kr and Xe where their lower position in the Periodic Table leads to a relatively lower ionization energy. On increasing the principal quantum number n the difference between the atom at the beginning of the series with the lowest ionization energy and the atom at the end with the highest becomes smaller. Nonetheless, the same pattern of starting the series with a reactive metal and finishing it with a noble and relatively unreactive gas is maintained for all the horizontal rows, except for the H and He row. The variation in ionization energies across these horizontal rows is much greater than that for the elements which make up the transition series, the lanthanides, and the actinides.

The large increase in ionization energies described above is reflected in the large increase in electronegativities of the elements. The increase in effective nuclear charge results in a reduction of the atomic radii across the series. The increased electronegativity combined with the deceased size leads to the transition from metallic to molecular in the properties of the elements. Specifically Na, Mg, and Al display typical metallic properties and their melting points and densities increase to the right as more electrons become involved in metal–metal bond formation. Silicon is a typical semi-conductor and has properties intermediate between metallic and molecular.

It adopts an infinite diamond-like structure rather than a close packed metallic structure and its conductivity is intermediate between those anticipated for a metal and an insulator. Phosphorus, sulfur, and chlorine form molecular structures in the solid state and utilize their valency electrons in such a way that the atoms generally achieve noble gas configurations. They each form $8 - N$ element–element bonds (N = number of s and p valence electrons) in the molecules shown in the margin.

Fig. 3.12 illustrates the relative binding energies of the second and third row elements. For both of the series of elements the maximum binding energies are observed for the Group 14 elements and the binding energies of the other elements progressively decrease from this maximum.

The shape of the curve is determined by the number of valence electrons associated with each of the atoms. For the s and p valence orbitals the maximum number of conventional covalent bonds which can be formed is 4, i.e. that which achieves the octet rule, and the Group 14 elements contribute the correct number of electrons to form such bonds. It is noteworthy that fluorine is the only second row element which has a lower binding energy than the corresponding third row element. The transition elements show a

P: $3s^2 3p^3$

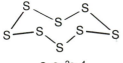

S: $3s^2 3p^4$

Cl——Cl

Cl: $3s^2 3p^5$

similar maximum in their enthalpies of atomization at approximately the centre of the transition series.

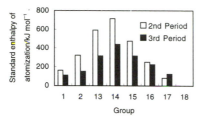

Fig. 3.12 The enthalpies of atomization (binding energies) of the s and p block elements of the second and third rows of the Periodic Table

> The binding energy of an element is a measure of the strength of the bond between two atoms of the element. It is equal to the standard enthalpy of atomization of the element

Isoelectronic relationships

The horizontal relationships developed above are complicated by the variations in valency which occur simultaneously across the Periodic Table. Some simplification of this problem may be obtained by comparing the properties of series of compounds which are isoelectronic and isostructural. For example, $[AlCl_4]^-$, $[SiCl_4]$, and $[PCl_4]^+$ are all tetrahedral and have a total of 32 valence electrons. Therefore, comparisons of their properties highlights the effect of changing the formal oxidation state of the central atom from Al^{III} to Si^{IV} to P^V. Fortunately, there are numerous series of compounds which share these relationships and they are discussed in some detail in Chapter 4.

Oxides, halides, and hydrides

Table 3.5 also traces the evolution of properties of the oxides, hydrides, and chlorides of the third row s and p block elements. The properties of the oxides and hydrides are also discussed in more detail below. On the left-hand side the large electronegativity difference between the metals and O, H, or Cl ensures that the compounds are essentially ionic. In Groups 13, 14, and 15 the bonding is polar-covalent, i.e. intermediate between ionic and covalent and in Groups 16 and 17 the oxygen compounds are predominantly covalent. These bonding differences are reflected in large differences in chemical properties, e.g. the Group 1 and 2 hydrides are hydridic reflecting the predominance of the M^+H^- and $M^{2+}2H^-$ bonding representations, the hydrides of Groups 14 and 15 are essentially neutral, and the Group 16 and 17 hydrides are protonic reflecting the bond polarities; $S^{\delta-}-H^{\delta+}$ and $Cl^{\delta-}-H^{\delta+}$. Similarly, the oxides progress from being basic, e.g. Na_2O, to amphoteric, e.g. Al_2O_3, to acidic, e.g. SO_3. These definitions reflect the relative tendency of these oxides to generate OH^- or H^+ ions in aqueous solution.

$$Na_2O \ + \ H_2O \ \rightleftharpoons \ 2Na^+ \ + \ 2OH^-$$
$$SO_3 \ + \ H_2O \ \rightleftharpoons \ 2H^+ \ + \ SO_4^{2-}$$

An insoluble oxide such as Al_2O_3 reacts with both acids and bases to form soluble salts and is therefore described as amphoteric:

$$Al_2O_3 + 6H^+ \rightleftharpoons 2Al^{3+}(aq) + 3H_2O$$

$$Al_2O_3 + 3H_2O + 2OH^- \rightleftharpoons 2[Al(OH)_4]^-$$

The ionic halides dissolve in water forming the corresponding solvated ions:

$$NaCl(s) \rightarrow Na^+(aq) + Cl^-(aq)$$

where $Na^+(aq)$ and $Cl^-(aq)$ represent the central ion surrounded by sheaths of water molecules. The covalent halides have polar-covalent E–Hal bonds which do not dissociate into $E^{n+}(aq)$ and $Hal^-(aq)$ ions because the bonds are not sufficiently polar. However, such compounds may react with water quite vigorously, e.g.

$$PCl_3(l) + 3H_2O(l) \rightarrow H_3PO_3(aq) + 3H^+(aq) + 3Cl^-(aq)$$

These molecular compounds do dissolve intact in organic solvents.

The halides also show differences which are reflected in their melting points and structural properties, as shown by the data in Table 3.6.

Table 3.6 Melting points and structures of chlorides of third row elements

	NaCl	MgCl$_2$	AlCl$_3$	SiCl$_4$	PCl$_3$	SCl$_2$	Cl$_2$
m.p./$^{\circ}$C	808	714	192*	−68	−92	−80	−101
Structure	6:6 infinite lattice	6:3 layer lattice (CdCl$_2$)	6:2 layer lattice (CrCl$_3$)	Molecular tetrahedral	Molecular trigonal pyramidal	Molecular angular	Molecular diatomic

* Under pressure since it sublimes as Al_2Cl_6 molecules

Sodium chloride has a typical ionic infinite structure and has a correspondingly high melting point. Magnesium and aluminium chlorides have layer structures with two-dimensional octahedral arrays of metal ions surrounded by chloride ions. These layers are held together by weaker van der Waals forces between chloride ions in adjacent layers. Such layer structures are often symptomatic of polar–covalent structures. $SiCl_4$, PCl_3, SCl_2, and Cl_2 have molecular structures and their melting points indicate the relative strengths of the van der Waals interactions. These decrease as the molecule becomes less polar and the polarizability of the molecule decreases. In this series of molecules the decrease in the number of chlorine atoms makes the molecule less polarizable.

For those elements with multiple oxidation states the degree of covalent character of the bonds increases as the oxidation state increases. For example, $PbCl_4$ is more covalent in its properties than $PbCl_2$. As a consequence $PbCl_2$ is soluble in hot water, but $PbCl_4$ hydrolyses according to the equation:

$$PbCl_4(l) + 2H_2O(l) \rightleftharpoons PbO_2(s) + 4H^+(aq) + 4Cl^-(aq)$$

Standard reduction potentials of the elements

The data summarized in Table 3.7 clearly suggest that the standard reduction potentials consistently become less negative across the series, i.e. the metals become less reducing. This trend may be interpreted using a Hess's law cycle similar to that shown on page 61. Across a series, the larger hydration enthalpies as n increases in M^{n+} do not compensate for the greater ionization enthalpies required to form $M^{n+}(g)$, and the larger enthalpies of atomization, i.e. $\Delta_{at}H^{\circ}$: Na < Mg < Al. The change also becomes more dramatic down a column of the Periodic Table, i.e. the change from Cs to Tl is 3.74 V, but only 1.05 V for Na to Al.

Table 3.7 Standard reduction potentials, E°/ V, for the ion/element couples of Groups 1, 2, and 13

$Li^+(aq)/Li$	$Be^{2+}(aq)/Be$	
−3.02	−1.85	
$Na^+(aq)/Na$	$Mg^{2+}(aq)/Mg$	$Al^{3+}(aq)/Al$
−2.71	−2.37	−1.66
$K^+(aq)/K$	$Ca^{2+}(aq)/Ca$	$Ga^{3+}(aq)/Ga$
−2.93	−2.87	−0.53
$Rb^+(aq)/Rb$	$Sr^{2+}(aq)/Sr$	$In^{3+}(aq)/In$
−2.99	−2.89	−0.34
$Cs^+(aq)/Cs$	$Ba^{2+}(aq)/Ba$	$Tl^{3+}(aq)/Tl$
−3.02	−2.90	+0.72

On the right hand side of the Periodic Table the Group 17 elements are strongly oxidizing and the elements become progressively less oxidizing from left to right as shown by the data of Table 3.8.

Table 3.8 Standard reduction potentials, E°/ V, for element / hydride couples of Groups 15 and 16, and element/ion couples of Group 17

$N_2(g)/NH_4^+(aq)$	$O_2(g)/H_2O$	$F_2(g)/F^-(aq)$
0.27	1.23	2.87
$P(s)/H_3P(g)$	$S(s)/H_2S(aq)$	$Cl_2(g)/Cl^-(aq)$
−0.04	0.14	1.36
$As(s)/H_3As(g)$	$Se(s)/H_2Se(g)$	$Br_2(l)/Br^-(aq)$
−0.38	−0.40	1.08
$Sb(s)/H_3Sb(g)$	$Te(s)/H_2Te(g)$	$I_2(s)/I^-(aq)$
−0.51	−0.50	0.62

3.5 Ionic–covalent transition

The transition from ionic to covalent bond formation is important for interpreting the physical and chemical properties of compounds. However, elementary textbooks do tend to oversimplify the issues. Specifically it is sometimes assumed that ionic compounds:

1.   Are involatile high m.p. solids.

2. Show poor conductivities in the solid state, but high conductivities when melted.
3. Possess a three dimensional lattice structure where all cations are surrounded by anions and *vice versa*.

Conversely, covalent compounds:

1. Are volatile low melting point solids which are soluble in organic solvents.
2. Show poor conductivities both as solids and liquids.
3. Possess crystal structures where individual and isolated molecules may be identified.

The limitations of these definitions may be appreciated by considering the properties of the compounds given in Table 3.9 which are isoelectronic, but where the electronegativity difference progressively decreases.

Table 3.9 Isoelectronic series of compounds with infinite structures

Compound / element	Lattice type and coordination	m.p. °C	$\Delta_{atom}H^{\ominus}$ /kJ mol^{-1}	Electronegativity difference
NaCl	infinite NaCl (6:6)	801	641	1.82
MgS	infinite NaCl (6:6)	2025	719	1.21
AlP	diamond (4:4)	1000 (decomp.)	796	0.59
Elemental Si	diamond (4:4)	1410	878*	0

*The value for elemental silicon has been multiplied by two to take into account the participating number of moles of atoms. The column labelled $\Delta_{atom}H^{\ominus}$ refers to the enthalpy change for the formation of the solid from the gaseous atoms in the gas phase:

$$MX(s) \rightarrow M(g) + X(g) \qquad \Delta_{atom}H^{\ominus} \quad \text{(for silicon M = X = Si)}$$

Although the bonding between atoms is clearly undergoing a transformation from ionic to covalent they all remain high melting point solids. Therefore, the melting point of a compound is not an accurate representation of the type of bonding. The melting point depends primarily on whether the compound forms a structure which is infinite, a layer, a linear chain, or a discrete molecule. Since all the compounds described above adopt infinite structures then they all have high melting points.

The data in Table 3.9 do not reinforce the popularly held misconception that ionic bonds are strong and covalent bonds are weak, indeed the bonding in all these compounds is strong and does not correlate directly with the electronegativity difference.

Ionic compounds based on simple monatomic cations and anions always have high melting points. However, the melting points of simple covalent compounds depend upon whether they retain a molecular structure in the solid state or adopt an infinite structure. The melting point of a compound with an infinite covalent structure depends primarily on the strength of the covalent bonds and the average number of bonds formed by each atom. Ionic compounds with polyatomic cations and anions may exhibit low melting points and Table 3.10 illustrates some specific examples. The larger cations or anions usually, but not always, result in a reduction in melting points because the centres of the ions are further apart and lattice enthalpies are

Table 3.10 Melting points of some molecular salts

	m.p. (°C)
[NMe$_4$]Br	>300
[NEt$_4$]Br	285 decomp.
[NPr$_4$]Br	270
[NBu$_4$]Br	102
[NBu$_4$]Br	102–104
[NBu$_4$]Br$_3$	71–73
[NBu$_4$]I	145–147
[NBu$_4$]IO$_4$	175 decomp.

reduced. Exceptions occur when the van der Waals forces between the cation and anion become significant, e.g. when there are phenyl substituents.

Since tables of ionic, covalent, and metallic radii are available it is useful to establish whether these data may be used as a criterion for distinguishing ionic and covalent bonding. The data for MgO, MgS, and MgSe given in Tables 3.11 and 3.12 clearly show that the same internuclear distances are calculated whether the ionic or covalent (metallic) radii are used and therefore they cannot be used to establish the type of bonding.

Table 3.11 Some ionic, metallic, and covalent radii

	Ionic			
	Mg^{2+}	O^{2-}	S^{2-}	Se^{2-}
Radii/pm	72	140	182	195
	Covalent (Mg, metallic)			
	Mg	O	S	Se
Radii/pm	148	66	104	117

Table 3.12 Internuclear distances in oxides, sulfides, and selenides

	Σionic radii/pm	Metallic + covalent radii/pm
MgO	212	214
MgS	254	252
MgSe	267	265

Both the ionic and covalent radii have been determined from experimental data and merely represent alternative ways of dividing the internuclear distance. The degree of ionic or covalent character can only be estimated by spectroscopic techniques which can evaluate more directly the residual charges in the cations and anions.

The data in Table 3.9 (p. 178) suggest that in this series of compounds the change in bond character leads to a preference for octahedral 6:6 coordination in the more ionic compounds and tetrahedral 4:4 coordination in the more covalent compounds. The compounds are more likely to conform to the effective atomic number rule and utilize tetrahedral sp^3 hybrids when the electronegativity difference is small and the bonds have a high degree of covalent character. This may lead to a preference for the 4:4 coordination and this topic is discussed in more detail in Chapter 4.

In a predominantly ionic compound the reduction in charges of the cation and anion associated with the progressive transfer of electron density from the isolated anions to the internuclear regions has the effect of reducing the lattice energy, since

$$\Delta_{latt}H^{\ominus} \propto z^+z^-/(r^+ + r^-) \propto z_{act}^2 \text{ (if the formula is MX)}$$

where z_{act}^2 represents the product of the actual charges on the cation and anion.

As the lattice enthalpies are reduced the covalent contribution to the binding enthalpy increases as more electron density is being shared by the atoms. The covalent bond order may tentatively be related to z_{act} by the following approximate relationship:

$$\text{Covalent bond order} \propto (1 - z_{act}^2)^{1/2}$$

so that it equals 0 when $z_{act} = 1$ and the bond is truly ionic and equals 1 when $z = 0$ and the bond is homopolar. If the covalent contribution to the total binding enthalpy is assumed to be proportional to the covalent bond order then the evolution of the ionic and covalent contributions to the total energy may be represented as in Fig. 3.13. In this schematic illustration it is assumed that at the extremes, i.e. 100% ionic or 100% covalent character the total binding enthalpies are equal—not an unreasonable assumption, given the data in Table 3.9.

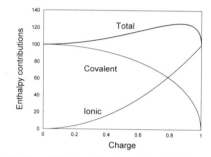

Fig. 3.13 The ionic, covalent, and total contributions to the binding enthalpy (scaled to 100) of binary compounds as a function of the charge separation between the two elements

$$\Delta_{latt}H^{\ominus} = \frac{N_A M z^+ z^- e^2}{4\pi\varepsilon_0 r_0}\left(1 - \frac{1}{n}\right)$$

where N_A is the Avogadro constant, M is the Madelung constant (values of M for common lattices are given below), e is the electronic charge, ε_0 is the permittivity of a vacuum, r_0 is the minimum interionic distance, and n is the Born exponent

	M
NaCl	1.748
CsCl	1.763
ZnS (wurtzite)	1.641
ZnS (sphalerite)	1.638
CaF$_2$	2.519
TiO$_2$ (rutile)	2.408

The changes in structure which occur for such series of compounds are discussed in Chapter 2.

Fig. 3.13 suggests that the different dependence on z_{act} of the ionic and covalent contributions may lead to situations where the covalent contribution may augment the ionic contribution for a significant range of z_{act} values.

When experimentally determined and computed lattice enthalpies derived from an electrostatic model are compared (see Table 3.13), it is found that the experimentally determined values often exceed the calculated values. In addition the difference reflects the anticipated degree of covalent character, i.e.

$$I > Br > Cl, F \quad \text{and} \quad Ag > Li > Na > K > Cs$$

Table 3.13 Experimental and calculated lattice energies (kJ mol^{-1})

	Exp.	Calc.*		Exp.	Calc.*
LiF	1033	1033	NaF	916	919
LiCl	845	840	NaCl	778	763
LiBr	799	794	NaBr	741	724
LiI	741	729	NaI	690	670
CsF	749	720	AgF	955	878
CsCl	653	626	AgCl	905	734
CsBr	632	600	AgBr	890	699
CsI	602	563	AgI	876	608

* Calculated from the Born–Landé equation (see marginal note)

If a series of compounds is compared where the charges on the cation and anion are not kept the same, the changes in physical and chemical properties are much more dramatic as shown by the data given in Table 3.14.

Table 3.14 Melting points (°C) for some horizontally related compounds, some of which adopt infinite lattices and the remainder are molecular compounds

Infinite lattices		Molecular compounds	
NaF	996	SiF_4	−96 (sublimes)
MgF_2	1263	PF_5	−83
AlF_3	1270 (sublimes)	SF_6	−51
		IF_7	5
		F_2	−219

However, in such a series not only is the bond polarity changing as a result of the electronegativity difference, but the number of anions is increasing and therefore the central atom is becoming progressively more coordinatively saturated, i.e. enough electron donation from the atoms surrounding it for the electroneutrality principle to be satisfied. Therefore, although NaF, MgF_2, and AlF_3 adopt infinite structures in order to fill the coordination spheres around the metal, the isolated NaF, MgF_2, and AlF_3 entities are sufficiently electrophilic and coordinatively unsaturated to attract lone pairs from the fluorine atoms of neighbouring molecules and thereby form infinite structures. SiF_4, PF_5, SF_6, and IF_7 do not polymerize because they are surrounded by sufficient fluorine atoms to complete their coordination environments. With several fluorines surrounding the central atom it becomes sterically less attractive to expand the coordination number by forming fluoride bridges to form chain, ring, or infinite structures. Therefore the compounds are molecular and their intermolecular interactions are limited to van der Waals forces. Consequently they have low melting points and boiling points.

A similar trend is observed for KCl, $CaCl_2$, $ScCl_3$, and $TiCl_4$ as indicated by the data given in Table 3.15.

Table 3.15 Structures and melting points for some chlorides

	Structure	m.p. (°C)
KCl	NaCl (6:6) infinite lattice	771
$CaCl_2$	TiO_2 (6:3) infinite lattice	782
$ScCl_3$	Layer lattice (6:2)	968
$TiCl_4$	Molecular – $TiCl_4$ molecules	−24 (normally a liquid)

3.6 Oxides

If the trends in ionic–covalent character are combined with a knowledge of the thermodynamically determined strengths of the bonds then significant insights into the chemical properties of the compounds may be derived. Fig. 3.14 illustrates the variations in the standard enthalpies of formation of the oxides of the Group 13–17 elements. There is a progressive decrease in the thermodynamic stabilities of the oxides across the series which reflects the decreasing electronegativity difference between the central atom and oxygen.

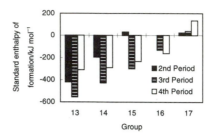

Fig. 3.14 Enthalpies of formation per oxygen atom for the series: (a) B_2O_3, CO_2, N_2O_3, and F_2O of the 2nd Period (b) Al_2O_3, SiO_2, P_4O_{10}, SO_3, and Cl_2O_7 of the 3rd Period (c) In_2O_3, SnO_2, Sb_4O_6, TeO_2, and I_2O_5 of the 4th Period

The value for F_2O appears anomalous, but it must be remembered that the bonds in this molecule have a reversed polarity, since fluorine is the only element with an electronegativity greater than oxygen. The thermodynamic data suggests that the oxidizing abilities of the oxides decrease from right to left across a row of the Periodic Table.

Table 3.16 Standard electrode potentials for oxo-anions

Redox couple	$E^{\ominus}/V$	
ClO_4^-/ClO_3^-	1.19	Most oxidizing
SO_4^{2-}/SO_3^{2-}	0.17	
PO_4^{3-}/HPO_3^{2-}	−1.12	Least oxidizing

This behaviour is reflected in the standard redox potentials for the oxo-anions of Cl, S, and P given in Table 3.16.

The oxides of Groups 17 and 18 are unstable with respect to the elements and therefore are thermodynamically unstable. Since there is not always a low energy kinetic pathway for this decomposition process, such compounds can behave in an unpredictable fashion. They will lie around for months seemingly stable and then detonate unexpectedly when subjected to pressure or bright light.

The important oxides of the p block elements are summarized in Table 3.17. The following general trends may be identified.

1. The oxides to the left form infinite polymeric structures, e.g. M_2O_3 (M = Group 13 element), and to the right the structures become more molecular. The transition does not occur at the same group for each period, and the dividing line is approximately diagonal. The reasons for this are discussed in more detail under Diagonal relationships (page 202).

2. The second row elements (C–F) form a wide range of volatile molecular oxides. The variety of compounds of this type arises from the ability of these elements to form multiply bonded species (see examples in the margin).

Table 3.17 Summary of the oxides of Groups 13–18

Group 13	Group 14	Group 15	Group 16	Group 17	Group 18
B_2O_3	CO	N_2O	O_2	F_2O	
	CO_2	NO	O_3	F_2O_2	
	C_3O_2	N_2O_3		F_2O_4	
		NO_2			
		(N_2O_4)			
Al_2O_3	SiO_2	P_4O_6	SO	Cl_2O	
		P_4O_7	SO_2	Cl_2O_3	
		P_4O_{10}	SO_3	ClO_2	
				Cl_2O_6	
				Cl_2O_7	
Ga_2O_3	GeO	As_4O_6	SeO_2	Br_2O	
Ga_2O	GeO_2	As_2O_5	SeO_3	BrO_2	
				BrO_3	
In_2O_3	SnO	Sb_2O_3	TeO_2	I_2O_4	XeO_3
In_2O	SnO_2	Sb_2O_5	TeO_3	I_4O_9	XeO_4
				I_2O_5	

3. The structures of the oxides can in general be interpreted using Lewis electron pair formalisms and the Valence Shell Electron Pair Theory. A full account of the applications of these models is provided in Chapter 4. Nitrogen, and to a lesser extent chlorine, are exceptional in forming oxides which are paramagnetic, because they have unpaired electrons, e.g. NO, NO_2, and ClO_2. The relative weakness of the nitrogen–nitrogen bond is an important factor in influencing the equilibrium constants of the following reactions.

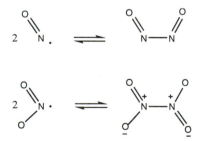

4. The oxides become progressively more oxidizing on moving to the right in the Periodic Table. The thermodynamic origins of this phenomenon are evident from Fig. 3.14.

5. In general the oxides become more acidic on moving to the right. The trend is most clearly visible for compounds where the central atom has the same formal oxidation state. The oxides are not in general as strong Lewis acids as the corresponding fluorides and chlorides. The ability of the oxygen atom to form multiple bonds and bridge several centres is responsible for this observation.

6. The oxides become more acidic as the oxidation state of the central atom is increased. For example, SO_2 is amphoteric, whereas SO_3 is a strong Lewis acid.

7. Some of the oxides can also function as Lewis bases in transition metal complexes. The molecules NO, CO, and SO_2 are particularly effective at stabilizing transition metal complexes in low oxidation states. There are other examples, e.g. P_4O_6, where the molecule functions as a Lewis base through the oxygen atoms.

8. The majority of the oxides form oxanions of the types, MO_4^{x-}, MO_3^{x-}, etc. which are the conjugate bases of the corresponding acids MO_4H_x and MO_3H_x. Iso-electronic series of the oxo-anions display an increased tendency to form polymeric structures on moving to the left of the Periodic Table. The increased negative charge on anions in such an isoelectronic series makes the oxygens more nucleophilic and more capable of attacking other molecules to generate oligomers and polymers. The corresponding acids generally become more acidic on moving to the right as shown by the data in Table 3.18.

Table 3.18 pK_a's of some oxo-acids

H_3PO_4	H_2SO_4	$HClO_4$
+2.1	−3	−10
H_3PO_3	H_2SO_3	$HClO_3$
+1.8	+1.9	−10

9. The mean bond enthalpies for single bonds to oxygen are particularly large for Si, P, and S and therefore these elements form particularly stable oxo-anion salts with a wide range of interesting chain and polymeric structures. The thermodynamic stability of these oxides also means that they are not strongly oxidizing unless they are associated with a very high oxidation state of the element. The single bond enthalpies follow the order: N–O < P–O > S–O.

3.7 Hydrides

Fig. 3.15 provides a periodic classification of the hydride compounds of the elements. The intermediate electronegativity of hydrogen (2.2) means that the polarity of the E–H bond varies across the Periodic Table. The s block elements form hydrides with infinite structures and chemically they show considerable hydridic (H^-) character and therefore they are classified as ionic or saline hydrides. Thermodynamically the formation of the hydride ion in the gas phase is much less favourable than that for the halogens and consequently ionic hydrides are only formed by the most electropositive metals. The alkali metal hydrides all adopt the NaCl lattice and therefore the metal and hydride are both octahedrally coordinated.

The alkaline earth metal hydrides adopt either the rutile (MgH_2) or a hexagonal close packed structure. The apparent radius of the H^- ion is surprisingly large and lies between that for Cl^- and F^-.

Chemically the ionic hydrides are very reactive towards even weak proton donors and react very rapidly with water. They are strong reducing agents and on an electrochemical scale they would be only slightly less reducing than the alkali metals. Unfortunately they cannot be readily dissolved in non-protic solvents and this has limited their usefulness as reagents. In organic chemistry LiAlH$_4$ and NaBH$_4$ which are soluble in organic solvents are more commonly used as hydridic reducing agents.

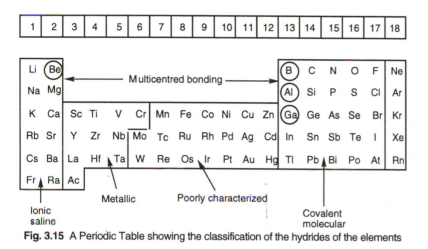

Fig. 3.15 A Periodic Table showing the classification of the hydrides of the elements

The more electronegative p block elements form a wider range of molecular hydrides. In these compounds the hydrides retain some of their hydridic character, for example for boron and aluminium, but become electroneutral for carbon or nitrogen, and are protonic (H$^+$) for the electronegative halogens.

The p block hydrides form essentially covalent bonds and their stoichiometries provide a good illustration of the applicability of the Effective Atomic Number Rule. The hydrides of Group 14–17 form conventional Lewis two-centre two-electron bonds and their shapes can be interpreted using the Valence Shell Electron Pair Theory, as shown in Fig. 3.16. These ideas are developed fully in Chapter 4.

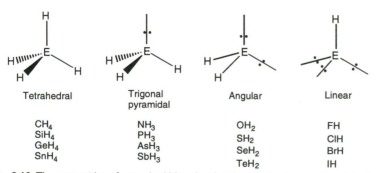

Fig. 3.16 The geometries of some hydrides showing the effects of lone pairs of electrons

The Group 13 hydrides such as B_2H_6 are sometimes described as electron deficient because their bonding cannot be immediately appreciated in terms of the Lewis electron pair bond theory. Specifically, B_2H_6 has a total of 12 valence electrons and yet its structure which is shown in Fig. 3.17 implies the presence of 8 bonds between boron and hydrogen.

However, the molecule can be seen to conform to the Effective Atomic Number Rule by forming coordinate (dative) bonds from the B–H bonds to the empty orbital on boron as indicated in the diagrams of Fig. 3.18.

The delocalization results in the delocalization of the B–H bond over the three atoms B–H–B and therefore the bond is described as a three-centre two-electron bond. This type of bonding is frequently represented as shown in Fig. 3.19. A molecular orbital description of the bonding in such molecules is given in Chapter 4.

Table 3.19 summarizes the mean bond enthalpies for the hydrides of Groups 1, 2, 13–17. In contrast to the trends noted above for the oxides there is generally an increase in the strengths of the bonds across the series (there is a slight anomaly at nitrogen). This behaviour is reflected in the standard Gibbs energies of the hydrides, which are summarized in Table 3.20.

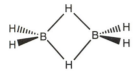

Fig. 3.17 The structure of the diborane molecule

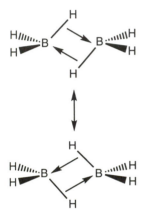

Fig. 3.18 Electronic arrangements of the diborane molecule which conform with the EAN rule

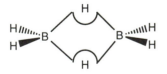

Fig. 3.19 A representation of the three-centre two-electron bonding in diborane

Table 3.19 Mean bond enthalpies (kJ mol^{-1}) for hydrides of the s and p block elements

LiH	BeH$_2$	BH$_3$	CH$_4$	NH$_3$	OH$_2$	FH
238	297	373	416	391	463	571
NaH	MgH$_2$	AlH$_3$	SiH$_4$	PH$_3$	SH$_2$	ClH
186	211	287	322	322	367	432
KH	CaH$_2$	GaH$_3$	GeH$_4$	AsH$_3$	SeH$_2$	BrH
175	159	260	288	297	317	366
RbH		InH$_3$	SnH$_4$	SbH$_3$	TeH$_2$	IH
167		225	253	257	267	298

Table 3.20 Standard Gibbs energies of formation (kJ mol^{-1}) for hydrides

LiH(s)	BeH$_2$(s)	B$_2$H$_6$	CH$_4$(g)	NH$_3$(g)	H$_2$O(g)	FH(g)
−68	+20	+87	−51	−17	−237	−273
NaH(s)	MgH$_2$(s)	AlH$_3$(s)	SiH$_4$(g)	PH$_3$(g)	SH$_2$(g)	ClH
−34	−36	−1	+57	+13	−34	−95
KH(s)	CaH$_2$(s)		GeH$_4$(g)	AsH$_3$(g)	SeH$_2$(g)	BrH
−36	−147		+113	+69	+16	−54
RbH(s)	SrH$_2$(s)		SnH$_4$(g)	SbH$_3$(g)	TeH$_2$(g)	IH(g)
			188	148		2

The following general trends emerge from these data.

1. For the p block elements the hydrides become thermodynamically less stable on moving to the left and down the Periodic Table. Consequently the hydride of gallium Ga_2H_6 (digallane) has only recently been isolated in a pure form and characterized. This synthesis required extraordinary precautions to be taken to exclude air and moisture and to ensure that the surfaces of the glass vessels were perfectly dry and free of oxygen.

2. The hydrides on the right hand side show little tendency to decompose with the formation of the element and dihydrogen. On moving to the left

this tendency increases and for example a well known test for arsine is its decomposition on heating to give a metallic film of arsenic.

3. The strength of the E–H bonds has a marked effect on the coordination chemistries of the hydrides. There are hundreds of examples of aqua (OH_2) and ammonia (NH_3) complexes, but the corresponding SH_2 and PH_3 complexes are much rarer. Once formed such complexes are much more reactive because of the weaker E–H bonds and consequently they decompose more readily to polynuclear compounds with sulfide or phosphide bridges.

4. The hydrides which have +ve Gibbs energies of formation may be synthesized either by the reaction of a metal chloride, or oxide with a hydridic metal hydride:

$$LiAlH_4 \; + \; SiCl_4 \;\; \rightarrow \;\; AlCl_3 \; + LiCl \; + \; SiH_4$$

$$BH_4^- \; + \; GeO_2 \; + H_2O \; \rightarrow \;\; GeH_4 \; + \; H_2BO_3^-$$

or the reaction of the salt of the appropriate anion with an acid:

$$Ca_3P_2 \; + \; 6H_2O \; \rightarrow \; 3Ca(OH)_2 \; + \; 2PH_3$$

The sections above have illustrated how the chemical properties of oxides and hydrides are influenced by their mean bond enthalpies. Analogous thermodynamic data for halides, sulfides, nitrides, etc. may similarly be used to interpret their chemical properties. Table 3.21, for example, presents the relevant data for p block fluorides. It is noteworthy that the trends in the mean bond enthalpies follow those for oxides more closely than the hydrides and therefore the trends in chemical properties for fluorides should follow more closely those noted above for oxides, e.g. the fluorides become stronger fluorinating agents on moving to the right in the Periodic Table.

Table 3.21 Mean bond enthalpies (kJ mol^{-1}) for fluorides of the the s and p block elements

BF_3	CF_4	NF_3	OF_2	FF
644	485	272	190	158
AlF_3	SiF_4	PF_3	SF_2	ClF
583	565	490	326	251
GaF_3	GeF_4	AsF_3	SeF_2	BrF
469	452	484	285	246
InF_3	SnF_4	SbF_3	TeF_2	IF
444	414	440	335	278

3.8 Common oxidizing and reducing agents

The discussion of the properties of oxides and hydrides developed above leads naturally into a more general discussion of reagents which are commonly used as oxidizing and reducing agents. Practical chemical experience leads to a hierarchy of preferred reagents for specific applications. The ultimate choice depends not only on the oxidizing or reducing ability of the compound, but also on its availability, its solubility in the solvent being used for the reaction, its speed of reaction, and the ease with which it can be separated from the product after the reaction. The cost of the reagent is also important especially if the reactions forms part of a commercial process.

For oxidation processes, dioxygen is itself an important and readily available oxidizing agent. However, the standard electrode potentials summarized in Table 3.22 suggest that there are alternative and more powerful oxidizing agents available for use in aqueous chemistry. The following reagents are commonly used in inorganic chemistry: $KMnO_4$, $K_2Cr_2O_7$, CrO_3, $NaOCl$, HNO_3, PbO_2, H_2O_2, Na_2O_2, $KClO_3$, KNO_3, BrO_3^-, H_5IO_6, and $S_2O_8^{2-}$ as strong oxidizing agents. With the exception of PbO_2 they are water soluble and may be used for homogeneous oxidation processes. They are all readily available and do not pose great separation problems.

Table 3.22 Commonly used oxidizing agents in aqueous solutions
(a) Standard reduction potentials for $a_{H^+} = 1$

Oxidized form		Reduced form	$E^{\ominus}$ /V
$O_3(g) + 2H^+ + 2e$	$\leftrightarrows$	$O_2(g) + H_2O$	2.1
$S_2O_8^{2-} + 2e$	$\leftrightarrows$	$2S_2O_4^{2-}$	2.0
$H_2O_2 + 2H^+ + 2e$	$\leftrightarrows$	$2H_2O$	1.77
$2HOCl + 2H^+ + 2e$	$\leftrightarrows$	$Cl_2(g) + 2H_2O$	1.63
$H_5IO_6 + H^+ + 2e$	$\leftrightarrows$	$IO_3^- + 3H_2O$	1.6
$MnO_4^- + 8H^+ + 5e$	$\leftrightarrows$	$Mn^{2+} + 4H_2O$	1.51
$2BrO_3^- + 8H^+ + 5e$	$\leftrightarrows$	$Br_2 + 6H_2O$	1.5
$Cl_2(g) + 2e$	$\leftrightarrows$	$2Cl^-$	1.36
$Cr_2O_7^{2-} + 14H^+ + 6e$	$\leftrightarrows$	$2Cr^{3+} + 2H_2O$	1.33
$O_2 + H^+ + 4e$	$\leftrightarrows$	$2H_2O$	1.23
$Br_2(l) + 2e$	$\leftrightarrows$	$2Br^-$	1.08
$Fe^{3+} + e$	$\leftrightarrows$	Fe^{2+}	0.77
$I_2 + 2e$	$\leftrightarrows$	$2I^-$	0.62

(b) Standard reduction potentials for $a_{OH^-} = 1$

Oxidized form		Reduced form	$E^{\ominus}$ /V
$O_3(g) + H_2O + 2e$	$\leftrightarrows$	$O_2 + 2OH^-$	1.4
$HO_2^- + H_2O + 2e$	$\leftrightarrows$	$3OH^-$	0.87
$ClO^- + H_2 + 2e$	$\leftrightarrows$	$Cl^- + 2OH^-$	0.94
$H_3IO_6^{2-} + 2e$	$\leftrightarrows$	$IO_3^- + 3OH^-$	0.70
$BrO_3^- + 3H_2O + 6e$	$\leftrightarrows$	$Br^- + 6OH^-$	0.61
$CrO_4^{2-} + 4H_2O + 3e$	$\leftrightarrows$	$Cr(OH)^{3+} + 5OH^-$	−0.12
$O_2 + 2H_2O + 4e$	$\leftrightarrows$	$4OH^-$	0.40

Ozone, which is not so readily available, is formed *in situ* by passing an electric discharge through dry dioxygen and is not only a very powerful oxidizing agent ($E^{\ominus} = 2.1$ V) but has the great advantage that when it reacts it forms only O_2 and H_2O. Therefore, it is an ideal reagent for sterilizing swimming baths. NaOCl, which is also used for this purpose produces NaCl, which is assimilated in the water, because of its high solubility, but NaOCl has the effect of converting some organic compounds into chlorinated derivatives which are lachrymatory.

Although the standard redox potentials in Table 3.22 are useful for underlining the thermodynamic basis for the relative oxidizing abilities of reagents, the data of course only relate to standard conditions and provides no guidance on the kinetics of the redox reactions and hence do not distinguish between slow and fast redox reactions.

Some of the powerful oxidizing agents in Table 3.22 are notoriously slow reagents. For example, peroxydisulfate ($S_2O_8^{2-}$) which has $E^{\ominus} = 2.0$ V only functions sufficiently quickly for practical purposes if a catalyst is added. Ag^+ is the most commonly used catalyst for this purpose. The speed of such slow oxidation reactions may also be enhanced by raising the temperature.

For all those reactions listed in Table 3.22, where the half reaction includes H^+, the electrode potential is influenced significantly by the pH of the solution. For those oxidation processes with H^+ on the left hand side increasing the H^+ ion concentration generally reduces $E^{\ominus}$ and therefore such reagents are frequently used in alkaline solutions.

Table 3.22 also provides some examples of oxidizing agents which are milder than O_2, e.g. Fe^{3+}(aq). The redox potentials of such oxidizing agents may be fine tuned by adding ligands, which are able to complex selectively to Fe^{3+} or Fe^{2+}. This topic is discussed further in Chapter 5, but briefly the addition of a ligand, e.g. NH_3, CN^-, etc., to an equimolar solution of Fe^{3+} and Fe^{2+} alters the electrode potential for the Fe^{3+}/Fe^{2+} couple. Specifically, if the ligand bonds more strongly to Fe^{3+} than Fe^{2+} the higher oxidation state is stabilized relative to the lower oxidation state and the Fe^{3+}/Fe^{2+} couple is a less effective oxidizing couple. Conversely, if the ligand coordinates more strongly to Fe^{2+} than Fe^{3+} the Fe^{3+}/Fe^{2+} couple becomes a less strongly oxidizing couple. The following data illustrates the general effect:

$$Fe^{3+}(aq) + e \rightleftharpoons Fe^{2+} \qquad E^{\ominus} = 0.77 \text{ V}$$

$$[Fe(CN)_6]^{3-} + e \rightleftharpoons [Fe(CN)_6]^{4-} \qquad E^{\ominus} = 0.36 \text{ V}$$

CN^- coordinates more strongly to Fe^{3+} than Fe^{2+}

$$[Fe(dipy)_3]^{3+} + e \rightleftharpoons [Fe(dipy)_3]^{2+} \qquad E^{\ominus} = 1.1 \text{ V}$$

dipy coordinates more strongly to Fe^{2+} than Fe^{3+}

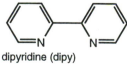

dipyridine (dipy)

If the reagent being added causes precipitation of one or more of the components of the couple, this also influences the redox potential. For example, addition of OH^- to Co^{2+}(aq) and Co^{3+}(aq) results in the precipitation of $Co(OH)_2$ and $Co(OH)_3$ respectively. However, the latter is much less soluble than the former and in effect preferentially stabilizes Co^{3+} relative to Co^{2+} making the couple a thermodynamically less effective oxidizing reaction. This conclusion is supported by the following electrode potentials.

$$Co^{3+}(aq) + e \rightleftharpoons Co^{2+}(aq) \qquad E^{\ominus} = 1.84 \text{ V}$$
$$Co(OH)_3 + e \rightleftharpoons Co(OH)_2 \qquad E^{\ominus} = -0.2 \text{ V}$$

These electrode potentials may be calculated from the standard electrode potentials using the Nernst equation:

R is the gas constant, T the absolute temperature, n the number of electrons transferred, and F is the Faraday constant

$$E = E^{\ominus} - \frac{RT}{nF} \ln \frac{[\text{reduced form}]}{[\text{oxidized form}]}$$

if the relevant equilibrium constants for the complexes or the solubility products of the precipitated compounds are known.

The halogens F_2, Cl_2, Br_2, and I_2 are very versatile oxidizing agents which span the whole range of standard electrode potentials and, apart from F_2 which requires specialized equipment, are readily available and convenient to use. The factors responsible for this large range of $E^{\ominus}$ values are discussed in Chapter 2. These reagents have the advantage that they can be used equally well in aqueous and non-aqueous solvents. Many halides become more sensitive to hydrolysis as the oxidation state of the central atom is increased. Consequently, the synthesis of such compounds in aqueous solutions using the halogens as oxidizing agents may lead to the hydrolysis of the oxidized product. This may be avoided by performing the oxidation in a non-aqueous solvent such as C_6H_6 or CCl_4. Many organic and organometallic compounds are insoluble in water and some are susceptible to hydrolysis and therefore their redox reactions are more effectively executed using reagents which are soluble in organic solvents. OsO_4 provides an example of such a reagent which is soluble in organic solvents and is a powerful oxidizing agent which is widely used for converting olefins into diols. $[Fe(\eta\text{-}C_5H_5)_2]^+$ is a much milder one–electron oxidizing agent which may be used to oxidize organometallic compounds in organic solvents and thereby represents an organically soluble analogue of the $Fe^{3+}(aq)$ reagent.

The strong oxidizing agents listed in Table 3.22 have certain common features which are noteworthy and relate to the thermodynamic properties of the bonds involving oxygen atoms discussed above. They are either based on the peroxide fragment, e.g. H_2O_2, Na_2O_2, or $S_2O_8^{2-}$ or are oxo-anions of the later p block elements and the group 16 and 17 transition metals.

The O–O single bond has a low mean bond enthalpy relative to the O=O double bond and the OH bond and therefore the thermodynamics greatly favour the conversion of the O–O bond into either of these alternative bonds.

For oxygen containing compounds of the p block elements the mean bond enthalpies generally decrease on moving from left to right and this affects their enthalpies of formation (see Fig. 3.19). Therefore, in general terms they become stronger oxidizing agents on moving to the right.

Table 3.23 summarizes the standard reduction potentials for commonly used reducing agents in aqueous inorganic chemistry. The table contains data for $a_{H^+} = 1$ and $a_{OH^-} = 1$ in order to emphasize the important effect of pH changes on the reducing abilities of the species. Of course, for these reagents kinetic factors can mask the thermodynamically favourable processes. For example, the direct reduction of $Cu^{2+}(aq)$ and many other metal cations by hydrogen (H_2) gas is thermodynamically favourable, but the mechanistic pathway resulting in the reduction is associated with large activation energies.

Table 3.23 Commonly used reducing agents in aqueous solutions
(a) Standard reduction potentials for $a_{H^+} = 1$

Oxidized form		Reduced form	$E^{\ominus}/$ V
$Zn^{2+} + 2e$	$\leftrightarrows$	$Zn(s)$	-0.76
$H_3PO_3 + 2H^+ + 2e$	$\leftrightarrows$	$H_3PO_2 + H_2$	-0.59
$Cr^{3+} + e$	$\leftrightarrows$	Cr^{2+}	-0.41
$Eu^{3+} + e$	$\leftrightarrows$	Eu^{2+}	-0.35
$H_3PO_4 + 2H^+ + 2e$	$\leftrightarrows$	$H_3PO_3 + H_2O$	-0.28
$N_2 + 5H^+ + 4e$	$\leftrightarrows$	$N_2H_5^+$	-0.23
$V^{3+} + e$	$\leftrightarrows$	V^{2+}	-0.20
$Sn^{2+} + 2e$	$\leftrightarrows$	$Sn(s)$	-0.14
$Ti^{4+} + e$	$\leftrightarrows$	Ti^{3+}	-0.05
$2H^+ + 2e$	$\leftrightarrows$	$H_2(g)$	0
$Sn^{4+} + 2e$	$\leftrightarrows$	Sn^{2+}	0.14

(b) Standard electrode potentials for $a_{OH^-} = 1$

Oxidized form		Reduced form	$E^{\ominus}/$ V
$ZnO_2^- + 2H_2O + 2e$	$\leftrightarrows$	$Zn + 4OH^-$	-1.21
$HPO_3^{2-} + 2H_2O + 2e$	$\leftrightarrows$	$H_2PO_2^- + 3OH^-$	-1.65
$CrO_2^- + 2H_2O + 3e$	$\leftrightarrows$	$Cr + 4OH^-$	-1.2
$PO_4^{3-} + 2H_2O + 2e$	$\leftrightarrows$	HPO_3OH^-	-1.05
$2H_2O + 2e$	$\leftrightarrows$	$H_2 + 2OH^-$	-0.83
$Sn(OH)_6^{2-} + 2e$	$\leftrightarrows$	$HSnO_2^- + 3OH^- + H_2O$	-0.96

The strongest quoted reducing agent in Table 3.23 is zinc ($E^{\ominus} = -0.76$ V). There are stronger reducing agents, but they are either impractically slow, e.g. Al and Mg, because of their oxide coating or are so strongly reducing that they reduce the solvent (water) rather than the substrate, e.g. the alkali metals. More strongly reducing conditions may be achieved using the following strategies. Firstly, reagents such as $NaBH_4$ may be employed which are capable of reducing water, but their reaction is sufficiently slow that they persist in water for sufficient period of time to affect the reaction. For example, sodium borohydride evolves hydrogen in water, but at a sufficiently slow rate for the BH_4^- anion to attack the substrate. Alternatively, a change of solvent may enable the use of a more reactive metal. For example, the alkali metals dissolve in liquid ammonia to give highly coloured solutions which involve the solvated M^+ cation and solvated

In organic solvents the approximate oxidizing abilities are:
Ce^{4+} > NO^+ > $[Ru(phen)_3]^{3+}$ > Ag^+ > Br_2 > I_2 > $[Fe(\eta-C_5H_5)_2]^+$
and reducing abilities are:
$C_{10}H_8^-$ > Na > Li > benzophenone > $Co(\eta-C_5Me_5)_2$ > $Co(\eta-C_5H_5)_2$ > hydrazine

For a more detailed discussion of standard electrode potentials in non-aqueous solvents see N. G. Connelly and W. E. Geiger, *Chem. Rev.*, 1996, **96**, 877.

electron. Such solutions are thermodynamically unstable with respect to the following reaction:

$$M(s) + NH_3(l) \leftrightarrows M^+(solv) + NH_2(solv)^- + \tfrac{1}{2}H_2(g)$$

but the kinetics of the reaction are sufficiently slow for this reaction to be disregarded until a catalyst such as Fe^{2+} is added. Therefore, this represents a very effective medium for the reduction of a wide range of inorganic compounds which are soluble in liquid ammonia. Similarly, many reactions involving alkali metals may be undertaken in donor organic solvents such as tetrahydrofuran. There are also organometallic compounds which are soluble in non-polar organic solvents, such as benzene, which are reasonably strong reducing agents, e.g. $Co(\eta-C_5H_5)_2$ and $Ti(\eta-C_6H_6)_2$.

The reducing agents which are summarized in Table 3.23 fall into the following categories:

1. Metals with medium electropositive character, e.g. Zn and Sn.
2. Metal ions with two adjacent oxidation states, e.g. Cr^{3+}/Cr^{2+}, Eu^{3+}/Eu^{2+}, V^{3+}/V^{2+}.
3. Lower oxidation state oxyanions of Group 15 and 16, e.g. HPO_3^{2-}, $S_2O_4^{2-}$, SO_3^{2-}. For these elements, the group oxidation state is particularly stable with oxygen ligands and therefore oxyanions in lower oxidation states serve as effective reducing agents.
4. Hydrazine N_2H_4 and hydroxylamine NH_2OH, which on oxidation form N_2, are effective agents because of the large difference in bond enthalpies for the N–N single bonds versus the N=N and N≡N multiple bonds.
5. On moving to the left in the Periodic Table the M–H bond enthalpy decreases and the hydrides become more hydridic, consequently the hydrides become more reducing. However, they cannot be used in aqueous solutions in general because of their sensitivity to hydrolysis. However, NaH, R_3SiH, and $LiAlH_4$ are routinely used in organic solvents and particularly ethers.

Explosives and rocket propellants

The standard enthalpies of formation of the oxides given in Fig 3.14 identifies a group of compounds which are thermodynamically unstable with respect to the elements, but which are kinetically inert. Such compounds have an activation barrier towards decomposition which is sufficiently large to prevent spontaneous decomposition. These compounds which occur to the right hand side of the Periodic Table decompose to liberate considerable volumes of gases—dioxygen primarily, but also at times other gases are produced, such as hydrogen, fluorine, and chlorine. The coincidence of processes which are highly exothermic and which also liberate gases in the decomposition reaction leads literally to a potentially explosive situation, because the gases which are liberated are caused to expand rapidly as a result of the large and rapid temperature rise. For such compounds a spark or the exertion of a slight pressure is often sufficient to initiate the autocatalytic reaction which results in the compounds exploding. Examples of oxygen containing compounds which are potentially explosive are summarized below.

XeO_3	NH_4ClO_3	H_2O_2	OF_2
HIO_4	NH_4ClO_4	$F_2O_6S_2$	Cl_2O
I_2O_5	NH_4NO_3		
Mn_2O_7	$NO^+NO_3^-$		

Besides the high oxidation state oxides of the group 17 and 18 elements and the transition metals the examples include ammonium amine salts where the liberation of N_2 and H_2O (steam) contribute to the gas output, peroxides and some low oxidation states oxyhalides. The major driving force in such reactions is the very favourable bond enthalpy associated with the O_2 double bond. Nitrogen compounds which liberate N_2 provide additional examples of explosive compounds.

BrN_3	NBr_3	N_4S_4	N_2H_4
HN_3	NI_3	N_4Se_4	N_2F_2
$Br_2P(O)N_3$	Hg_3N	N_4Te_4	
$Cd(N_3)_2$			

The rate with which the gases are formed in such reactions and the rate of their expansion is awesome and for example in the microseconds following a car accident a small solid sample of sodium azide can be detonated and the resulting explosive liberation of gases is sufficiently fast to inflate the air-bag and prevent the driver's head hitting the steering wheel.

Carbon containing compounds which are potentially explosive are less numerous, but there are examples of cyanates, e.g. BrNCO, and ethynides e.g. Au_2C_2, which are explosive as a result of the thermodynamically favourable release of either CO or C_2H_2.

Explosive reactions may also result from the bringing together of compounds which are strong reducing and oxidizing agents. The Chinese recognised this more than two thousand years ago in the invention of gunpowder and fireworks which require the combination of sulfur and carbon with potassium nitrate. The combination or hydrocarbons and sugars with $KClO_3$ or KNO_3 represent more modern examples of this principle, which result in rather unpredictable and unsafe explosive mixtures. The booster rockets in the Space Shuttle contain a mixture of aluminium powder and NH_4ClO_4 and once such reactions are initiated they cannot be stopped. The Challenger Space Shuttle tragedy provided a salient reminder of the power and potentially uncontrollable nature of these explosive combinations. Solid boron hydrides such as $B_{10}H_{14}$ have also been used as propellants for smaller hand-held anti-tank rockets.

Liquid propellants are safer to handle in general and are more controllable because the source of the propellant can be isolated and the quantity released can readily be controlled by valves. The most commonly used combinations of reductants and oxidants are H_2/O_2, kerosene/O_2, and methyl hydrazines/O_2.

If the oxidizing and reducing functions are combined within one molecule, some highly unstable and unpredictable explosives can result. For example, in inorganic chemistry, it is not a good idea (and unfortunately in some cases has been lethal) to combine perchlorate anions with inorganic cations which have organically based ligands. However, commercially based organic explosives have been developed by the introduction of nitro groups into organic aromatic compounds, cellulose, etc. leading for example to TNT (trinitrotoluene) and nitro-glycerine.

Combinations of reducing metal powders and metal oxides may be used to weld metals together because of the high temperatures achieved in such reactions. For example, the *thermite reaction* consists of the combination of either chromium oxide or iron oxide powder with aluminium powders which is initiated by using a burning magnesium ribbon. In a finely divided form metal powders are frequently explosive. In these reactions gases are not being formed, but being consumed to form a solid oxide, but the reactions are highly exothermic and if the metal particles are widely spread in the gas, the heat liberated during the reaction rapidly expands the gases and causes the explosion.

3.9 Catenation

Table 3.24 summarizes the bond enthalpy terms for homonuclear bonds of the Groups 14–17 elements of the Periodic Table.

Table 3.24 The bond enthalpy terms (mean bond energies) for single homonuclear covalent bonds for the elements of Groups 14–17 of the Periodic Table

Group 14	Group 15	Group 16	Group 17
C–C	N–N	O–O	F–F
356	167	144	158
Si–Si	P–P	S–S	Cl–Cl
226	209	226	242
Ge–Ge	As–As	Se–Se	Br–Br
188	180	172	193
Sn–Sn	Sb–Sb	Te–Te	I–I
151	142	149	151

For the elements C to F there is a large decrease in the strengths of the bonds across the series. However, for the subsequent rows the bond enthalpy terms stay remarkably constant. The large decrease from C–C to N–N has been attributed to secondary electron–electron repulsion energies between the lone pairs on the nitrogen atoms. The O–O and F–F bonds are also weakened by such effects. These effects diminish down the column because the large atoms place the lone pairs farther apart and therefore the repulsions which they generate are reduced. For the second row elements B to F catenation, i.e. the ability to form compounds with element–element bonds, is important only for B and C, which form a wide range of polyhedral molecules. For the third row elements catenation is important for Si, P, and S, but becomes progressively less important for the heavier elements. The data in Table 3.25 which summarizes the important classes of catenated compounds for these elements in general reflects the trends in bond enthalpy terms noted above.

Table 3.25 Important classes of catenated compounds

Boron	Carbon	Nitrogen	Oxygen		
Deltahedral boranes $closo$–$B_nH_n{}^{2-}$, $nido$–B_nH_{n+4}, $arachno$–B_nH_{n+6}, icosahedral boron polyhedra in element	Polyhedra C_{60}, C_{70}, etc. Wide range of chain, ring and polyhedral hydrocarbons	N_2H_4, N_2F_4, $N_3{}^-$, N_2O_4 (very weak bond)	O_3, H_2O_2		O_2F_2
	Silicon	**Phosphorus**	**Sulfur**		
	Si_nH_{2n+2} (n = 1–10) $(Me_2Si)_n$ (n = 5–100) $Si_4{}^{4-}$ tetrahedron	Phosphorus tetrahedra in white phosphorus (P–P bonds in other allotropes leading to 2-dimensional structures) $P_2I_4(RP)_n$ (n = 4–6) rings phosphorus sulfides $(PO_2)_6{}^{6-}$	S_n rings in element, $S_n{}^{2-}$ chains (n = 3–5)RS_2R (R = halide or alkyl) $S_2O_4{}^{2-}$, $S_2O_6{}^{2-}$, $S_4O_6{}^{2-}$, $S_8{}^{2+}$		
	Germanium	**Arsenic**	**Selenium**		
	Ge_nH_{2n+2} (n = 1–10) Ge_2Cl_6	As_4 polyhedra and As–As bonds in other allotropes $(RAs)_n$ rings (n = 4–6)	Se–Se bonds in element, Se_2X_2 (X = Cl or Br) $Se_4{}^{2+}$, $Se_8{}^{2+}$		
	Tin	**Antimony**	**Tellurium**		
	Sn_2H_6, Sn_nMe_{2n+2} (n = 1–6) $(SnR)_6$ rings	Very few catenated derivatives	Te–Te bonds in element, very few catenated derivatives		

3.10 Lone pairs and empty orbitals

A comparison of the reactivities of a series of molecules of an analogous type from Groups 13, 14, and 15 provides a basis for evaluating the effect of empty orbitals and lone pairs on chemical reactivity. The organometallic compounds Me_3In, Me_4Sn, and $SbMe_3$ provide such a series. All these compounds have roughly comparable metal–carbon bond strengths and they are thermodynamically unstable with respect to both oxidation and hydrolysis, because the resulting oxides and hydroxides are thermodynamically more stable. However, they exhibit remarkably different properties towards air and water, as indicated in Table 3.26.

Table 3.26 Reactivities to air and water of some organometallic compounds

Compound	Reactivity to air	Reactivity to water
Me_3In	Pyrophoric	Readily hydrolysed
Me_4Sn	Inert	Inert
Me_3Sb	Pyrophoric	Inert

The reactivities of these compounds are governed by the activation energies for the oxidation and hydrolysis processes. Hydrolysis requires an empty orbital on the central metal atom for the initial nucleophilic attack, which presumably is rate determining. Oxidation, must require pre-coordination of the dioxygen molecule, which since it has both filled and empty frontier orbitals (i.e. HOMO and LUMO), can be achieved either if the molecule has a low lying empty acceptor orbital or a filled lone pair orbital. The molecular structures of the three organometallic molecules and their relevant structures are illustrated in Fig. 3.20.

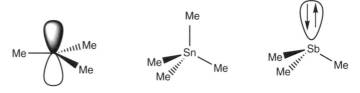

Fig. 3.20 The structures of three organometallic compounds showing the influences of vacant and lone pair orbitals

Me_3In has an empty p orbital perpendicular to the plane, which can function as an effective Lewis acid site towards either O_2 or OH_2 and consequently it is highly reactive towards both air and water. The compound Me_4Sn has neither an empty p orbital nor a lone pair and therefore it is inert towards both air and water. The tin atom does have empty 4d orbitals, but presumably they are of too high an energy to function as effective Lewis acid acceptor sites. Me_3Sb has no empty p orbitals and it is therefore inert towards nucleophilic attack by water molecules, but the lone pair is able to function as a Lewis base donor towards the dioxygen molecule.

The arguments developed above can be applied to other series of molecules which differ in the number of Lewis acid sites. In other situations the simplified analysis is complicated by dimerization processes involving lone pairs on halogens and chalcogenides which reduce the Lewis acid character of the central atom. Such compounds only behave as Lewis acids if the incoming Lewis base competes effectively with the lone pairs responsible for the bridge bonds.

3.11 Comparison of elements of Groups *N* and (*N* +10)

Elements belonging to the same period, but separated by 10 places in their periodic groups show similarities in their chemical properties and particularly in their highest oxidation state compounds. Elements which are connected in this fashion are summarized in Table 3.27.

The post-transition elements in the (*N*+10) groups have valence electrons in the s and p subshells, outside a filled d shell which is effectively core-like, because the ionization energies of the d electrons are large. The effective number of valence electrons which these atoms have is therefore identical to those of the atoms in group *N*, i.e. ($x + y$) in Table 3.27. The ($x + y$) valence electrons, nevertheless, occupy different valence orbitals, i.e. s and d for Groups *N* and s and p for groups (*N*+10). If these elements form compounds resulting either from the complete ionization of these electrons or their complete participation in covalent bond formation then the compounds have identical stoichiometries. Since oxygen and fluorine are particularly effective at bringing out the highest oxidation states and valencies then these similarities should be most obvious in the fluorides and oxides of those elements which form compounds in oxidation states 5–8. Table 3.28 provides specific examples of such analogous compounds.

Table 3. 27 Elements which belong to Groups N and $(N+10)$

Group (N)	$ns^x(n-1)d^ynp$	Group $(N+1)$	$(n-1)d^{10}ns^xnp^y$	$(x+y)$
1	K, Rb, Cs	11	Cu, Ag, Au	1
2	Ca, Sr, Ba	12	Zn, Cd, Hg	2
3	Sc, Y, La	13	Ga, In, Tl	3
4	Ti, Zr, Hf	14	Ge, Sn, Pb	4
5	V, Nb, Ta	15	As, Sb, Bi	5
6	Cr, Mo, W	16	Se, Te, Po	6
7	Mn, Tc, Re	17	Cl, Br, I	7
8	Fe, Ru, Os	18	Ar, Kr, Xe	8

Table 3.28 Analogous compounds from Groups N and $(N+10)$

Group N	Group $(N+10)$
KCl, RbCl, CsCl	CuCl, AgCl, AuCl
$CaCl_2$, $SrCl_2$, $BaCl_2$	$ZnCl_2$, $CdCl_2$, $HgCl_2$
$ScCl_3$, YCl_3, $LaCl_3$	$GaCl_3$, $InCl_3$, $TlCl_3$
$TiCl_4$, $ZrCl_4$, $HfCl_4$	$GeCl_4$, $SnCl_4$, $PbCl_4$
VF_5, $VOCl_3$, VO_4^{3-}	AsF_5, $AsOCl_3$, AsO_4^{3-}
CrF_6, CrO_2F_2, CrO_4^{2-}	SeF_6, SeO_2F_2, SeO_4^{2}
MnO_4^-, ReF_7	BrO_4^-, IF_7
OsO_4	XeO_4

The compounds in Groups 1–3 and 11–13 have a significant degree of ionic character and therefore the compounds have a formal oxidation state which is reasonably realistic. This formal oxidation state is achieved by ionizing all the valence electrons from the Group 1–3 atom and the s and p valence electrons for the Group 11–13 atoms. This leaves either a noble gas core or a core with a filled and stable d subshell. For the Groups 4–8 and 14–18 the compounds have a high degree of covalent character and the valence electrons occupy hybrid orbitals which are sufficiently closely related to result in similar bond formation. For example, in tetrahedral $TiCl_4$ the four valence electrons of Ti probably occupy tetrahedrally pointing sd^3 hybrid orbitals. Tetrahedral $GeCl_4$ forms analogous sp^3 tetrahedral hybrids.

The sp^3 and sd^3 hybrids may both be used to describe the bonds in tetrahedral molecules because the tetrahedron does not have a centre of symmetry. The gerade nature of the d orbitals and the ungerade nature of the p orbitals is lost and they can both mix effectively with s. For molecules with a centre of symmetry, the gerade and ungerade nature of the orbitals is retained and the hybrids are no longer necessarily equivalent. For example, sd hybrids make an angle of 90° whereas sp hybrids make an angle of 180°.

Sometimes the elements in Groups N and $(N+10)$ form analogous species in negative oxidation states. For example, gold has the ability to form the Au^- species which are analogous to the anions formed by the alkali metals. In both cases the stability of the anionic species may be related to the completion of the subshell structures [noble gas]s^2 for the alkali metals and [noble gas]$5d^{10}6s^2$ for gold.

The similarities in the formulae of the compounds formed by elements in Groups N and $(N+10)$ can be most useful if one is seeking isostructural analogues in the Periodic Table, but fails to emphasize the important differences in the properties of the elements in other oxidation states.

Atoms belonging to the same row of the Periodic Table, but connected by the $N/(N+10)$ relationship, have the following important differences in properties.

1. The $(N+10)$ atom has a much smaller size, because the presence of the d shell results in a significant increase in the effective nuclear charge experienced by its s and p valence electrons.

2. The larger effective nuclear charge also increases the ionization energies of the atom and makes it more electronegative.

3. Although the N and $(N+10)$ atoms both have $(x+y)$ identical the fact that they are occupying different valence orbitals can have a significant impact on the properties of compounds which have these valence orbitals partially occupied.

These differences in atomic properties have the following effects on the chemical properties of these elements and their compounds.

(a) The $(N+10)$ elements are more noble than the N elements, i.e. they are less reducing. The standard electrode potential data in Table 3.29 provides numerical data to support this observation.

Table 3.29 Standard reduction potentials, E°/V, for the ion/element couples of Groups N and $(N+10)$

Group N		Group $(N+10)$	
$K^+(aq)/K$	$Ca^{2+}(aq)/Ca$	$Cu^+(aq)/Cu$	$Zn^{2+}(aq)/Zn$
−2.93	−2.87	0.52	−0.76
$Rb^+(aq)/Rb$	$Sr^{2+}(aq)/Sr$	$Ag^+(aq)/Ag$	$Cd^{2+}(aq)/Cd$
−2.99	−2.89	2.00	−0.40
$Cs^+(aq)/Cs$	$Ba^{2+}(aq)/Ba$	$Au^+(aq)/Au$	$Hg^{2+}(aq)/Hg$
−3.02	−2.90	1.68	0.95

(b) Although the compounds may share a common formula they can at times have different solid state structures and physical and chemical properties. For example, KCl and RbCl have the sodium chloride structure, but CsCl has a body-centred cubic structure with 8-coordination. CuCl and AgCl have the wurtzite structure, but AuCl has a linear chain structure. The alkali metal chlorides are also much more soluble in water than the coinage metal chlorides. K_2O is a colourless soluble, basic oxide, whereas Cu_2O is a red insoluble oxide with significant covalent character. The smaller sizes of the $(N+10)$ metal ions are a major contributing factor to the lower coordination numbers, but the greater covalency of the bonds is also significant.

(c) The $(N+10)$ metal ions form much more stable complexes than the N metal ions. For example, K^+ shows no tendency to form complexes in aqueous solutions with cyanide ion, whereas Cu^+, Ag^+, and Au^+ all form very stable complexes with CN^- with the general formula $[M(CN)_2]^-$. Even with neutral ligands such as NH_3, both $[Ag(NH_3)_2]^+$ and $[Au(NH_3)_2]^+$ are known, but corresponding alkali metal complexes are not formed in aqueous solutions.

(d) The $(N+10)$ metal ions show more pronounced **soft** behaviour as Lewis acids. Therefore, the difference in stability of oxides and sulfides in the solid state and complexes with oxygen and sulfur ligands lead to the classification of metals such as silver, gold, and mercury as prototypical soft metal ions, and to calcium and potassium as typical hard metal ions.

(e) Although the N and $(N+10)$ metals may form compounds with similar formulae their kinetic labilities may be very different. For example, $TiMe_4$ and $GeMe_4$ may be anticipated to have similar chemical properties, because their metal–carbon bond enthalpies are not too dissimilar. However, the latter boils without decomposition at 43°C, whereas the former decomposes above −78°C.

(f) In intermediate oxidation states the compounds have very different properties probably because of differences in the orbitals occupied. For example, Ti^{III} has the electronic configuration $3d^1 4s^0 4p^0$ and forms a wide range of stable octahedral complexes, e.g. $[Ti(OH_2)_6]^{3+}$. These stable radical species exist because the radial distribution function of the 3d orbitals is rather contracted and the orbitals are well shielded within the octahedral ligand environment. The complexes are paramagnetic and show electronic transition which involve the promotion of electrons within the d shell. In contrast Ge^{III} has the electronic configuration $3d^{10} 4s^1 4p^0$ and stable molecules of Ge^{III} are rather rare. The unpaired electron occupies the s orbital which extends effectively from the nucleus and also can hybridize effectively with the 4p orbitals to form a radical species with a sp^3 hybridized orbital singly occupied. Such radical species may readily dimerize to form metal–metal bonded compounds. Indeed it is only possible to form stable Ge^{III} compounds if very sterically crowded ligands are coordinated to it.

(g) Although the d shells behave in a core-like manner for the elements in Groups 12–18 in Group 12 they are sufficiently available for the elements to form compounds with partially filled d shells. For example, copper forms a very wide range of Cu^{II} d^9 complexes, silver forms Ag^{II} d^9 and Ag^{III} d^8 complexes, and gold forms a very wide range of stable Au^{III} d^8 complexes, and is known in the +5 oxidation state in

AuF_5 (d^6). These compounds, of course, have no analogues in Group 1 chemistry.

(h)   The higher electronegativities of the Groups 11 and 12 metals makes their compounds generally more highly coloured than the corresponding compounds of the Groups 1 and 2 metals, i.e. the ligand to metal charge transfer energy is lowered if the electronegativity of the central atom is increased.

(i)   The (N +10) metal ions do not form hydrated metal ions in solution as readily. Their hydration enthalpies are generally not sufficiently favourable to overcome their high lattice energies. For example, $HgCl_2$ persists in aqueous solution in a molecular form rather than as $Hg(OH_2)_x^{2+}$.

(j)   The (N +10) metal ions show a greater tendency to disproportionate in a way which is not possible for the N metal ions. For example, as shown in the margin, disproportionation possibilities exist for the Group 11 metals.

$$2Cu^+ \rightleftharpoons Cu + Cu^{2+}$$
$$K = 10^6$$

$$2Ag^+ \rightleftharpoons Ag + Ag^{2+}$$
$$K = 10^{-20}$$

$$2Au^+ \rightleftharpoons Au + Au^{3+}$$
$$K = 10^9$$

The differences between N and (N +10) elements can be appreciated in more specific detail by a comparison of the fluoride, oxide, and carbonyl compounds of vanadium and arsenic provided in Table 3.30.

The similarities in the highest oxidation state compounds and the divergence in properties for the lower oxidation state compounds are particularly noteworthy. The following differences are particularly important.

1.   The differences in the colours associated with the compounds.
2.   The difference in magnetic properties and, in some instances, electrical conductivity properties.
3.   The strong preference for octahedral coordination in the lower oxidation state vanadium compounds.
4.   The greater range of oxidation states displayed for vanadium, and the formation of many complexes by the metal in these oxidation states.
5.   The complete absence of carbonyl compounds for arsenic.
6.   The absence of V^{3-} anions except in the presence of CO.
7.   The absence of simple hydrides for vanadium.

The comparison has been made on the basis of a limited range of compounds, but it could be extended to other classes of compounds, e.g. organometallic compounds, sulfides, multiply bonded compounds, in order to provide a more detailed analysis of the differences between N and (N +10) groups of elements.

Besides the horizontal relationships discussed above, there are other horizontal relationships which originate from subshell structures. For example, Tl^I compounds are frequently considered as pseudo-alkali metal compounds because the oxidation state similarity and also because thallium(I) compounds are reasonably ionic. The subshell electronic configuration $5d^{10}6s^26p^0$ for thallium(I) and the inert pair effect which stabilizes Tl^I relative to Tl^{III} provide a basis for this horizontal relationship.

Table 3.30 Comparison of some vanadium and arsenic compounds

Ox. No.	Vanadium	Arsenic
+5	VF_5 colourless solid/liquid (m.p. 20°C). Monomeric in gas phase. Powerful fluorinating agent, readily hydrolysed. Forms octahedral VF_6^- with F^-.	AsF_5 colourless gas (m.p. -80°C). Readily hydrolysed. Strong Lewis acid. Readily forms AsF_6^-.
	V_2O_5 (m.p. 677°C) chain structure based on either trigonal bipyramids or square-pyramids. Amphoteric mild oxidizing agent. Catalyst for industrial oxidation of SO_2 to SO_3. Slightly soluble in water.	As_2O_5 colourless glassy solid based on chains of AsO_6 and AsO_4 octahedra and tetrahedra. Oxidation catalyst, very soluble in H_2O. Thermally unstable and strong oxidizing agent.
	VO_4^{3-} exists only in this form at high pH in solution and in solid state. At lower pH's a series of iso-polyvanadates are formed, $V_2O_7^{2-}$, $V_3O_9^{3-}$, and $HV_{10}O_{28}^{5-}$, etc. based on tetrahedral and octahedral vanadium-oxide polyhedra.	H_3AsO_4 formed when As_2O_5 is dissolved in water. Tribasic acid which forms salts such as Na_3AsO_4 with strong alkalis. These readily dehydrate to $NaAsO_3$, which has a polymeric structure. $KAsO_3$ has cyclic $As_3O_9^{3-}$ cyclic trimers similar to iso-polyvanadates. Arsenates show less tendency to polymerize than phosphates or vanadates.
+4	VF_4 lime green solid (m.p. 325°C) which has a layer structure based on octahedral coordination centres. Paramagnetic $\mu_{eff} = 1.86\mu_B$ (one unpaired electron), readily disproportionates on heating to VF_5 and VF_3. Forms complexes with Py and NH_3.	Corresponding compound not known.
	VO_2 blue-black amphoteric solid (m.p. 197°C). Distorted rutile structure based on octahedral vanadium centres, with metal–metal bonds at low temperatures. At higher temperatures a phase transition results in breaking of metal–metal bonds and increase in conductivity and paramagnetism. VO^{2+} fragment is very frequently observed in square-pyramidal complexes of vanadium.	Corresponding compound not known.
+3	VF_3 yellow green solid (m.p. 800°C) based on infinite structure with octahedral vanadium centres. Paramagnetic $\mu_{eff} = 2.55\mu_B$ (corresponding to two unpaired electrons). Forms a wide range of octahedral complexes, e.g. VF_6^{3-}.	AsF_3 colourless gas/liquid (m.p. $- 6^\circ$C) which is pyramidal in the gas phase and has an infinite polymeric structure base on tricapped trigonal prisms in the solid. Fumes in air and acts as a mild fluorinating agent.
	V_2O_3 black solid with a corundum structure based on octahedral vanadium centres. (m.p. 1970°C). A basic oxide.	As_2O_3 colourless powder which forms sheet structure based on pyramidal AsO_3 units in the monoclinic form. Forms As_4O_6 molecules in the gas phase which persist in the cubic phase of the compound in the solid state. Main species when dissolved in water is probably $As(OH)_3$, which is a very weak acid.
	$V(OH)_3$ brown green solid with octahedral vanadium centres. Paramagnetic.	
	VH_3 not known as a stable compound.	AsH_3 colourless gas (m.p. -116°C) which is highly poisonous and unstable. Reducing agent forming metallic arsenic film on heating and arsenic oxides in the presence of oxidizing agents.
+2	VF_2 blue solid which has a tetragonal rutile structure in the solid state with octahedral vanadium centres. Antiferromagnetic in solid state resulting from coupling of three unpaired spins on each vanadium.	AsF_2 does not exist as a stable isolable species.
	VO grey non-stoichiometric metallic solid with NaCl lattice.	AsO does not exist as a simple isolable species.

Table 3.30 (Cont.) Comparison of some vanadium and arsenic compounds

Ox. No.	Vanadium	Arsenic
0	With π-acid ligands, e.g. $V(CO)_6$	Does not form stable low oxidation compounds.
−1	$V(CO)_6^-$	Not known.
−3	$V(CO)_5^{3-}$ only—not possible to obtain compounds with V^{3-} entity when coordinated to ligands such as CO.	NiAs, AlAs are prototypical of a wide range of compounds which formally have the As^{3-} anion in combination with a M^{3+} cation. These compounds generally adopt either a hexagonal close packed structure with the metal in octahedral sites or the ZnS structures based on tetrahedral coordination.

3.12 Diagonal relationships

The previous sections have defined and illustrated vertical and horizontal relationships associated with the Periodic Table. An atom's size and electronegativity and the strengths of the bonds it makes with other elements have a particularly important influence on the properties of the element and its compounds. Since the electronegativity of atoms generally increases on moving to the right and decreases on moving down a column two atoms which are diagonally related could have similar electronegativities. The radii of atoms decrease across a row and increase down a column and therefore diagonally related atoms may also have similar radii. Therefore, there are situations where the comparable electronegativities and radii may lead to similarities in chemical properties which are apparent even though the two elements concerned have different valencies.

Metals, metalloids, and non-metals

For Groups 13–17 the elements become more metallic down the groups. The demarcation line between metal and non-metal behaviour in this part of the Periodic Table is neither a vertical nor horizontal line but a diagonal line. As noted earlier in this chapter, the distinction between metal and non-metal can be defined in terms of physical or chemical properties and the exact location of the demarcation line depends upon the precise property being used as a criterion, nonetheless the following elements lie close to the borderline; B, Si, Ge, As, Sb, Te, and Po and may therefore be described as semi-metals or metalloids.

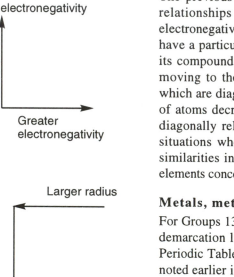

For a discussion of metallic bonding see: R. Hoffmann, *Solids and surfaces—A Chemist's view of Bonding*, VCH-Weinheim, 1988 and D. Pettifor, *Bonding and Structure of Solids*, Oxford University Press, 1995, and L. C. Allen and J. K. Burdett, *Angew. Chem. Int. Ed.*, 1995, **34**, 2003.

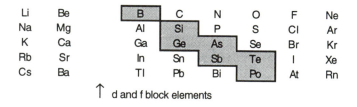

Fig. 3.21 An indication of the distinction between the metals and non-metals of the main group elements

The electrical conductivity of an element is related to the ease with which its electrons migrate to neighbouring atoms. The polarizability of an atom and its electronegativity are therefore important parameters. The electronegativity of an atom measures its ability to attract an electron and its polarizability reflects its ability to respond to the attractive field exerted by adjacent atoms. In general the polarizability of an atom is related to its size and increases down a group of the Periodic Table.

The number of nearest neighbours in the structure and whether they are close packed around the atom also influences the conductivity. The migration of electrons is facilitated by a large number of closely packed neighbouring atoms. Indeed the conductivities of all elements increase with pressure and even solid hydrogen becomes a conductor if a sufficiently high pressure is applied.

Given the factors outlined above, Edwards and Sienko have suggested that the metal to non-metal transition for elements may be defined by values of the following ratio:

$$\frac{R}{V} < 1 \ \text{for a non-metal}; \quad \frac{R}{V} \geq 1 \ \text{for a metal}$$

where R is the molar refractivity of the isolated atom and V is the molar volume of the element in the condensed phase. The molar refractivity is closely related to the polarizability of the atom and the molar volume is intimately connected to the coordination number of the element and the extent of close packing of the atoms. The R/V plot shown in Fig. 3.22 demonstrates the validity of this simple ratio in the interpretation of the non-metal–metal classification.

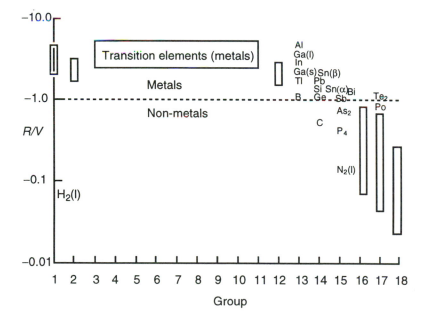

Qualitative explanation of the differences between metals, metalloids, and non-metals

The definition of a metal which is based on its electrical conductivity properties suggests that the relative ease of transmission of electrons through the solid is the primary issue. For an infinite three-dimensional array of molecules, e.g. P_4 molecules in white phosphorus, the relative ease of transmission of electrons from one end of the solid to the other depends to a large extent on a balance of the strength of the bonding within the individual molecules and between adjacent molecules. Strong element–element bonding within a molecule means that the electrons reside in a deep potential well and it is difficult for them to experience an effective attractive attraction from neighbouring molecules. Therefore, they have no great inclination to migrate and the solid is an insulator. Since, in general, the element–element bond enthalpies decrease down a column of the Periodic Table the depth of the potential well also decreases and it becomes easier for an electron to be disengaged from its parent molecule. (Continued on next page)

Fig. 3.22 A graph of the *R/V* values for the elements (adapted from P. P. Edwards and M. J. Sienko, *Chem. Brit.*, 1983, 39)

(Continued from previous page)

The strength of the interaction between molecules is influenced primarily by their polarizabilities, i.e. by van der Waals forces. For an isostructural series of elements the van der Waals interactions increase down the column and the following consequences result. The ratio of the strengths of the element–element bonds and the van der Waals interactions decrease and therefore the environment experienced by an electron located within a molecule or one placed between molecules becomes smaller. Furthermore, the electron within a molecule experiences a greater attraction from neighbouring molecules. The stronger intermolecular interactions may also increase the number of atoms from neighbouring molecules which are within van der Waals contact. These enhanced van der Waals interactions facilitate the transmission of electrons from molecule to molecule and when taken together with the decreased element–element bond strengths suggest increased metallic character down a particular column of elements.

The equalization of intra- and inter-molecular interactions and the increased numbers of neighbouring atoms leads progressively to more closed packed structures and eventually the complete loss of distinction between intra- and inter-bonding, so that in the resulting metallic lattice very facile electrical conduction occurs.

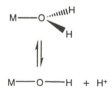

Both the polarizability and volume of an atom increase down a column of the Periodic Table and the ratio R/V gradually increases leading to the observed non-metal–metal transition down columns of Groups 13–16. Across the series the polarizabilities of the atoms decrease and it is necessary to descend further down the column before R/V becomes greater than unity. This leads to the observed diagonal demarcation line.

The diagonal behaviour noted above for the metal to non-metal transition is reproduced in other physical and chemical properties of the elements. For example, when the basic, amphoteric, or acidic classification of oxides is superimposed on the Periodic Table the diagonal demarcation line shown in Fig. 3.23 is apparent.

Water soluble acidic oxides lower the pH of the pure water which is used to dissolve them by creating more solvated hydrogen ions in solution, e.g.

$$SO_3(g) + H_2O(l) \leftrightarrows SO_4^{2-}(aq) + 2H^+(aq)$$

Water insoluble oxides are described as acidic if they react with bases to form salts, e.g.

$$SiO_2(s) + Na_2O(s) \leftrightarrows Na_2SiO_3(s)$$

Water soluble basic oxides raise the pH of water by increasing the concentration of OH^- ions.

$$K_2O(s) + H_2O(l) \leftrightarrows 2K^+(aq) + 2OH^-(aq)$$

Water insoluble basic oxides react with acids to form salts, e.g.

$$Ca(OH)_2(s) + 2HCl(aq) \leftrightarrows Ca^{2+}(aq) + 2Cl^-(aq) + 2H_2O(l)$$

Amphoteric oxides exhibit both acidic and basic properties, e.g.

$$Al(OH)_3(s) + 3HCl(aq) \leftrightarrows Al^{3+}(aq) + 3Cl^-(aq) + 3H_2O(l)$$

$$Al(OH)_3(s) + OH^-(aq) \leftrightarrows Al(OH)_4^-(aq)$$

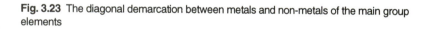

Fig. 3.23 The diagonal demarcation between metals and non-metals of the main group elements

In general, whether an oxide is acidic or basic is determined by the electropositive character of the central atom. The more electropositive (or less electronegative) the central atom the more basic the oxide. It follows that as the central atom becomes more electronegative the oxide becomes more acidic. Qualitatively, this variation may be represented by the following diagram which emphasizes the way in which the electron withdrawing ability of the central atom can influence the release of protons from the coordinated water molecules. More electronegative atoms, M, promote proton production and more electropositive atoms discourage the proton dissociation process shown in the margin.

Since electropositive character increases towards the left and down the Periodic Table, the resultant borderline between basic and acidic oxides occurs along a diagonal.

The diagonal relationship carries over to the structures of the oxides in the solid state. The basic and acidic properties of oxides depends mainly on the atom's electronegativity and size. Since the electronegativity increases across the Periodic Table and size decreases, but down the table electronegativity decreases and the size increases, it is not surprising that the oligomeric polar-covalent oxides lie along a diagonal. The classification of oxides according to their structures is given in Table 3.31. The structures of the hydrides and fluorides, summarized in Tables 3.32 and 3.33, show a similar trend with the transition from ionic to molecular covalent occurring along a diagonal.

More electropositive

More electropositive

Oligomers are polymers consisting of a small number of the basic repeat unit which is a monomeric molecular fragment

Table 3.31 Classification of oxides according to structure

Group 1	2	13	14	15	16	17	18
Li	Be	B	C	N	O	F	
Na	Mg	Al	Si	P	S	Cl	
K	Ca	Ga	Ge	As	Se	Br	
Rb	Sr	In	Sn	Sb	Te	I	Xe
Cs	Ba	Tl	Pb	Bi	Po	At	
Ionic infinite, ionic character increases to the left.					Oligomeric polar-covalent		Molecular covalent

Table 3.32 Classification of hydrides

Group 1	2	12	13	14	15	16	17
LiH	BeH_2		BH_3	CH_4	NH_3	H_2O	HF
NaH	MgH_2		AlH_3	SiH_4	PH_3	H_2S	HCl
KH	CaH_2	ZnH_2	GaH_3	GeH_4	AsH_3	H_2Se	HBr
RbH	SrH_2	CdH_2	InH_3	SnH_4	SbH_3	H_2Te	HI
CsH	BaH_2			PbH_4	BiH_3		
Ionic infinite		Intermediate		Molecular covalent			

Table 3.33 Fluorides of the elements in their group oxidation states

Group 1	2	12	13	14	15	16	17
LiF	BeF_2		BF_3	CF_4			
NaF	MgF_2		AlF_3	SiF_4	PF_5	SF_6	
KF	CaF_2	ZnF_2	GaF_3	GeF_4	AsF_5	SeF_6	
RbF	SrF_2	CdF_2	InF_3	SnF_4	SbF_5	TeF_6	IF_7
CsF	BaF_2	HgF_2	TlF_3	PbF_4	BiF_5		
	Ionic infinite		Polymeric			Molecular covalent	

The diagonal demarcation lines noted above for the changes in structure from molecular to polymeric to infinite have their origins in the Lewis acidities of the putative molecular compounds. Specifically, if a molecular compound MX_n does not have a high Lewis acidity it will not attempt to expand its coordination number by accepting electron pairs from a second molecule, i.e. it is coordinatively saturated.

A detailed discussion of this topic has also been given in R. J. Puddephatt and P. K. Monaghan, *The Periodic Table of the Elements*, Clarendon Press, Oxford, 1986.

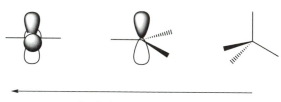

Coordinative unsaturation

The coordinative unsaturation at the metal centre depends on the positive charge at the metal centre, the availability of empty orbitals on the metal atom and the available space around the metal atom. In general the orbital availability and size effects dominate. If the coordinative saturation is relieved by the formation of one or two additional dative bonds then the resulting compound is either oligomeric or polymeric, but if all the lone pairs on the anion become involved in dative bond formation then the resulting structure is an infinite polymeric structure. The way in which donation from lone pairs may result in the formation of polymeric chains is illustrated below.

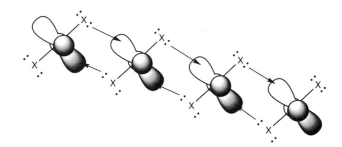

Coordinative
unsaturation

Coordinative
unsaturation

In general the coordinative unsaturation of the central atom increases down the column of the Periodic Table because the atom's radius is increasing and it is able to accommodate more atoms around it and thereby increase its coordination number. For example, CF_4, SiF_4, and GeF_4 are monomeric and have tetrahedral monomeric structures, whereas SnF_4 and PbF_4 have infinite polymeric structures based on octahedral coordination centres.

Similarly moving to the left in the Periodic Table leads to an increase in coordinative unsaturation. Although the formal oxidation state on the central metal atom decreases on moving to the left the number of orbitals available at the metal centre and the space around the metal atom increase and the latter effects dominate. Therefore, the coordinative unsaturation increases. For example, PF_5 and SiF_4 are molecular but AlF_3 is an infinite solid with octahedral geometries about aluminium. The trends in coordinative unsaturation summarized below account for the observed diagonal demarcation lines.

3.13 Diagonal relationships in rows 2 and 3

In descending a column of the Periodic Table the element's electronegativity generally decreases (electropositive character increases) and the atomic radius increases. These differences are particularly dramatic between rows 2 and 3 of the Periodic Table and lead to the elements Li–F showing distinctly anomalous physical and chemical properties. Diagonal relationships are frequently observed between the elements of rows 2 and 3 as shown below.

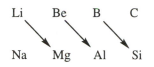

Lithium and magnesium

Table 3.34 Enthalpies of vaporization, metallic radii, and electronegativity values of some elements

	Li	Na	Mg
Enthalpy of vaporization (kJ mol^{-1})	135	98	132
Metallic radius/pm	155	190	160
Electronegativity coefficient	1.15	1.0	1.25

1. Lithium has an enthalpy of vaporization, metallic radius, and electronegativity coefficient which are more similar to those of magnesium than sodium and therefore there are many similarities in the chemistries of lithium and magnesium. Lithium has an anomalously high enthalpy of vaporization, melting point, boiling point, density, and hardness.
2. LiOH and Li_2CO_3 are much less soluble in water than the corresponding Na and K compounds. Their fluorides, carbonates, and phosphates are insoluble.
3. The oxo-anion salts of Li and Mg (CO_3^{2-}, NO_3^-, and SO_4^{2-}) decompose quite readily on heating to give the corresponding oxides.
4. Lithium and magnesium are unusual in reacting directly with N_2 to form the nitrides Li_3N and Mg_3N_2.
5. Neither lithium or magnesium form stable peroxides when burned in an atmosphere of dioxygen which remains in excess.
6. Lithium and magnesium both form organometallic compounds which have considerable covalent character and are therefore soluble in organic solvents. Organolithium, LiR, and Grignard reagents, RMgX, are used in an analagous fashion in organic syntheses.

Beryllium and aluminium

1. The enthalpy of atomization of beryllium is much higher than that of magnesium and therefore it has higher melting and boiling points, enthalpy of fusion, and density and it is much harder.

Table 3.35 Physical properties of beryllium, magnesium, and aluminium

	Be	Mg	Al
Melting point ($^\circ$C)	2970	1107	2467
Enthalpy of vaporization /kJ mol^{-1}	310	132	284
Metallic radius/pm	112	160	143
Electronegativity coefficient	1.5	1.25	1.45

2. The bonds in compounds of beryllium and aluminium are highly covalent because of the high electronegativities of the elements. In many of the more covalent compounds Be and Al adopt sp^3 hybridized tetrahedral geometries. The larger size of Al^{3+} compared to Be^{2+} (50 pm versus 31 pm) leads to some octahedral ions for the former, e.g. AlF$_6$$^{3-}$ and [Al(OH$_2$)$_6$]$^{3+}$, and AlCl$_3$ and AlF$_3$ in the solid state. The organometallic and halide compounds of Be and Al are more soluble in organic solvents than those of the other Group 2 elements.

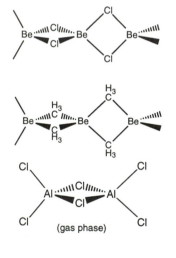

3. Beryllium and aluminium have similar standard reduction potentials:

	Be	Mg	Al
E°/V	−1.85	−2.37	−1.66

Despite the similarities in their standard reduction potentials, neither of the metals react with water because they both form an impermeable oxide film. Unlike the other alkaline earth metals and the alkali metals beryllium and magnesium do not form coloured solutions in liquid ammonia.

4. Beryllium and aluminium both form amphoteric oxides. Both dissolve in alkali to evolve H$_2$.

5. The halides of beryllium are hygroscopic and fume in air and in water AlCl$_3$ hydrolyzes to yield [Al(OH$_2$)$_6$]$^{3+}$. Positive ions of both beryllium and aluminium have high hydration enthalpies (see Table 3.36) and this leads to highly soluble salts particularly for beryllium.

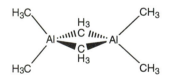

Table 3.36 Hydration enthalpies (kJ mol^{-1})

	Be^{2+}	Mg^{2+}	Al^{3+}
$\Delta_{hyd}H^\circ$	−2455	−1900	−4630

$$[\text{Be(OH}_2)_4]^{2+} \underset{\text{H}_3\text{O}^+}{\overset{\text{H}_2\text{O}}{\rightleftarrows}} [(\text{H}_2\text{O})_3\text{Be-O-Be(OH}_2)_3]^{2+} \overset{\text{OH}^-}{\rightleftarrows} \text{Be(OH)}_2$$

$$\downarrow\uparrow \text{ OH}^-$$

$$[\text{Be(OH)}_4]^{2-}$$

$$[\text{Al(OH}_2)_6]^{3+} \underset{\text{H}_3\text{O}^+}{\overset{\text{H}_2\text{O}}{\rightleftarrows}} [\text{Al(OH)(OH}_2)_5]^{2+} \overset{\text{OH}^-}{\rightleftarrows} \text{Al(OH)}_3$$

$$\downarrow\uparrow \text{ OH}^-$$

$$[\text{Al(OH)}_4]^-$$

Although Be^{2+} is smaller than Al^{3+}, the larger positive charge of the latter leads to a much more negative value for its hydration enthalpy.

6. Both beryllium and aluminium form reasonably stable complexes particularly with oxygen and nitrogen based ligands.
7. Unlike magnesium, but similar to aluminium, beryllium forms the ethynide BeC_2 rather than the carbide Be_2C.

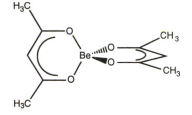

Lithium and beryllium are significantly different from the heavier elements of their groups. Specific points follow.

1. They exhibit lower coordination numbers because of their small radii. For example, Be in $BeCl_2$ and BeO is four coordinate, whereas Ca in $CaCl_2$ and CaO is octahedral.
2. The large hydration enthalpies of Li^+ and Be^{2+} lead to many of their salts having hydrated metal cations.
3. Their salts with oxoanions, e.g. NO_3^-, CO_3^{2-}, SO_4^{2-}, are much less stable than those of the heavier elements of Groups 1 and 2. Lithium also does not form stable bicarbonates, I_3^-, SH^-, and O_2^- salts and $BeCO_3$ and $BeSO_4$ decompose more readily on heating than the heavier Group 2 salts.
4. The organometallic compounds of lithium and beryllium are more covalent and consequently more soluble in organic solvents and less reactive.
5. Their complexes are more stable and numerous.
6. They both react more slowly with water. Lithium reacts more slowly with cold water than the heavier alkali metals and beryllium only reacts with steam.

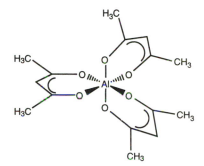

Boron and silicon

Table 3.37 summarizes the elemental radii, enthalpies of vaporization, and electronegativity coefficients of B, Al, and Si. The diagonal similarities are less pronounced than those discussed above. Boron and silicon do have the following properties in common.

Table 3.37 Physical properties of boron, aluminium, and silicon

	B	Al	Si
Enthalpy of vaporization / kJ mol^{-1}	536	284	297
Metallic / covalent radius / pm	98	143	132
Electronegativity coefficient	2.0	1.45	1.74

For both elements the exceptional stabilities of the bonds to oxygen lead to a particularly wide range of borates and silicates. Also the similarities in bond strengths for B–H and Si–H bonds lead to hydrides with analogous properties.

The following similarities may be identified for the elements.

1. Boron is semi-metallic and silicon is a semi-conductor. Boron is a black, very hard solid with a high melting point and is chemically inert. Silicon has a metallic appearance and is lustrous, relatively unreactive, and has a melting point of 1410°C (cf. boron 2300°C).

2. Boron and silicon do not exist as aquated ions, $[B(OH_2)_4]^{3+}$ and $[Si(OH_2)_6]^{4+}$. Both form a wide range of oligomers based on boron–oxygen and silicon–oxygen bonds. In borates, BO_3 triangles and BO_4 tetrahedra link through oxygen bridges to form chains and rings. In silicates, SiO_4 tetrahedra build up into chains, multiple chains, rings, sheets, and 3-dimensional networks. Both boric oxide, B_2O_3, and silicon dioxide, SiO_2, form glassy materials.

3. The halides of boron and silicon are molecular covalent solids or liquids which readily hydrolyse and behave as Lewis acids.

$$BCl_3 + 3H_2O \rightleftharpoons B(OH)_3 + 3HCl$$

$$SiCl_4 + 4H_2O \rightleftharpoons Si(OH)_4 + 4HCl$$

4. The hydrides of boron and silicon are volatile, flammable covalent compounds. They are both formed by similar hydrolysis reactions.

$$Mg_3B_2 \xrightarrow[\text{acid}]{\text{dilute}} B_nH_{n+4}, B_nH_{n+6}$$

$$Mg_2Si \xrightarrow[\text{acid}]{\text{dilute}} Si_nH_{2n+2}$$

Summary
Diagonal demarcation lines in the Periodic Table separating either chemical or structural properties occur when the atomic characteristic responsible for the changes in the properties occurs with approximately equal weight across and down a column of the Periodic Table. When the changes in atomic properties are much more dramatic down a column than those between elements which occur in a common row, the traditional Mendeleev vertical trends predominate.

Further reading

D. F. Shriver, P. W. Atkins, and C. H. Langford, *Inorganic Chemistry*, 2nd Ed., OUP, 1994.
F. A. Cotton, G. Wilkinson, and P. L. Gaus, *Basic Inorganic Chemistry*, 3rd Ed., John Wiley & Sons, 1995.

4 Isoelectronic and isostoichiometric relationships

4.1 Introduction

The previous two chapters showed that the properties of compounds depend upon several variables. Therefore, it is not always possible to relate a trend to a single parameter. Two methods have been developed to reduce the number of variables. One depends on identifying series of molecules which are isoelectronic, i.e. have the same number of valence electrons. Isoelectronic series of molecules are usually isostructural and if their structures can be rationalized then the generalization is applicable to the whole series of molecules. The second method depends on identifying a class of compounds with the same basic formula, i.e. an isostoichiometric series. By examining the structural variations which result when the total number of valence electrons is increased it becomes possible to characterize the type of orbital which is being occupied. The ions and molecules of the series H_2^+, H_2, and H_2^-, are isostoichiometric; the number of valence electrons increases from 1 to 3. In molecular orbital terms the cation has a single electron occupying the bonding molecular orbital and has a formal bond order of $\frac{1}{2}$. On adding a second electron to this molecular orbital the bond order is increased to 1. The bond length in H_2 therefore shortens relative to that in H_2^+ because of the increased bond order. When a third electron is added to form H_2^- it has to enter an antibonding molecular orbital and therefore the H–H bond order decreases to $\frac{1}{2}$ and the bond length increases. The relationship between formal bond order, bond lengths, and bond dissociation enthalpies is illustrated in the margin for hydrogen and helium diatomics.

This chapter extends these isoelectronic and isostoichiometric relationships to more complex systems and develops some important general bonding principles.

Lewis structures

The great majority of inorganic molecules may be described by Lewis structures where each atom achieves a closed shell electronic configuration. The total number of bonds (x) in such molecules is defined by the Effective Atomic Number Rule (EAN) as follows:

$$x = [2h + 8p - \text{total number of valence electrons}] \div 2$$

where h is the number of hydrogen atoms (for which the EAN rule is 2) and p is the number of p-block atoms (for which the EAN rule is 8).

A summary of bond lengths and bond dissociation energies for simple diatomics with 1–4 electrons. i.e. with formal bond orders of $\frac{1}{2}$, 1, $\frac{1}{2}$, and zero.

	X–X /pm	D(X–X) /kJ mol^{-1}
H_2^+	106	256
H_2	74	436
He_2^+	108	230
He_2	no covalent bond	

For a recent characterization of Xe_2^+ in the solid state, which has demonstrated that it has an exceptionally long bond (309 pm), see K. Seppelt *et al.*, *Angew. Chem. Int. Edn*, 1997, **36**, 273.

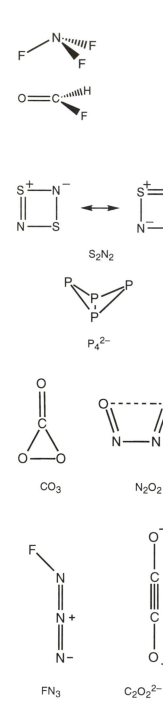

S_2N_2

$P_4{}^{2-}$

CO_3

N_2O_2

FN_3

$C_2O_2{}^{2-}$

For example, NF_3 has 26 valence electrons, $h = 0$ and $p = 4$

$$x = [2 \times 0 + 8 \times 4 - 26] \div 2 = 3 \text{ bonds}$$

HCOF possesses 18 valence electrons, $h = 1$, and $p = 3$

$$x = [2 \times 1 + 8 \times 3 - 18] \div 2 = 4 \text{ bonds}$$

Isoelectronic molecules are often also isostructural particularly if the molecule has only a single centre. For more complex chain and ring molecules the formula specifies the total number of bonds, but does not define the relative locations of single and multiple bonds. The molecules and ions shown in the margin all have 22 valence electrons, but their actual structures depend on the combinations of atoms present.

For example, N_2O_2 possesses 22 valence electrons, $h = 0$, and $p = 4$

$$x = [2 \times 0 + 8 \times 4 - 22] \div 2 = 5 \text{ bonds}$$

All the molecules and ions shown in the margin may also be represented by Lewis structures formed from five bonds. Although it is at times necessary to introduce resonance between alternative, but equivalent, structures in order to account for the observed high symmetry of the molecule.

S_2N_2 and isoelectronic $S_4{}^{2+}$ have ring structures with equivalent bonds and the electrons in the π-system perpendicular to the plane form a delocalized aromatic π electron system by means of the resonance forms shown in the margin. The ion, $P_4{}^{2-}$, in contrast has a butterfly structure and no multiple bonding. In N_2O_2 and P_2O_2 the central bond occurs between the Group 15 atoms and the N–N bond is particularly weak in the former. In the solid state $C_2O_2{}^{2-}$ has a linear structure with a C–C bond length similar to that in ethynides. The molecules CO_3 and CS_3 have only been isolated in inert gas matrices and infrared studies have suggested a Y-shaped geometry. The structure of N_3F has been established by microwave spectroscopy in the gas phase.

Clearly some additional criteria are required for ring and chain compounds in order to locate the positions of the atoms and bonds. The following generalizations are helpful in this regard.

1. Hydrogen atoms are always terminally located unless the molecule has three-centre two-electron bonds, e.g. B_2H_6, or three-centre four-electron bonds, e.g. F–H–F⁻.

2. Structures which utilize the conventional valencies of the atom are generally preferred to those which lead to charged atoms. This generally means that the halogens are located terminally and atoms from Groups 14 and 15 are generally located internally. This rule may be overturned if a ring compound with a cyclic delocalized system is formed, e.g. N_2S_2 versus N_2O_2. The utilization of conventional valencies also results in the location of lone pairs on internal atoms which can influence the

observed geometry markedly. For example, $C_2O_2^{2-}$ is linear whereas N_2O_2 is angular.

3. The less electronegative atom generally is located in central positions and the more electronegative atoms in outer positions.

The last criterion is useful for locating the central atoms in interhalogen and related compounds, e.g. ClF_4^- has a central chlorine atom.

In compounds where the EAN rule is not obeyed the formula for deriving the number of bonds presented above does not specify the atom connectivities. It defines the number of bonds in the resonance structure (examples are shown in the margin) which is based on the EAN rule.

$$XeF_2 \qquad x = [24 - 22] \div 2 = 1$$
$$XeF_4 \qquad x = [40 - 36] \div 2 = 2$$
$$XeF_6 \qquad x = [56 - 50] \div 2 = 3$$

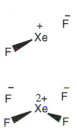

The bonding in such compounds is discussed in more detail in Section 4.5.

There are examples of compounds which are isoelectronic but which show different degrees of aggregation (oligomerization). For example, $C_2O_4^{2-}$ and N_2O_4 are isoelectronic, but the former has a C–C bond, whereas the latter exists in equilibrium between monomeric NO_2 and N_2O_4 dimers. In the solid state at $-11°C$ it exists exclusively as the dimer and in the gas phase primarily as a monomer.

4.2 Isoelectronic molecules and ions

Groups 13–18 provide many examples of molecules and ions which are isoelectronic and isostructural and some illustrative examples are summarized below. The relevant structural types are illustrated in the margins.

Diatomic molecules Y_2—10 valence electrons

$$N_2 \quad NO^+ \quad CO \quad CN^- \quad C_2^{2-} \quad BF$$

Linear molecules XY_2—16 valence electrons

CO_2	N_2O	N_3^-
CH_2N_2	NO_2^+	FCN
$H_2C=C=CH_2$	$H_2C=C=O$	$HNCO$
H_3BCN^-	H_3CCN	NCN^{2-}
H_3BCO	BO_2^-	$H_3CC{\equiv}CH$

Angular triatomic molecules XY_2—18 valence electrons

CF_2	O_3	
SO_2	SiF_2	ClO_2^+
$SnCl_2$		

Angular

Trigonal planar

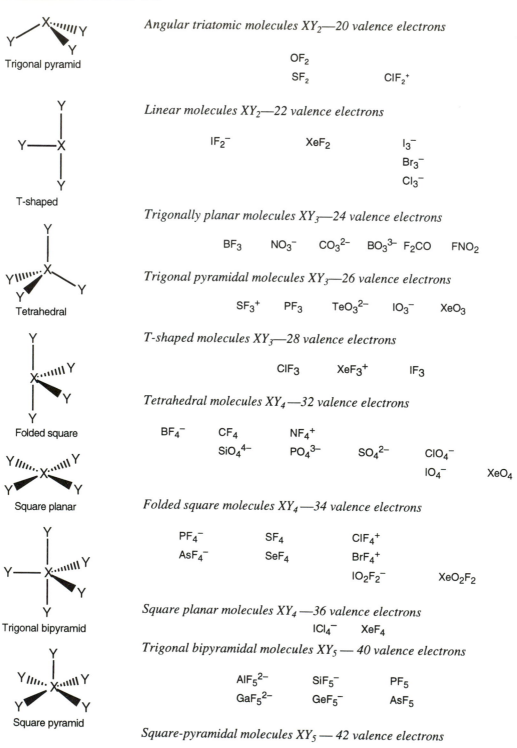

Trigonal pyramid

Angular triatomic molecules XY₂—20 valence electrons

$$OF_2$$
$$SF_2 \qquad ClF_2^+$$

T-shaped

Linear molecules XY₂—22 valence electrons

$$IF_2^- \qquad\qquad XeF_2 \qquad\qquad I_3^-$$
$$Br_3^-$$
$$Cl_3^-$$

Tetrahedral

Trigonally planar molecules XY₃—24 valence electrons

$$BF_3 \qquad NO_3^- \qquad CO_3^{2-} \qquad BO_3^{3-} \quad F_2CO \qquad FNO_2$$

Trigonal pyramidal molecules XY₃—26 valence electrons

$$SF_3^+ \qquad PF_3 \qquad TeO_3^{2-} \qquad IO_3^- \qquad XeO_3$$

Folded square

T-shaped molecules XY₃—28 valence electrons

$$ClF_3 \qquad XeF_3^+ \qquad IF_3$$

Tetrahedral molecules XY₄—32 valence electrons

$$BF_4^- \qquad CF_4 \qquad NF_4^+$$
$$SiO_4^{4-} \qquad PO_4^{3-} \qquad SO_4^{2-} \qquad ClO_4^-$$
$$IO_4^- \qquad XeO_4$$

Square planar

Folded square molecules XY₄—34 valence electrons

$$PF_4^- \qquad SF_4 \qquad ClF_4^+$$
$$AsF_4^- \qquad SeF_4 \qquad BrF_4^+$$
$$IO_2F_2^- \qquad XeO_2F_2$$

Trigonal bipyramid

Square planar molecules XY₄—36 valence electrons

$$ICl_4^- \qquad XeF_4$$

Trigonal bipyramidal molecules XY₅—40 valence electrons

$$AlF_5^{2-} \qquad SiF_5^- \qquad PF_5$$
$$GaF_5^{2-} \qquad GeF_5^- \qquad AsF_5$$

Square pyramid

Square-pyramidal molecules XY₅—42 valence electrons

$$IF_5^+ \qquad XeF_5^+$$

Octahedral molecules XY$_6$ — 48 and 50 valence electrons

$$AlF_6^{3-} \qquad SiF_6^{2-} \qquad PF_6^{-} \qquad SF_6$$

$$SnF_6^{2-} \qquad Sn(OH)_6^{2-}$$

$$SbF_6^{3-} \qquad TeF_6^{2-} \qquad IF_6^{-} \qquad XeF_6$$

Octahedral

Pentagonal bipyramidal molecules XY$_7$ — 56 valence electrons

$$IF_7$$

Square antiprismatic molecules XY$_8$ — 64 and 66 valence electrons

$$IF_8^{-} \qquad XeF_8^{2-}$$

Pentagonal pyramid

The large range of compounds and ions given above have been organized in isoelectronic series, but sometimes the term is used more loosely to describe molecules, where although the total number of valence electrons is different, the number of valence electrons around the central atom is equal. For example, $SiCl_4$, SiH_4, and $SiMe_4$ may be described as isoelectronic, because H, Cl, and Me all are one electron donors to the valence orbitals of silicon. Similarly, SiF_4, $SiCl_4$, $SiBr_4$, and SiI_4 are not strictly isoelectronic, but they do have the same number of valence electrons.

Square antiprism

> The most important relationship connecting the individual compounds in such isoelectronic series is a structural one. The compounds share a common geometry about the central atom and the formal bond orders are identical. However, the polarities of the bonds and the charges on individual atoms differ greatly and therefore their chemical properties can be very different.

Bond length and angle variations

The isoelectronic relationship provides a basis for comparing structural data for series of compounds where the only important variable is the central atom. For example, the bond lengths given in Table 4.1 for the series of octahedral molecules and ions illustrate the contraction in the sizes of the atoms across the Periodic Table.

Table 4.1 Bond lengths for some octahedral species

	AlF_6^{3-}	SiF_6^{2-}	PF_6^{-}	SF_6
d(X–F)/pm	180	171	160	156
	GaF_6^{2-}	GeF_6^{2-}	AsF_6^{-}	SeF_6
d(X–F)/pm	188	172	167	169
	InF_6^{3-}	SnF_6^{2-}	SbF_6^{-}	TeF_6
d(X–F)/pm	204	197	184	182

The bond length in SeF_6 is the only one which appears to be out of line, since the remainder illustrate the anticipated contraction in covalent radii across the series.

Isoelectronic molecules with multiple bonds also show the anticipated contraction on moving to the right-hand side:

	C_2^{2-}	CN^-	CO	NO^+
Bond length/pm	120	117	113	106

In the less symmetrical square pyramidal molecules the structural data, given in Table 4.2, provides a basis for evaluating the effect of varying the central atom on the relative lengths of the axial and basal bonds.

Table 4.2 Bond lengths and angles for some square pyramidal species

	SbF_5^{2-}	TeF_5^-	IF_5	XeF_5^+
d(X–F), *axial* /pm	192	186	184	179
d(X–F), *basal* /pm	208	195	181	179
F(*axial*)–X–F(*basal*) angle	79°	78°	81°	79°

The X–F (*axial*) bond length is generally shorter than the X–F (*basal*) bond, but the difference becomes progressively smaller across the series. Such a generalization could be explored further and perhaps even be explained by more detailed valence theory calculations. Across the series, the angle between the axial and basal fluorine atoms does not vary significantly and this removes an additional variable from such a detailed analysis. However, in the trigonal pyramidal anion XO_3^{x-} (X = Te, I, or Xe) the angle does vary in a regular manner as given in Table 4.3.

Table 4.3 Bond lengths and angles for some trigonal pyramidal species

	TeO_3^{2-}	IO_3^-	XeO_3
d(X–O)/pm	188	182	176
O–X–O angle	94.6°	99.0°	103°

A theoretical analysis would therefore have to account for both the variation in bond lengths and bond angles across the series.

Similar trends are apparent for the triatomic XH_2 molecules and ions:

	NH_2^-	OH_2	FH_2^+
H–X–H angle	104°	105°	118°

	CH_2^-	NH_2	OH_2^+
H–X–H angle	99°	103°	111°

	BH_2^-	CH_2	NH_2^+
H–X–H angle	102°	110°	115°

For completeness, some other bond length comparisons in trigonal and tetrahedral molecules are summarized in Table 4.4.

Table 4.4 Bond lengths for some trigonal and tetrahedral species

	BO_3^{3-}	CO_3^{2-}	NO_3^-	
d(X–O)/pm	138	129	122	
	SiO_4^{4-}	PO_4^{3-}	SO_4^{2-}	ClO_4^-
d(X–O) /pm	163	154	149	146
		AsO_4^{3-}	SeO_4^{2-}	BrO_4^-
d(X–O) /pm		176	161	161

Geometric generalizations

The extensive range of isoelectronic molecules discussed above suggests that the geometries of these molecules are determined primarily by the total number of valence electrons rather than the polarities of the bonds and their multiple bond character. The XY_n molecules with $8n$ valence electrons have the symmetrical structures indicated in Table 4.5.

Table 4.5 The dependence of the geometry of XY_n molecules on the number of valence electrons

Molecule	No. of valence electrons	Geometry
XY_2	16	Linear
XY_3	24	Trigonal planar
XY_4	32	Tetrahedral
XY_5	40	Trigonal bipyramid
XY_6	48	Octahedral
XY_7	56	Pentagonal bipyramid /capped octahedron
XY_8	64	Dodecahedron square-antiprism

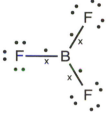

In these molecules the electrons contributed by the central atom complete the octets of electrons of the peripheral atoms by forming conventional covalent bonds and no residual unshared electron pairs remain on the central atom. For example, in BF_3 (see the diagram in the margin) the three valence electrons of boron complete the octets of the more electronegative fluorine atoms leading to a total of 24 ($8n$ with $n = 3$) valence electrons. A similar pattern is observed for hydrides and the molecules XH_n with $2n$ valence electrons adopt the symmetrical structures summarized in Table 4.6. Therefore, for both XY_n and XH_n the attainment of closed shell electronic configurations for the peripheral atoms is a significant factor.

The XY_n molecules with $8n + 2$ valence electrons have a residual lone pair of electrons located on the central atom. This lone pair is 'stereochemically active' and the molecules with $8n + 2$ electrons have characteristic geometries. Specifically the lone pair exerts its stereochemical effect by occupying a ligand position in the coordination sphere. The polyhedra defined by the ligands and the lone pair are identical to those described above for XY_n with $8n$ valence electrons.

If the lone pair is indicated in the formula by E the series of molecular shapes, shown in Table 4.7, is generated.

Table 4.6 The geometry of XH_n molecules

Molecule	No. of valence electrons	Geometry
XH_2	4	Linear
XH_3	6	Trigonal planar
XH_4	8	Tetrahedral

Table 4.7 The dependence of shape on the number of valence electrons when one lone pair is present

	Number of valence electrons	Molecular shape	
EXY$_2$	18	Angular	
EXY$_3$	26	Trigonal pyramid	
EXY$_4$	34	Folded square	
EXY$_5$	42	Square pyramid	

The non-regularity of the structures appears to depend in part on the size of the central atom. For example, the extent of distortion from regular octahedral symmetry is: IF_6^- > SeF_6^{2-} > BrF_6^- and the coordinated anions, e.g. $SeCl_6^{2-}$, $SeBr_6^{2-}$, TeX_6^{2-} (X = Cl, Br, or I) are all symmetrical. (K. Seppelt *et al.*, *Angew. Chem. Int. Ed.*, 1996, **35**, 398; 1995, **34**, 1586)

Each of the molecular structures illustrated on the right hand side of the table is related to the symmetrical coordination polyhedra described for XY$_n$, but one of the Y atoms has been replaced by a lone pair of electrons.

The next member of the series, EXY$_6$, should by analogy have a pentagonal bipyramidal structure with a lone pair occupying one of the ligand positions. In practice, such molecules have either a pentagonal pyramidal structure (e.g. $Sb(C_2O_4)_3^{3-}$ and $[XeOF_5]^-$), a distorted octahedral structure based on a face capped octahedron (e.g. XeF$_6$) or a regular octahedral structure, where the lone pair no longer appears to be stereochemically active, e.g. $TeCl_6^{2-}$.

Those molecules with $8n + 4$ valence electrons have two lone pairs located on the central atom which are both stereochemically active. If the lone pairs are again represented by E the structures of these E$_2$XY$_4$ molecules may be rationalized also in terms of the symmetrical coordination polyhedra with lone pairs occupying two positions previously occupied by the atoms Y. Table 4.8 gives specific examples of these shapes.

Table 4.8 Shapes of molecules with two lone pairs

Molecule	No. of valence electrons	Shape	Derived from:
E$_2$XY$_2$	20	Angular	Tetrahedron
E$_2$XY$_3$	28	T-shaped	Trigonal bipyramid
E$_2$XY$_4$	36	Square planar	Octahedron
E$_2$XY$_5$	44	Pentagonal planar	Pentagonal bipyramid

In summary, the shapes of molecules of the Groups 13–18 elements conform to a rather simple and useful general pattern, which appears to be decided primarily by the total number of valence electrons. Specifically, the completion of octets around the peripheral atoms and the stereochemical activities of the lone pairs, which remain on the central atom, decide which basic coordination polyhedron is preferred. Secondly, stereochemically active lone pairs and the Y atoms are interchangeable. This wealth of structural data may therefore be economically summarized by the structure matrix shown in Fig. 4.1. The molecules with $8n$, $8n + 2$, $8n + 4$, and $8n + 6$ valence electrons lie successively on diagonals of this matrix.

XY_6

(SF_6)

The derivation of the matrix does not require any knowledge of the bond orders and polarities of the bonds in the molecule—only the total number of valence electrons. For example, no distinction is made between CO_2 and BeF_2, although the former is usually written with double bonds between C and O, because they are isoelectronic.

This structure matrix has been derived using the minimum of theory and the underlying electronic reasons for the adoption of such a simple structural paradigm represents an intriguing and important problem for valence theory. The simplest accepted theoretical interpretation is described as the Valence Shell Electron Pair Repulsion Theory (VSEPR).

This is a semi-classical theory which views the bonding and lone pairs in the Lewis description of chemical bonds as localized regions of electron density which repel each other sufficiently strongly to represent the prime electronic factor responsible for determining the geometries of main group molecules. VSEPR gives a geometric dimension to the Lewis theory by proposing that the most favourable molecular shape is that which minimizes the repulsion between sigma-bonding electron pairs around specific atoms.

The coordination polyhedra which minimize these electron pair repulsions in a molecule AB_n with no non-bonding pairs of electrons have shapes which are summarized in Table 4.9 and are illustrated along the diagonals of the matrix shown in Fig. 4.1.

If the atoms Y are replaced by lone pairs the polyhedra are retained and an electron pair occupies the vertex which was previously occupied by Y. For example, the following molecules SF_6, BrF_5, and XeF_4 all have six σ-electron pairs and have geometries based on the octahedron.

The series XY_5, EXY_4, E_2XY_3, and E_3XY_2; XY_4, EXY_3, and E_2XY_2; and XY_3 and EXY_2, display a similar structural patterns and are illustrated in rows of the structure matrix in Fig. 4.1.

More detailed aspects of the shapes are rationalized by proposing that bonding pairs and lone pairs do not generate the same extent of electron–electron repulsion. Specifically the following order of repulsions rationalizes the observed geometries:

lone pair–lone pair > lone pair–bond pair > bond pair–bond pair

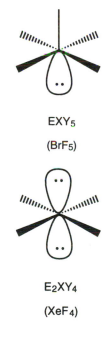

EXY_5

(BrF_5)

E_2XY_4

(XeF_4)

In its most widely used form Valence Shell Electron Pair Theory has been developed by R. J. Gillespie, *Angew. Chem. Int. Ed.*, 1996, **35**, 495.

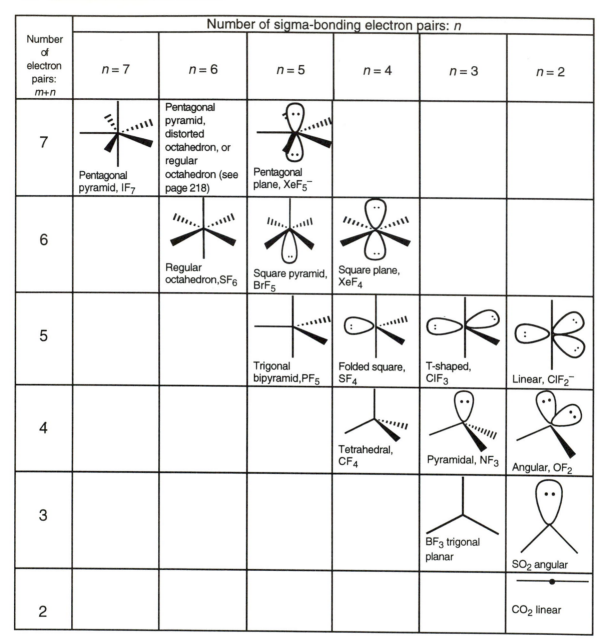

Fig. 4.1 Structure matrix for E_mXY_n main group molecules

Table 4.9 Polyhedra arising from the numbers of σ-bonding electron pairs

XY_n	8	7	6	5	4	3	2
Polyhedron	Square antiprism	Pentagonal bipyramid	Octahedron	Trigonal bipyramid	Tetrahedron	Trigonal plane	Linear

The molecule adopts its shape, or adjusts its shape to minimize these repulsions. For example, XeF_4 adopts a *trans*-arrangement of lone pairs rather than a *cis*-arrangement because the lone pair–lone pair repulsions are minimized in the former.

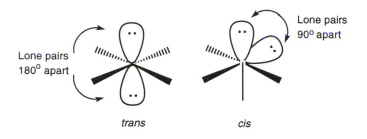

The manner in which the molecular shape is adjusted to minimize repulsion may be illustrated by the following series which have geometries based on the tetrahedron.

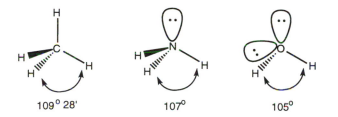

In ammonia and water the H–N–H and H–O–H bond angles decrease to reduce the lone pair–bond pair repulsions. The energy gain compensates for the increased bond pair–bond pair repulsions.

Finally, VSEPR theory suggest that multiple bonds exert a stronger repulsion than single bonds and the molecule adjusts its geometry to reflect this. The following molecules illustrate specific applications of this principle:

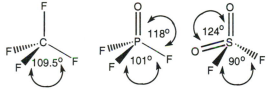

Exceptions

The simple theory developed above clearly is able to rationalize a wide range of structures of molecules, but there are exceptions and some of the more important are summarized below.

1. Alternative geometries are at times observed for XY_n with 5, 7, and 8 coordination because they have similar energies to those predicted by VSEPR. Examples, $[SbPh_5]$, $[BiPh_5]$, and $[InCl_5]^{2-}$ have square-pyramidal not trigonal bipyramidal geometries; $[NbF_7]^{2-}$ (capped trigonal prismatic) and $[NbOF_6]^{3-}$ (capped octahedral) rather than pentagonal

The association of lone pairs with specific locations in the coordination sphere has important implications regarding the stereochemistries of many of the compounds formed by the parent molecule. Specifically, making a coordinate bond or a hydrogen bond which involves donation of electron pairs from the parent molecule to the Lewis acid or H–A occurs essentially in the lone pair direction(s), and only relatively minor changes occur in the bond angles involving the other atoms linked to the central atom. For molecules with more than one location this may lead to alternative geometries for the adducts.

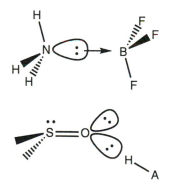

bipyramidal. For 8-coordination the square antiprism is almost as favourable as the dodecahedron

2. The lone pair in EXY_n is not stereochemically active in the following EXY_n compounds $[TeCl_6]^{2-}$ and $[SbBr_6]^{3-}$ ions which retain a regular octahedral geometry. Similarly although PbO has an infinite layer lattice based on XY_4E square pyramids with E in the axial position PbS has a sodium chloride structure with regular octahedra.

3. Steric effects resulting from bulky ligands can overwhelm lone pair effects. For example, both $SnCl_2$ and $Sn(C_5H_5)_2$ have angular geometries but $Sn(C_5Ph_5)_2$ has a linear geometry.

4. Strong multiple bonding effects can lead to delocalization of the lone pair and consequently it is no longer stereochemically active. Examples, $C(NO_2)_3^-$ is trigonal planar not pyramidal, $[O(RuCl_5)_2]^{4-}$ is linear rather than angular about oxygen because of $p_\pi - d_\pi$ bonding.

5. Coordination compounds with incomplete d shells frequently have geometries which do not conform to VSEPR and it is necessary to use alternative arguments to rationalize their structures.

6. The alkaline earth halides MX_2 in the gas phase have linear geometries for M = Be and Ca, but angular geometries for the heavier metals.

The structures of these molecules are discussed further using models which are based more fully on modern quantum mechanical ideas later in this chapter.

4.3 Reactivity consequences of the structure matrix

The structure matrix shown in Fig. 4.1 can be used to provide some insight into the stereochemical changes which accompany the reactions of main group molecules. The structural changes which accompany Lewis acid/base, autoionization, and oxidative addition and reductive elimination reactions can be related to the structure matrix. The changes associated with their reactions are indicated on the modified structure matrix shown in Fig. 4.2.

Lewis acid and base reactions

Addition or loss of halide ion Y^-, or more generally a Lewis base, from the molecules XY_n (Y = F, Cl, Br, or I), as shown in Fig. 4.3, does not influence the numbers of lone pairs on the central atom. Such reactions therefore only connect molecules which are diagonally related in the matrix. Examples of reactions of this type are shown in the margin.

N(SiH$_3$)$_3$ is planar rather than pyramidal. Whether this results from effective $p_\pi - d_\pi$ bonding or steric effects is still a matter of debate. NPr$^i_3$ is planar and the absence of d-valence orbitals on carbon precludes a π-bonding explanation.

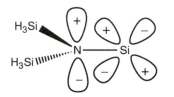

$$BF_3 + F^- \rightleftharpoons BF_4^-$$

$$XY_3 \qquad\qquad XY_4^-$$

$$CO_2 + O^{2-} \rightleftharpoons CO_3^{2-}$$

$$XY_2 \qquad\qquad XY_3^-$$

$$IF_5 + F^- \rightleftharpoons IF_6^-$$

$$EXY_5 \qquad\qquad EXY_6^-$$

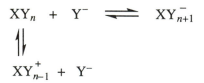

Fig. 4.3 General equations showing the addition or loss of a halide ion, Y^- from the molecule XY_n

Autoionization

Autoionization reactions connect three members of a diagonal series of Fig. 4.2

$$2PCl_5 \rightleftharpoons PCl_4^+ + PCl_6^-$$

$$2XY_5 \qquad XY_4^+ \quad XY_6^-$$

Autoionization reactions are very important for understanding the properties of non-aqueous solvents such as BrF_3, IF_5, and $POCl_3$. Halide abstraction reactions also do not involve the transfer of lone pairs and therefore the Lewis acid and base which are involved move one up and one down, respectively, along the diagonal.

$$SF_4 + BF_3 \rightleftharpoons SF_3^+ + BF_4^-$$

$$EXY_4 \quad XY_3 \qquad EXY_3^+ \quad XY_4^-$$

This reaction is widely used to generate cationic ions and some additional examples are given below:

$$AsF_2^+SbF_6^- \qquad SF_3^+SbF_6^- \qquad ClF_2^+AsF_6^- \qquad XeF^+AsF_6^-$$
$$AsCl_2^+AsF_6^- \qquad\qquad\qquad IF_4^+AsF_6^- \qquad XeF_3^+AsF_6^-$$
$$\qquad\qquad\qquad\qquad\qquad IF_6^+AsF_6^-$$

From the examples above it is apparent that AsF_5 and SbF_5 are particularly effective Lewis acids for forming fluoro- and chloro-cations.

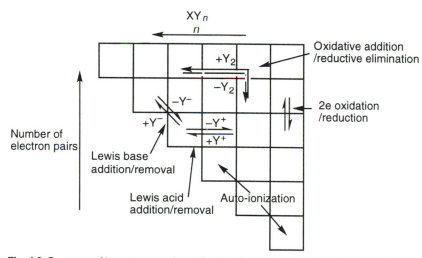

Fig. 4.2 Summary of important reactions of molecular species of the p block elements superimposed upon the structure matrix

The molecules shown in Table 4.10 are good halide acceptors because they can expand their coordination numbers readily.

Table 4.10 Examples of some good halide acceptors, Y = Cl, Br, I

XY_3	XY_4	XY_5		EXY_3	EXY_4	EXY_5	E_2XY_2	E_2XY_3	E_3XY
BY_3	SiY_4	PY_5	>	PY_3	SF_4	BrF_5	SeY_2	ClF_3	ClF
	GeY_4	AsY_5	>	AsY_3	SeY_4	IF_5	TeY_2	BrF_3	BrF
	SnY_4	$Sb\,Y_5$	>	SbY_3	TeY_4			IF_3	IF

With respect to the base NMe_3 the Lewis acidity orders are:

$$BF_3 \; > \; SiF_4 \; < \; PF_5$$
$$PF_5 \; < \; AsF_5 \; < \; SbF_5$$

It is noteworthy that apart from BF_3, which has an empty p orbital which enables it to function as a Lewis acid, all the examples are drawn from the subsequent rows of the Periodic Table. Clearly these larger atoms are able to expand their coordination spheres much more readily than the second row atoms. It is a matter of controversy whether the higher energy empty nd orbitals of these atoms are sufficiently low lying for them to function effectively as the Lewis acid acceptor orbitals. The relative Lewis acidities of BF_3, BCl_3, BBr_3, and BI_3 are discussed in Section 4.5.

Redox reactions

Redox processes which maintain the formula of the compound, but alter the number of valence electrons, result in vertical connections in the structure matrix as the following examples show.

$$NO_2^+ \; + \; 2e \; \rightleftharpoons \; NO_2^-$$
$$XY_2 \qquad\qquad\qquad EXY_2$$

$$I_3^+ \; + \; 2e \; \rightleftharpoons \; I_3^-$$
$$E_2XY_2 \qquad\qquad\qquad E_3XY_2$$

However, such transformations are rarely observed in aqueous solutions because the Lewis acid and base character of the two ions are usually so extremely different that they cannot coexist in solution without one or other of them reacting with the solvent.

Single electron processes which lead to intermediate species are more common and may be studied using electron spin resonance spectroscopy or matrix isolation techniques.

$$ClO_2(g) \; + \; e \rightarrow ClO_2^-$$

$$SO_2(g) \; + \; e \rightarrow SO_2^-$$

$$NO^+ \; + \; e \rightarrow NO$$

If E_mXY_n acts as a Lewis base the lone pair on the central atom is donated to the Lewis acid and the structural type changes from E_mXY_n to $E_{m-1}XY_{n+1}$ and this corresponds to a single movement to the left in the structure matrix.

$$
\begin{array}{lll}
NH_3 & + \quad H^+ & \rightarrow \quad NH_4^+ \\
PCl_3 & + \quad S & \rightarrow \quad PCl_3S \\
EXY_3 & & \quad \quad \quad XY_4
\end{array}
$$

Oxidative addition and reductive elimination reactions

Oxidative addition and reductive elimination reactions involve the destruction or creation of lone pairs and therefore are associated with complex movements in the structure matrix. Formally, an oxidative addition reaction which results in the formation of two additional bonds at the metal centre may be viewed as either the sequential addition of Y^+ and Y^- or the concerted addition of Y_2 across the XY_n molecule.

An example of an oxidative addition reaction is:

$$
\begin{array}{ccc}
ICl_2^- \; + \; Cl_2 & \rightleftharpoons & ICl_4^- \\
E_3XY_2 & & E_2XY_4
\end{array}
$$

and generally such reactions can be written as:

$$
E_mXY_n \; + \; Y_2 \; \rightleftharpoons \; E_{m-1}XY_{n+2}
$$

An example of a reductive elimination reaction is:

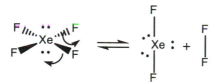

and generally such reactions can be written as:

$$
E_mXY_n \; \rightleftharpoons \; E_{m+1}XY_{n-2} \; + \; Y_2
$$

The movements in the matrix (Fig. 4.2, p. 223) therefore resemble those of a knight on a chess board, i.e. one position up and two positions across to the left for oxidative addition and two positions to the left and one position down for a reductive elimination.

4.4 Reactivity patterns for isoelectronic molecules

Although the compounds in an isoelectronic series are structurally related, their stabilities, Lewis acid and base behaviour, and redox properties differ greatly. The residual charge on the central atom, the polarities and strengths of the bonds, and the kinetic lability of the molecules significantly influence the observed reactivites of the molecules.

For example, in the following isoelectronic series: C_2^{2-}, CO, N_2, and NO^+ all the species have a triple bond, but their reactivities are remarkably different. The C_2^{2-} ion is instantly protonated by water to produce ethyne, whereas N_2 and CO are totally unreactive and NO^+ is rapidly hydrolysed to HNO_2, the initial step involving nucleophilic attack at nitrogen. The species also become progressively more oxidizing and less reducing, e.g. NO^+ may be used to dissolve gold in CH_3CN by oxidizing the metal to $[Au(NCCH_3)_2]^+$, whereas CO is widely used as a high temperature reducing agent in many metallurgical process requiring the reduction of the metal oxide to the metal. CO, N_2, and NO^+ all form a wide range of complexes with the transition metals in low oxidation states, but the ease of formation of the complexes and their reactivities differ greatly.

It has been demonstrated that HCO^+ is only formed in significant quantities in the superacid HF/SbF_5 if the CO pressure is 200 atm. (P. J. F. de Rage, J. A. Gladysz and I. T. Horvàth, *Science*, 1997, **276**, 776)

Similarly NO_2^+ and CO_2 are isostructural and isoelectronic, but the residual positive charge in the former makes it a much stronger Lewis acid and it is widely used in organic chemistry as a nitrating agent for aromatic molecules.

In general for an isoelectronic series the molecule becomes a stronger Lewis acid as the residual charge becomes more positive and the atoms are replaced by others further to the right in the Periodic Table.

The compounds also become stronger oxidizing agents when these changes are made. For example, XeF_2 and XeF_4 are stronger oxidizing agents than IF_2^- and IF_4^-. The following standard reduction potentials illustrate the general trend.

									$E^{\ominus}/V$
ClO_4^-	+	$2H^+$	+	$2e$	$\rightarrow$	ClO_3^-	+	H_2O	1.19
SO_4^{2-}	+	H_2O	+	$2e$	$\rightarrow$	SO_3^-	+	$2OH^-$	−0.93
PO_4^{3-}	+	$2H_2O$	+	$2e$	$\rightarrow$	HPO_3^-	+	$3OH^-$	−1.12
XeO_3	+	$6H^+$	+	$6e$	$\rightarrow$	$Xe(g)$	+	$3H_2O$	1.0
IO_3^-	+	$3H_2O$	+	$6e$	$\rightarrow$	I^-	+	$6OH^-$	0.26
H_4XeO_6	+	$2H^+$	+	$2e$	$\rightarrow$	XeO_3	+	$3H_2O$	3.0
H_5IO_6	+	H^+	+	$2e$	$\rightarrow$	IO_3^-	+	$3H_2O$	1.6

The pK_a's of a isoelectronic series of main group hydrides also show systematic variations.

The dominant influence across the period is the increasing electronegativity of the central atom which leads to a more polar bond and encourages the following dissociation:

$$EH_n \rightleftharpoons EH_{n-1}^-(aq) + H^+(aq) \qquad (n = 4\text{–}1)$$

Table 4.11 gives a comparison of the mean bond dissociation enthalpies and the pK_a values in aqueous solution for the isoelectronic series of hydrides for the 2nd and 3rd row elements. The bond enthalpy rather than the electronegativity differences are more important in influencing the pK_a's of hydrides belonging to the same column of the Periodic Table, e.g. HCl is a stronger acid than HF.

Table 4.11 Values of pK_a's and bond enthalpies of some 2nd row hydrides

	CH_4	NH_3	OH_2	FH
pK_a	58	35	14	3
$E(EH)/kJ\ mol^{-1}$	414	389	464	565
	SiH_4	PH_3	SH_2	ClH
pK_a		27	7	7
$E(EH)/kJ\ mol^{-1}$			368	431

The solvated metal ions $[M(H_2O)_6]^{n+}$ also represent an isoelectronic series and their pK_a values vary as follows:

	$[Na(OH_2)_6]^+$	$[Ca(OH_2)_6]_2^{2+}$	$[Al(OH_2)_6]^{3+}$
pK_a	>20	13	5

As the charge on the central metal ion increases its electron-withdrawing effect on the coordinated H_2O molecule increases and the following dissociation process is encouraged:

Indeed, the polarizing power of metal ions with a charge of +4 or more is so great that their aquo-ions are not observed and they are generally isolated as hydroxides or oxides, e.g. $Si(OH)_4$ rather than $[Si(OH_2)_6]^{4+}$; $SO_2(OH)_2$ rather than $[S(OH)_6]^{6+}$. The size of the ion is also important and for example, unlike $[Al(OH_2)_6]^{3+}$, the corresponding boron compound exists as $B(OH)_3$ rather than $[B(OH_2)_m]^{3+}$ (m = 4 or 6).

The ability of molecules to react with themselves to form polymeric structures also increases with charge. As the negative charge on the anions ClO_4^-, SO_4^{2-}, PO_4^{3-}, and SiO_4^{4-} increases the ions show a more pronounced tendency to form oxo-polyanions as indicated in Table 4.12.

$$2SiO_4^{4-} + 2H^+ \rightleftharpoons 2SiO_3(OH)^{3-}$$

$$\rightarrow Si_2O_7^{6-} + H_2O \rightarrow etc.$$

Table 4.12 Examples of some oxo–polyanions

SO_4^{2-}	$S_2O_7^{2-}$				
PO_4^{3-}	$P_2O_7^{4-}$	$P_3O_{10}^{5-}$			
SiO_4^{4-}	$Si_2O_7^{6-}$	$Si_3O_9^{6-}$	$[SiO_3]_n^{2n-}$	$[Si_2O_5]_n^{2n-}$	SiO_2
	(Linear)	(Cyclic)	(Polymer chains)	(Polymer sheets)	3D structure

The polymerization process is more favourable for the more highly charged anions because the oxygen atoms are more negatively charged and therefore are able to act more effectively as nucleophiles towards the central atom of a second MO_4^{x-}. In addition the polymerization process reduces the number of terminal O^{2-} ions and replaces them with bridging oxygen atoms and the resultant condensed species bear a smaller formal charge.

4.5 Octet rule, hybridization schemes, and multiple bonding

It is instructive to consider in more detail the role of the octet rule on the structural generalizations developed above. In the structure matrix illustrated on page 220 only those compounds which lie on the horizontal row with 4 sigma electron pairs conform to the octet rule, i.e. CF_4, NF_3, and OF_2.

Below this line the molecules are electronically unsaturated, i.e. they have fewer than 8 valence electrons involved in sigma bonding. This coordinative unsaturation is generally relieved either internally by multiple bond formation or externally by accepting an electron pair from a Lewis base. Specifically in a molecule such as CO_2 the formation of π-bonds between carbon and oxygen results in the attainment of the octet configuration. In terms of valence bond theory, this is represented by the resonance between the following canonical forms:

$$^+O\equiv C-O^- \longleftrightarrow O=C=O \longleftrightarrow {}^-O-C\equiv O^+$$

For a recent account of hybridization schemes see Lin Zhenyang and D. M. P. Mingos, *Struct. Bond.*, 1990, **72**, 73.

In orbital terms these Lewis structures are represented by sp hybridization at carbon and p_π–p_π multiple bond formation between carbon and oxygen as shown below. The different canonical forms represent the alternative π-bonding possibilities.

The generation of the appropriate sp hybridized state for carbon requires a preliminary promotion of an electron from 2s to 2p. The orbital interaction diagram for CO_2 is shown in Fig. 4.4.

In BeF_2 which is isoelectronic with CO_2 the coordinative unsaturation is more evident because the sigma-bonding framework leads to two p orbitals on Be which are empty. However, coordinate (dative) π-bonds between the filled p orbitals on F and these empty orbitals relieve the electronic unsaturation as indicated in the margin.

The orbital representation of the bonding in BF_3, shown in Fig. 4.5, is based on sp^2 hybrids at boron which are directed at 120° which leaves an empty p_z orbital on the boron atom. The Lewis structures, shown in the margin, result in electron donation to the empty p orbital perpendicular to the BF_3 plane and similarly reduce the degree of electronic unsaturation.

It is noteworthy that although there are three filled p orbitals on the fluorine atoms which have the correct symmetry to donate to the boron $2p_z$ orbital, only one interaction can be represented at a time by the canonical forms shown in the margin and the orbital box diagram above. The resonance formalism illustrated in the margin represents the valence bond way of illustrating the shared role of the fluorine donor orbitals. In no sense is the dative bond moving from atom to atom, but the resonance formalism is a shorthand way of saying that the wave function for the molecule is best represented by taking a linear combination of the wave functions for each of the three canonical forms.

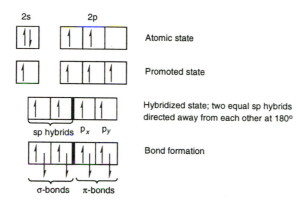

Fig. 4.4 Orbital interaction diagram for the CO_2 molecule

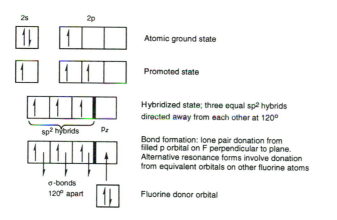

Fig. 4.5 Orbital diagram for the BF_3 molecule

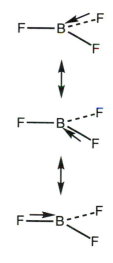

The relative strengths of multiple bonds for a column of elements is discussed in Chapter 2

The extent to which intramolecular donation from filled orbitals on the peripheral atoms relieves the electronic unsaturation on the central atom depends on the electronegativity difference between the atoms involved and the extent of overlap between their orbitals. In general the donation becomes less effective if the atoms have very different electronegativities and the overlap becomes less effective if the atoms do not belong to the same row of the periodic table. Also $2p_\pi$–$2p_\pi$ overlaps are generally much larger than those for $3p_\pi$–$3p_\pi$, $4p_\pi$–$4p_\pi$, etc.

Molecules such as BF_3, BeF_2, and CO_2 are able to function as Lewis acids because an alternative way of making up for the electron unsaturation is provided by the donation of an electron pair from an external source, i.e. a Lewis base.

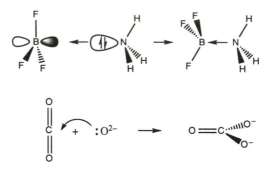

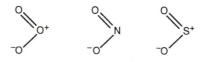

Therefore, the Lewis acidity of unsaturated molecules depends on the polarity of the bonds, the extent of π-bonding and the total charge on the species, e.g. the Lewis acidities decrease in the order: $NO_2^+ > CO_2 > N_2O > N_3^-$. Indeed, the azide ion is better known as a Lewis base than a Lewis acid. Although the aluminium halides show the relative Lewis acidities:

$$AlCl_3 \quad > \quad AlBr_3 \quad > \quad AlI_3$$

which reflect the decreasing electronegativities of the halide substituents, for the boron halides the relative Lewis acidities towards a base such as NMe_3 is:

$$BI_3 \quad > \quad BBr_3 \quad > \quad BCl_3 \quad > \quad BF_3$$

The electron withdrawal through the sigma bonds is controlled by electronegativity effects, and therefore the ability to form good π-bonds between B and the halides diminishes in the order: $F > Cl > Br > I$, because the overlap between orbitals with matching sizes is better than that between mismatched orbitals. The latter effect predominates and accounts for the observed trend.

Molecules such as O_3, NO_2^-, and SO_2 are angular and the bonding may be described as an adaptation of the orbital picture developed above for BF_3. Specifically, sp^2 hybridization is invoked at the central atom, but in contrast to BF_3 one of the hybrids is occupied by a lone pair of electrons.

Lewis acid

Lewis base

Since such molecules have both a lone pair and are electronically unsaturated they are amphoteric, i.e. they function as Lewis bases and Lewis acids. Examples of this duality for SO_2 are illustrated in the margin.

Compounds in the structural matrix illustrated in Fig. 4.1 which lie above the $(m + n) = 4$ row exceed the octet rule if they are described in terms of conventional two-centre two-electron bonds. Two major proposals have been proposed to account for the occurrence of such compounds. The first recognizes that for the elements in rows 3, 4, 5, and 6 there are empty nd ($n = 3$–6) atomic orbitals lying above the ns and np subshells, which may be utilized for bonding if they can be accessed by electron promotion processes which do not require too much of an energy input. For example, for SF_6 the orbital interaction diagram shown in Fig. 4.6 may be proposed.

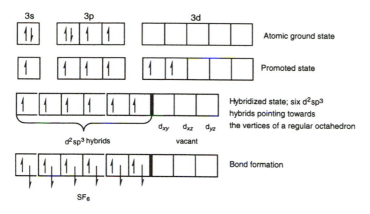

Fig. 4.6 Orbital interaction diagram for the SF_6 molecule

The resulting d^2sp^3 hybrid orbitals are directed towards the vertices of an octahedron and therefore could rationalize the observed geometry. However, it has been pointed out that the promotion energy required to promote electrons from the 3p subshell to the 3d orbitals is prohibitively large (800 kJ mol^{-1} for sulfur) and the energy could not be regained by the energy released when the additional S–F bonds are formed. In addition, accurate molecular orbital calculations which have been completed on SF_6 have not demonstrated a significant participation of the 3d orbitals in the S–F bonds.

An alternative interpretation retains the concept of the octet rule, but accommodates the excess of electrons on the peripheral atoms by invoking the concept of the three-centre four-electron bond. These bonds utilize only the s and p valence orbitals of the central atom and therefore do not require the involvement of the d orbitals of the central atom.

Three-centre four-electron bonds

The orbital interactions between a p orbital and two hydrogen orbitals positioned 180° apart either side of the p orbital are illustrated schematically in Fig. 4.7.

This arrangement of atoms and orbitals can accommodate four electrons; two electrons in the in-phase bonding combination which is bonding between all three atoms and two electrons in the non-bonding orbital which is localized on the peripheral atoms. The remaining orbital is antibonding and is not utilized in bonding. In consequence, of the initial four electrons only two are utilized directly in bonding and two remain localized on the peripheral atoms. This bonding mode may be represented by the resonance combination of the following valence bond canonical forms:

$$F^- \quad Cl\text{–}F \quad \leftrightarrow \quad F\text{–}Cl \quad F^-$$

(S) indicates that the linear combination is symmetric to a mirror plane perpendicular to the bond axes. (A) indicates that the linear combination is antisymmetric to the mirror plane operation.

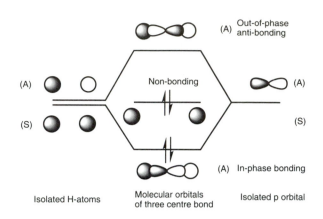

Fig. 4.7 A molecular orbital diagram showing the formation of a three-centre four-electron bond

In terms of orbital box diagrams the 3-centre 4-electron bond is represented by the following formalism:

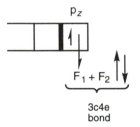

This indicates the orbital of the central atom involved in the bonding and the four electrons associated with the 3c4e bond.

Note that as a consequence of the resonance between these equivalent canonical forms the central atom has only a single electron pair bond shared between both atoms. Furthermore, the fluorine atoms share an electron pair which is not involved in bonding. Therefore, in a molecule which has n electron pairs in excess of those required by the octet rule, n three-centre four-electron bonds are formed.

It should be emphasized that the bond order resulting from the three-centre interaction is not 1/2, as intuition may suggest, but $1/\sqrt{2}$ (i.e. 0.717). The bond order is defined as the product of the coefficients of the orbitals contributing to the bond times the number of electrons occupying the orbital. In the three-centre bonding orbitals the coefficients of the external and central atom are 1/2 and $1/\sqrt{2}$ and since two electrons occupy this orbital the bond order is : $2 \times 1/2 \times 1/\sqrt{2} = 1/\sqrt{2}$.

In ions such as F–Cl–F⁻, I_3^-, etc. the components F–Cl and F⁻, and I_2 and I⁻ can exist in their own right as independent stable species. Therefore, molecules which have three-centre four-electron bonds show a continuum of structures with varying bond lengths. The structural data in Table 4.13 illustrate that the bond lengths in I_3^- and Br_3^- depend on the counter-ion present.

Table 4.13 Comparison of bond lengths in Y_3^- ions and the parent Y_2 molecules

Ion	Compound	Bond lengths/pm	
I_3^-	Me_4NI_3	2.90	2.90
I_3^-	CsI_3	2.83	3.04
	I_2	2.66	
Br_3^-	Me_3NHBr_3	2.53	2.53
Br_3^-	$CsBr_3$	2.44	2.77
	Br_2	2.38	

In the salts with the large organic cations the Y_3^- ions show symmetrical structures and the bond lengths are somewhat longer than those in the parent Y_2 molecule, The latter reflects the lower formal bond order in the three-centre four-electron bond. However, in inorganic salts the bond lengths diverge, with one approaching the length observed in the parent diatomic molecule as the other bond increases in length. In the limit the separated Y^- ion and Y_2 molecule would be separated by such long distances that they could be considered as isolated entities. The range of structures is illustrated schematically in Fig. 4.8.

The relevant Lewis structures, which serve as a starting point for the bonding description of the molecules in the structure matrix are illustrated in Fig. 4.9. The bonds in such molecules cannot all be described by conventional two-centre two-electron bonds if only the valence s and p orbitals of the central atom are used.

Br—Br—Br⁻

Br₃⁻ symmetric

Br—Br-----Br⁻

Br₃⁻ asymmetric

Br—Br   Br⁻
Br₂ and Br⁻ isolated

Fig. 4.8 Continuum of structures observed for ions with three-centre four-electron bonds

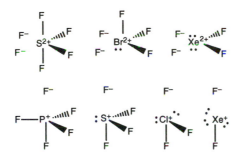

Fig. 4.9 Some Lewis structures for molecules where three-centre four-electron bonding is involved.

Note that in these neutral molecules the central atom bears a single positive charge in the five-coordinate structures and a double positive charge in the six-coordinate structures. It is noteworthy that in both series the octet rule limits the total number of bonds and lone pairs which can be used in a specific canonical form to four.

The number of bonds in these canonical forms is defined by the formula given on page 211, i.e. $x = [2h + 8p - \text{total number of valence electrons}] + 2$.

Therefore, although the three-centre four-electron bond leads to a reduction in the formal covalent bond order an additional stabilization is achieved by electrostatic interaction terms between the positive charge on the central atom and the negative charges on the fluorine atoms. *Three-centre four-electron bonding is particularly favoured if the peripheral atoms are highly electronegative and the central atom has a reasonably low ionization energy.*

The structures shown in Fig. 4.10 suggest a certain unequivalence between the bonds, however resonance between the alternative but equal canonical forms leads to a structure which has high symmetry, i.e. D_{4h} for square planar, D_{3h} for five coordination, and O_h for six coordination.

When the Lewis structures are translated into orbital interaction diagrams not only are the positive charges retained for a series of molecules which lie on a horizontal row of the structure matrix, but also a common hybridization scheme persists across the series (see Fig. 4.11).

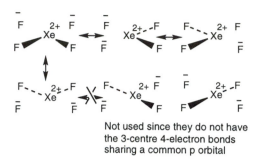

Not used since they do not have the 3-centre 4-electron bonds sharing a common p orbital

Fig. 4.10 Resonance structures for XeF_4

Geometric consequences of the hybridization model

In the valence shells of the main group atoms, the energies of the ns electrons are always lower than those of the np electrons and therefore an electron pair which resides in a sp, sp^2, or sp^3 hybrid orbital is more stable than one which resides in a pure p orbital. Furthermore, the atom's ability to form hybridized orbitals depends on the abilities of the s and p orbitals to overlap with the orbitals of the atoms bonded to the central atom. As discussed in Chapter 1 the radial distribution functions of the ns and np orbitals of the main group atoms differ significantly except for the second row atoms. Specifically the np orbital is more diffuse than the ns orbital and overlaps more effectively with the ligand orbitals at distances which are usually associated with normal single bonds.

The following structural generalizations follow from the important differences in the energies and radial distribution functions of the ns and np orbitals:

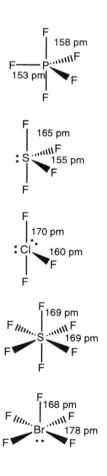

1. The three-centre four-electron bonds are generally longer than the two-centre two-electron bonds within the same molecule. The three-centre four-electron bonds involve pure p orbitals on the central atom whereas the two-centre two-electron bonds involve sp hybrid orbitals. The former overlap more effectively at longer distances, i.e. r_{max} for np is larger than r_{max} for ns. In addition the formal covalent bond order in the three-centre four-electron bond is smaller because only a single bonding pair is involved directly in bonding. The second electron pair is in a non-bonding orbital localized on the ligands. This bonding relationship is reflected in the pattern of bond lengths for these series of molecules which lie along horizontal rows of the structure matrix.

    The bonds involving the three-centre four-electron bonds are consistently longer than the localized bonds and the difference becomes larger as the number of lone pairs is increased (see structures in margin).

2. More electronegative substituents favour the sites involving the three-centre four-electron bonds and less electronegative substituents favour the sp^x hybridized sites. The three-centre four-electron bonds are associated

with a build-up of electron density on the outer atoms (see Fig. 4.7) and consequently electronegative substituents which are more able to attract electron density favour occupying these sites. The following molecules illustrate this substitutional preference.

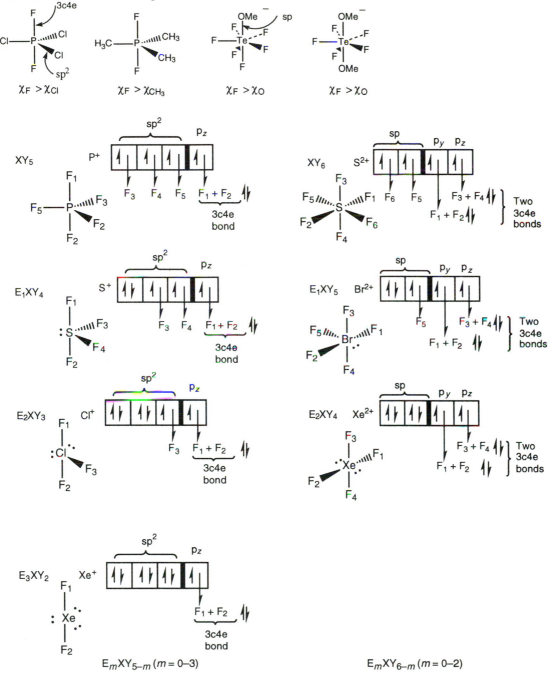

Fig. 4.11 Valence bond orbital box diagrams for E_mXY_n molecules

3. Atoms which are capable of forming multiple bonds are generally less electronegative and consequently they favour the sp^x hybridized sites. Also the short bond lengths which result from multiple bonding can improve the overlap with the ns contribution to the hybrid orbitals. Examples, of such molecules include:

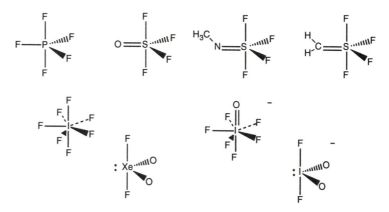

4. The lone pairs of electrons in E$_m$XY$_n$ molecules favour occupation of the sp^x hybridized sites because of the relative energies of the ns and np orbitals, i.e. ns is much more stable than np. In trigonal bipyramidal molecules the sp^2 hybrids lie in the equatorial plane and therefore the lone pair occupies an equatorial site. This preference is retained even when three are several lone pairs on the central atom. The following structures illustrate this generalization.

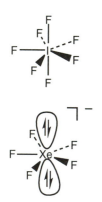

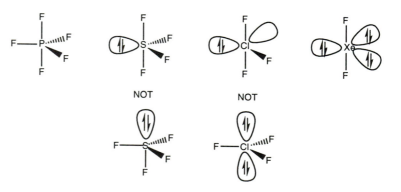

In pentagonal bipyramidal molecules the sp hybrids point towards the poles and therefore the lone pairs occupy apical rather than equatorial sites (see structures in margin).

5. The bond angles in molecules containing lone pairs decrease as the central atom is chosen from further down the column of the Periodic Table. In the series NH$_3$, PH$_3$, AsH$_3$, and SbH$_3$ the hydrogen atoms overlap more effectively with np relative to ns as the column is descended, because the r_{max} values for the np orbitals are larger (see page 25). It follows that the X–H bonds have a higher proportion of p orbital character and the hybrid orbital which contains the lone pair of electrons

has a higher proportion of s orbital character. Therefore, the H–X–H bond angles become closer to 90° and deviate increasingly from that expected for sp³ hybridized orbitals, i.e. 109.5°.

Similarly in the series SiF_4, POF_3, and SOF_2 the F–X–F angles diminish because the percentage s character in the X–F bonds decreases and that in the X–O bonds increases. As the X–O bond shortens as a result of multiple bonding the X-O overlaps involving the s orbital improves at the expense of that involving p. The percentage of s character increasing has the effect of opening up the O–X–F bond angle. Conversely, the increasing p orbital character in the F–X–F bonds results in a closing up of this angle.

> The arguments developed above illustrate that the important geometric features of main group molecules may be satisfactorily interpreted using ideas from valence bond theory. These orbitally based ideas have also proved useful for accounting for the differences in bond lengths in these molecules and the broad trends in their reactivities.

Fig. 4.12 A Lewis structure for the IF_7 molecule

Extrapolation of the bonding scheme developed above for SiF_4, PF_5, and SF_6 suggests that the appropriate Lewis structure for IF_7, shown in Fig. 4.12, should be based on I^{3+} and involve three three-centre four-electron bonds.

However, the relevant orbital diagram, shown in Fig. 4.13(a), leaves a single electron in an s orbital which is non-directional and can in principle interact equally well with each of the seven fluorine atoms.

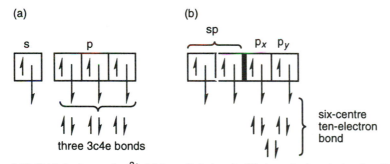

Fig. 4.13 Orbital schemes for I^{3+}: (a) three 3c4e bonds, (b) six-centre ten-electron bonding

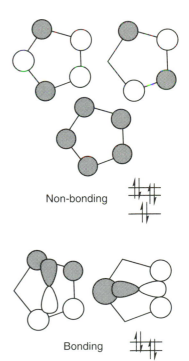

Fig. 4.14 A six-centre ten-electron bonding scheme for the IF_7 molecule

Alternatively the bonding may be described in terms of sp hybrids in the axial direction which are used to form two conventional two-centre bonds as shown in Fig. 4.13(b). This leaves two p orbitals and five fluorine orbitals in the planes and this cannot be readily accommodated within the three-centre four-electron scheme. It is possible to devise a delocalized six-centre ten-electron system of molecular orbitals such as that shown in Fig. 4.14 to account for the bonding. This partitioning of localized bonds along the axis and delocalized bonding in the plane satisfactorily accounts for the shorter I–F axial bonds in this molecule.

For the related EXY_6 molecules, e.g. XeF_6, $TeCl_6^{2-}$, etc. similar dual possibilities exist. The orbital interaction diagram which involves axial hybridization suggests a pentagonal pyramidal structure and this geometry is indeed observed in $[Sb(C_2O_4)_3]^{3-}$ and $[XeOF_5]^-$. Alternatively the molecule may adopt a regular octahedral structure and locate the lone pair of electrons in the s orbital as shown in Fig. 4.15.

Unlike PF_3, SF_4, and BrF_5 the electron pair is not occupying a hybrid orbital with directional character, but is localized in an s orbital which interacts equally with all the six peripheral atoms. The interaction is repulsive since the s orbital and the orbitals on fluorine are both filled. The octahedron is ideally set up to form three orthogonal three-centre four-electron bonds, one along each of the x, y, and z axes and therefore this favourable overlap leads to a geometry which competes with the pentagonal pyramid, which has the lone pair in a more energetically favourable sp hybrid, but has a less favourable arrangement of fluorine from the point of view of overlap. The ions and molecules shown in Table 4.14 have a regular octahedral structure. Their bond lengths indicate that the bonds to the halides are significantly longer than those in the related XY_6 molecules and ions which have two electrons fewer.

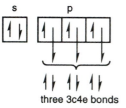

s   p

three 3c4e bonds

Fig. 4.15 Three 3c4e bonding scheme for Xe^{3+}

Table 4.14 Bond lengths for some regularly octahedral species

	$PbCl_6^{4-}$	$SbCl_6^{3-}$	$TeCl_6^{2-}$	IF_6^-	XeF_6
d(X–Y)/pm	293	265	254		189
	$PbBr_6^{4-}$	$SbBr_6^{3-}$	$TeBr_6^{2-}$		
d(X–Y)/pm	312	280	270		
		$BiCl_6^{3-}$	$PoCl_6^{2-}$		
d(X–Y)/pm		266	254		
		$BiBr_6^{3-}$	$PoBr_6^{2-}$		
d(X–Y)/pm		284	264		

IF_6^- and XeF_6 show slightly irregular structures which do not resemble either the octahedron or the pentagonal pyramid, but may be described as trigonally distorted octahedra. Therefore, they may be described as capped octahedra, a geometry which is almost energetically equal to the pentagonal bipyramid for ML_7, with the lone pair occupying the capping position.

The anion XeF_8^{2-}, which is expected to have an irregular structure with a lone pair localized on Xe, has a regular square antiprismatic structure. However, XeF_5^- has a pentagonal planar structure which resembles XeF_4 in having lone pair orbitals above and below the plane of the Xe and F ligands. Therefore, for Xe it appears that the stereochemical activity of the lone pairs depends on the coordination number. As the coordination number is increased the distortions of the structure necessary for a space to be generated for the lone pair become energetically less favoured and a regular geometry is observed. In such regular structures the lone pair of electrons no longer occupy a hybrid orbital, but are localized in the s orbital of the Xe atom.

This electron pair therefore repels all ligands equally and the resultant bond lengths tend to increase.

The relevant hybridization schemes for the molecules discussed above are summarized in the matrix shown in Fig. 4.16. From this matrix it is evident that electron precise molecules which conform to the effective atomic number rule (octet rule) only form a small part of the whole. They tend to be formed in compounds where the electronegativity differences are small. Therefore, hydrides and alkyls of the p block elements generally conform to the octet rule. Also compounds with element-element bonds, i.e. the electronegativity difference is zero, obey the rule.

The bonding model developed above suggests that for the p block elements the octet rule is a dominant influence in deciding the stoichiometries and structures of molecules. However, in order to achieve this it has been necessary to introduce the three-centre four-electron bonding model. It is important to explore whether the change in bond type from conventional two-centre two-electron bonding to three-centre four-electron bonding is reflected in the bond enthalpies. Table 4.15 summarizes the bond enthalpy terms for a range of fluorides and chlorides of the p block elements.

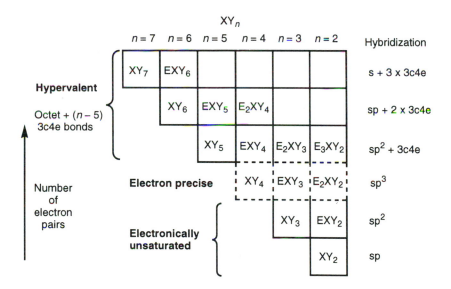

3c4e = three-centre four-electron bond (3cfe)

Fig. 4.16 A matrix showing hybridization schemes adopted by molecules

Particularly noteworthy is the general decline in bond enthalpy terms with valency, which is particularly marked when the central atom is relatively small, e.g. Cl versus Xe, and when the peripheral atoms are relatively large, cf. PF_3 and PF_5 versus PCl_3 and PCl_5. This decline in bond enthalpies is a feature of all compounds throughout the Periodic Table and cannot be attributed to a change in bond type. It is a chemical equivalent of the law of diminishing returns: the first bonds which are formed are the strongest because the parent atoms are in a neutral state. When the molecule which has formed forms additional bonds the atoms within the molecule have developed partial positive and negative charges and therefore it is more difficult to involve them in bonding. Furthermore, as more atoms are introduced around the central atom then they begin to repel each other and these effects reduce the bond enthalpy terms.

Hypervalency can also influence the rates of nucleophilic substitution reactions. For example, $[SiF_3PhMe]^-$ undergoes nucleophilic substitution reactions at an accelerated rate compared with $[SiF_2PhMe]$. See R. R. Holmes, *Chem. Rev.*, 1996, **96**, 927.

Table 4.15 Selected bond enthalpies for fluorides and chlorides of the p block elements (kJ mol^{-1}). Those in bold exceed the inert gas rule.

ClF	BrF	IF	PF$_3$	SF$_2$	XeF$_2$
255	250	281	490	367	133
ClF$_3$	**BrF$_3$**		**PF$_5$**	**SF$_4$**	**XeF$_4$**
174	202		323	339	131
ClF$_5$	**BrF$_5$**	**IF$_5$**	PCl$_3$	**SF$_6$**	**XeF$_6$**
151	187	268	461	329	126
		IF$_7$	PCl$_5$		
		232	257		

Table 4.15 indicates that there is no strong demarcation line between classical and three-centre four-electron molecules (which are shown in **bold**). Therefore, although the covalent bond order of a three-centre four-electron bond is $1/\sqrt{2}$ that of a conventional two-centre two-electron bond, the energy difference must be made up by electrostatic contributions to the bond enthalpy which arise from the resonance forms:

$$F–Xe^+ \ F^- \ \leftrightarrow \ F^- \ ^+Xe–F$$

Clearly these resonance forms will only make a significant contribution if the peripheral atom has either a high electron capture enthalpy or electronegativity and the central atom has a low ionization potential. Fluorine and oxygen as the most electronegative elements are most able to stabilize these charged resonance forms and therefore it is not surprising that these elements provide the majority of examples of such compounds. The ability to stabilize such compounds falls in the order: $F > Cl > Br > I$, and for the central atom the order is: $Xe > Kr > Ar > Ne$.

The following data for sulfur–oxygen compounds, where the steric effects are minimal, emphasize the way in which the decline in the bond enthalpy terms with valency is an inherent feature of bond formation for elements with multiple valencies. The resonance forms for these molecules which maintain the octet rule are illustrated in the margin.

Although the formal average multiple bond orders decrease in the series the bond enthalpy decreases less than may have been anticipated because the additional electrostatic interactions make a stabilizing contribution.

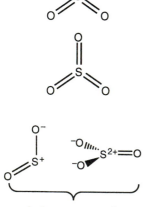

and other resonance forms

	SO$_2$	SO$_3$
Bond enthalpy (kJ mol^{-1})	537	474

The multiply bonded resonance forms shown in the margin require the participation of the 3d orbitals of sulfur. These resonance forms are unlikely to contribute significantly because of the large p–d promotion energy. The resonance forms for the SF_6 molecule, which are based on the EAN rule are shown in Fig. 4.17 and are more significant than that which implies the full participation of the 3d orbitals shown in the margin.

Fig. 4.17 Some resonance forms of the SF_6 molecule which conform to the octet rule

Hydrogen bonding

The bonding in an ion such as FHF^- may also be described in terms of three-centre four-electron bonding and the relevant orbital diagram is illustrated in Fig. 4.18.

For a detailed discussion of hydrogen bonding see P. L. Huyskens, W. A. P. Luck, and T. Zegers-Huyskens, *Intermolecular Forces*, Springer Verlag, Heidelberg, 1991.

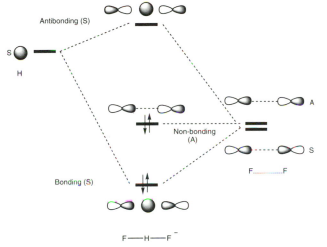

Fig 4.18 Three-centre four-electron bonding in F_2H^-

This differs from that shown in Fig. 4.7 in the following important ways. The electronegativity of H and F are very different and therefore the lines representing the energies of the isolated F atoms are placed lower on the energy scale than that for the H atom. Secondly the hydrogen 1s orbital is symmetric (S) to the mirror plane operation which reflects the fluorine atoms into themselves and therefore it forms bonding and antibonding combinations with the symmetric combination of fluorine orbitals. The antisymmetric combination (A) remains non-bonding because it does not overlap with the hydrogen 1s orbital. Occupation of the bonding and non-bonding molecular orbitals by electron pairs leads to the three-centre four-electron bond description for F–H–F⁻. This ion has probably the strongest known hydrogen bond and it has been estimated that the dissociation enthalpy for the process:

$$HF_2^- \rightleftharpoons HF + F^-$$

is 165 kJ mol^{-1}. The H–F distance in the ion is 113 pm which is somewhat longer than that in HF (92 pm) which is consistent with the bond lengthening noted previously for systems with three-centre four-electron bonds, e.g. Br_3^- and I_3^-. The strong three-centre four-electron hydrogen bond is favoured when the terminal atoms have high electronegativities and therefore the stability order is: $F_2H^- > Cl_2H^- > Br_2H^-$. Furthermore, such symmetrical strong hydrogen bonds are most commonly observed for the electronegative second row atoms F, O, and N.

One feature of three-centre four-electron bonding, noted previously, was that the symmetrical structure is not always observed, but instead a continuum of structures which range from the symmetric to the very asymmetric occur and a similar pattern is observed for three-centre four-electron bonds involving hydrogen. Some specific examples are shown in Fig. 4.19.

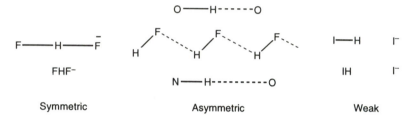

Fig 4.19 Examples of symmetric and asymmetric hydrogen bonding

Table 4.16 Some hydrogen bond donors and acceptors

Hydrogen bond donors, D–H	Acceptors, A
C–H	N
N–H	P
O–H	O
F–H	S
P–H	F
S–H	Cl
Cl–H	Br
I–H	I
	C=C
	C≡C
	aromatic rings

The broad range of groups where hydrogen bonding interactions have been identified and the large variation in their strengths has made it problematical to provide a single definition of a 'hydrogen bond'

In the gas phase the isolated H–F molecule has an internuclear distance of 92 pm and in the symmetrical [F–H–F]$^-$ anion the corresponding distances are lengthened to 113 pm. In the solid state HF adopts the linear chain structure shown in Fig. 4.19, where the H–F distances within the molecules are only slightly lengthened to 95 pm and only long intermolecular contacts of 156 pm are observed. Therefore, whereas [F–H–F]$^-$ may be viewed as a single molecular entity, the solid state structure HF may be interpreted in terms of H–F molecules with additional intermolecular H---F interactions. These differences are reflected in the relative energetics of the interaction—in solid H–F the intermolecular interactions contribute a stabilization of 30 kJ mol^{-1}, whereas in [F–H–F]$^-$ the mean bond enthalpy is 165 kJ mol^{-1}. In the asymmetric structure the covalent contributions from the three-centre four-electron bond are reduced and the interaction is replaced progressively by electrostatic attractive terms arising from the molecular dipoles.

$$\overset{\delta+\ \delta-}{H—F} \qquad \overset{\delta+\ \delta-}{H—F} \qquad \overset{\delta+\ \delta-}{H—F}$$

These three-centre interactions effectively bridge the conventional separation between covalent bonds and intermolecular interactions.

More generally, the type of interaction described above involving a D–H bond and a second atom which contributes an additional electron pair is so important and ubiquitous that it is given the title 'hydrogen bond'. Hydrogen bonds are observed in the gas, liquid, and solid state phases and the D–H and A groups shown in Table 4.16 have all been implicated in hydrogen bonding.

It is noteworthy that although the majority of the A groups are atoms which have lone pairs, multiple bonds may also participate as hydrogen bond acceptors.

The predominant nature of the electrostatic interaction in asymmetric hydrogen bonds leads to the strongest hydrogen bonds being formed when A is very electronegative, i.e. F, N, or O. Data for some more strongly bonded hydrogen bonded compounds and ions are given in Table 4.17.

Table 4.17 Examples of typical hydrogen bonding enthalpies

X–H ---Y	Example	Bond enthalpy /kJ mol^{-1}
F–H–F	KHF_2	165
$[HF]_n$		29
O–H...O	$(HCO_2H)_2$	30
O–H–O	$H_2O_5^+$	151
H_2N–H ...NH_2	NH_3	17
H_2S..H_2S		7
HOH...F$^-$	H_2OF^-	98

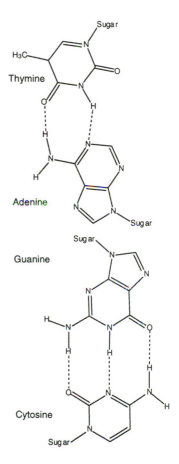

Complementary hydrogen bonding interactions between base pairs in DNA (Watson-Crick model).

The strong hydrogen bonds formed by compounds of F, O, and N have very important chemical and biological implications since hydrogen bonding interactions involving these atoms (10–50 kJ mol^{-1}) are stronger than those commonly associated with van der Waals forces (5–10 kJ mol^{-1}). The boiling points of hydrides of the second row elements are increased by up to 150°C, relative to those for comparable compounds of Cl, S, and P. The specific effect associated with hydrogen is underlined by the following boiling points: H_2O (100°C), F_2O (–114°C), and $(CH_3)_2O$ (–138°C). The importance of the proton acceptor group having a lone pair is underlined by the relative boiling points of CH_4 (–180°C) and NH_3 (–30°C).

Almost all biological molecules have groups which are capable of acting as proton donors and acceptors and the interactions between these groups and either the solvent or other groups within the molecule influences the mode of folding of proteins and the type of base pairing interaction which occur in the helical structures of DNA and RNA. Those parts of a protein chain which do not contain groups which are capable of forming strong hydrogen bonds are described as hydrophobic and these parts of the chain fold in such a way that these regions are located in the centre of the protein. The surface of the protein has groups which are capable of forming strong hydrogen bonds with the solvent water molecules. The complementary hydrogen bonding interactions between thymine and adenine, and cytosine and guanine, are very important in the replication processes associated with the helical structure in DNA.

Table 4.17 summarizes a wider range of hydrogen bond donors and acceptors. Apart from those involving F, O, and N the strength of the interactions are generally less than 20 kJ mol^{-1}. The weaker interactions therefore may not have a dramatic effect on the boiling and melting points of compounds containing these groups, but nonetheless may influence the way in which these molecules pack in the crystalline state. Since the hydrogen bond may be viewed as a perturbation of the D–H by the lone pair on A the D–H bond frequently lies in the same direction as that associated with the lone pair on A. As the hydrogen bonding becomes weaker then a wider range of D–H---A angles is observed and the D–H direction can no longer be associated with the lone pair position. The hydrogen bonds associated with multiple bonds generally occur in a direction which maximize the interaction with the π-electron density as illustrated by the two arrangements shown in the margin. Hydrogen bonds may also involve several centres leading to the bi- and tri-furcated interactions illustrated in the margin.

Unconventional hydrogen bonds have been observed between the proton donors N–H and OH and a range of borohydrides and metal hydrides which function as proton acceptors.

For a more detailed discussion see R. H. Crabtree *et al.*, *Acc. Chem. Res.*, 1996, **29**, 348, G. B. Aakeroy and K. R. Seddon, *Chem. Soc. Rev.*, 1993, **22**, 397; G. A. Jeffrey and W. Saenger, *Hydrogen Bonding in Biological Structures*, 2nd Edn., J. Wiley & Sons New York, 1994, and M. S. Gordon, *Acc. Chem. Res.*, 1996, **29**, 536.

$$\overset{\delta-\quad\delta+}{N\text{—}H} \text{-------------} \overset{\delta-\quad\delta+}{H\text{—}B}$$
$$O\text{—}H \text{------------} H\text{—}M$$

These interactions are thought to contribute 12–30 kJ mol^{-1}. The large difference in melting points between CH_3–CH_3 (m.p. −181°C) and H_3B–NH_3 (m.p. 104°C) has been attributed to the occurrence of such intermolecular interactions in the latter.

Three-centre two-electron bonding

In the compounds described above the octet electronic configuration about the central atom was retained by invoking three-centre four-electron bonds. There is another class of compound where the octet configuration is achieved around the central atom, but more than four bonds are made around the central atom. This is achieved by forming three-centre two-electron bonds.

Fig. 4.20 illustrates the generalized molecular orbital scheme for a linear H_3 molecule. The formation of bonding, non-bonding, and antibonding molecular orbitals will now be familiar. Clearly for such a three orbital interaction the non-bonding orbital may be occupied by an electron pair leading to a **three-centre four-electron bond** or remain empty leading to a **three-centre two-electron bond**. Since the non-bonding orbital is localized completely on the peripheral atoms then the strength of H–H bonding is entirely equivalent in the three-centre two-electron bond and the three-centre four-electron bond. However, the charge distributions in H_3^+ and H_3^- are quite different. In the former the terminal atoms bear a charge of $+\frac{1}{2}$ and in the latter a charge of $-\frac{1}{2}$. The three-centre four-electron bond is favoured by the presence of electronegative atoms on the outside and three-centre two-electron bonding by electropositive peripheral atoms.

Therefore, it is not surprising that F–H–F$^-$ has a three-centre four-electron bond and the B–H–B bond in B_2H_6 has a three-centre two-electron bond. The occurrence of three-centre two-electron bonds is very prevalent in the hydride and organometallic chemistry of Group 13 compounds and some illustrative examples are shown below.

The resonance forms corresponding to the molecular orbital diagrams in Fig. 4.20 are illustrated at the top of Fig. 4.22. Significantly, they also lead to charges of either $+\frac{1}{2}$ or $-\frac{1}{2}$ on the terminal atoms.

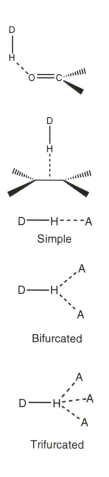

D——H----A

Simple

D——H （A, A）

Bifurcated

D——H （A, A, A）

Trifurcated

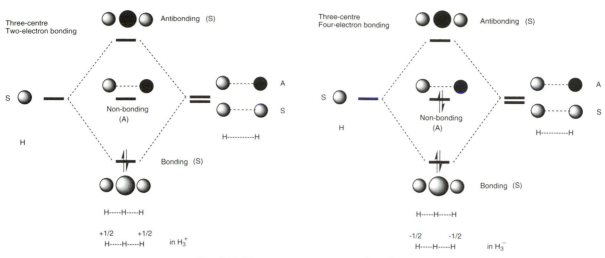

Fig. 4.20 Three centre bonding in H_3^+ and H_3^-

The bond lengths in H_2 (bond order 1), H_2^+ (bond order $= \frac{1}{2}$) and triangular H_3^+ are 74 pm, 106 pm, and 86 pm respectively, suggesting that the three-centre two-electron bond has a formal bond order closer to 1 than $\frac{1}{2}$.

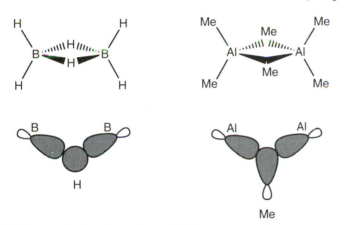

Fig 4.21 Three-centre two-electron bonds in B_2H_6 and Al_2Me_6. In these examples the three-centre two-electron bonds include two sp^3 hybrid orbitals on the boron or aluminium atoms, but the bonding pattern is not greatly altered from that in the H_3 species.

Fig. 4.22 also repeats the application of these resonance forms to molecules such as PF_5. The delocalized three-centre model also suggests that a related series of compounds should exist based on three-centre two-electron bonds. The relevant valence bond canonical forms for CX_5^+ are illustrated in the lower part of Fig. 4.22. Three-centre two-electron bond formation results in the development of positive charges on the peripheral atoms rather than the negative charges illustrated for PF_5. Such compounds would be favoured when the peripheral atoms have low electronegativities, i.e. they are electropositive. Furthermore, the atom X has to complete an electronic shell when forming the ion X^+, or as in the unique case of hydrogen form a completely empty shell structure.

Molecules of this type are exemplified by the series of molecules and ions shown in Fig. 4.23, each of which have 8 valence electrons around the carbon atom.

CH_5^+ was first observed mass spectrometrically and in the condensed phase in super-acid (FSO_3H–SbF_5, HF–SbF_5) matrices at $-180°C$. Computational studies suggest that it has the following structure:

This is based on a three-centre two-electron bond between H_2 and an sp^3 hybrid on carbon rather than the symmetrical trigonal bipyrimidal structure shown in Fig. 4.23.

For a detailed discussion see G. A. Olah and G. Rasul, *Acc. Chem. Res.*, 1997, **30**, 245.

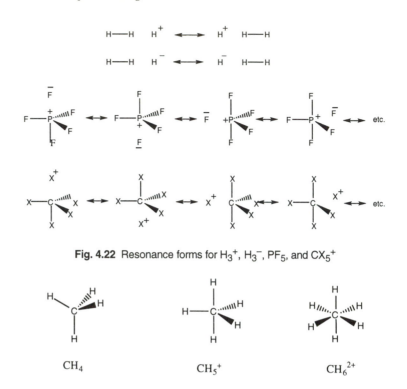

Fig. 4.22 Resonance forms for H_3^+, H_3^-, PF_5, and CX_5^+

Fig. 4.23 Hypothetical protonated forms of methane

For a more detailed discussion of these gold such compounds see H. Schmidbauer, *Chem. Soc. Rev.*, 1995, **24**, 391.

Protonated compounds may be detected for hydrocarbons in strongly acid media, but are not commonly observed as stable compounds which can be isolated and stored in containers at room temperature. However, the fragments R_3PAu and HgCl have bonding characteristics which closely resemble those of a hydrogen atom. Specifically they have a single out-pointing hybrid orbital with s, p, and d character which is occupied by a single electron and may be used to form bonds to other atoms. At its simplest both H and R_3PAu are univalent. Such fragments are described as *isolobal* and the symbolism below is used to illustrate their bonding abilities.

$$H \longleftrightarrow AuPR_3 \longleftrightarrow HgCl$$

The Au^+ and Hg^{2+} ions also have completed sub-shell structures, i.e. their electronic configurations are $5d^{10}6s^06p^0$. These fragments are much more electropositive than hydrogen and consequently are better candidates for forming three-centre two-electron bonding interactions. Schmidbaur and his co-workers have isolated and characterized as air stable crystalline solids many examples of hypervalent compounds which are electronically related to CH_4, CH_5^+, and CH_6^{2+}, but have the $AuPR_3$ fragment replacing the H atom. Some representative examples containing a range of p-block central atoms are illustrated in Fig. 4.24. The related compound $[C(HgCl)_4]$ has also been studied and included in the Fig. 4.24 since it clearly is structurally related. Gold is more effective than H at stabilizing the hypervalent compounds, because it is more electropositive and also is capable of forming weak gold–gold bonding interactions which stabilize the high coordination number compounds. Schmidbaur has proposed the term 'aurophilicity' to describe these weak gold–gold interactions.

The bonding in hypervalent electron deficient molecules and ions such as those illustrated in 4.24 may be described using the valence bond hybridization schemes illustrated in Fig. 4.25. The orbital box diagrams resemble those derived previously for the hypervalent molecules and ions shown in Fig. 4.23 except that 3c2e bonds are formed rather than 3c4e bonds.

Fig 4.24 Examples of high coordination number carbon, nitrogen, and oxygen compounds. The three compounds on the left conform to the octet rule.

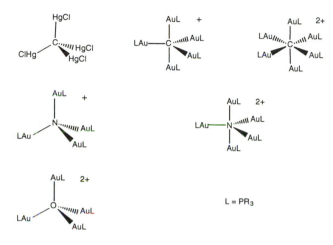

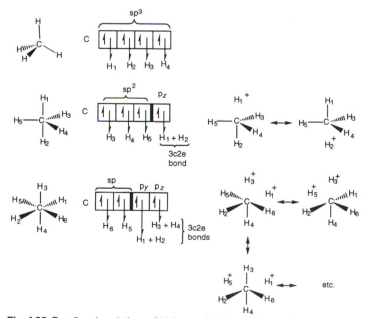

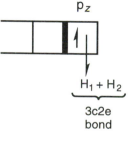

In terms of orbital box diagrams, the three-centre two-electron bond may be represented by the diagram below which clearly indicates the orbitals involved and the two electrons associated with the bond.

Fig. 4.25 Bonding descriptions of high coordination number carbon compounds

4.6 Molecular orbital analysis

The bonding models developed above have involved subtle combinations of ideas taken from valence bond theory and molecular orbital theory. The three-centre four-electron model is a simplified and symmetry based molecular orbital concept which has been re-expressed in a semi-localized form. This has then been combined with hybridization concepts derived from valence bond theory in order to emphasize the importance of the octet rule and provide a simple localized picture of the bonding which rationalizes the molecular structures.

One disadvantage of this methodology is that it becomes necessary to introduce an increasing number of resonance canonical forms to accurately describe the equivalent bonds observed in the structure (see for example Fig. 4.10). This carries with it the implication that the bonding in such molecules is highly delocalized. Such situations are more conveniently described in terms of molecular orbital theory which has as its starting point a delocalized view of the bonding.

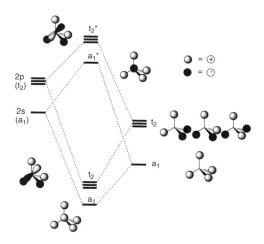

Fig. 4.26 Molecular orbital description of AB_4

The derivation of linear combinations of atomic orbitals from group theory is beyond the scope of this book, but is explained in detail in F. A. Cotton, *Chemical Aspects of Group Theory*, 3rd Edn., John Wiley & Sons, New York, 1990.

Fig. 4.26 illustrates the molecular orbital diagram for the tetrahedral XY_4 molecule. The orbitals of the Y atoms generate four linear combinations one of which is totally symmetric and matches the s orbital on the central atom. The symmetry designation of this orbital is a_1. The three remaining linear combinations are degenerate and match the p orbitals of the central atom. Their symmetry designation is t_2. The molecular orbitals of the XY_4 molecule are derived by taking in-phase and out-of-phase combinations of the Y linear combinations and the symmetry matching orbitals of the central atom.

This results in a_1 and t_2 bonding molecular orbitals and a_1* and t_2* antibonding molecular orbitals. It is apparent from Fig. 4.26 that filling of the bonding molecular orbitals requires 8 valence electrons and leads to a molecule which has all the bonding orbitals filled and a large energy gap between the highest occupied molecular orbital and the lowest unoccupied molecular orbital. Molecules which satisfy this closed shell requirement include BH_4^-, CH_4, and NH_4^+. The bonding and antibonding molecular orbitals in such molecules are equal in number and exactly half the molecular orbitals are filled. The bonding in such molecules can therefore, be equally well described either in terms of the molecular orbital diagram shown in Fig. 4.26 or in terms of four localized two-centre two-electron bonds.

If molecular orbital calculations are completed on a series of isoelectronic tetrahedral molecules, e.g. BH_4^-, CH_4, and NH_4^+, the charges on the atoms reflect the electronegativity differences and the total charge on the molecule. The results of such calculations are given in Table 4.18.

Table 4.18 Charges on the atoms of some tetrahedral isoelectronic species

	BH_4^-	CH_4	NH_4^+
Charge on central atom	-0.14	-0.13	+0.15
Charge on H's	-0.215	+0.0325	+0.21

The total charges on all the atoms add up to the residual charge on the molecule or ion. The covalent nature of the bonding ensures that the charge separations between the atoms remain quite small. It is noteworthy that in this series the charges on the four hydrogen atoms change more dramatically than the charge on the central atom and this corresponds to the observation that the hydrogen atoms in BH_4^- are hydridic and those in NH_4^+ protonic.

The increasing electronegativity of the central atom across the series results in a stabilization of the highest occupied molecular orbital (HOMO) and the lowest unoccupied molecular orbital (LUMO), both of which have t_2 symmetry. Data for BH_4^-, CH_4, and NH_4^+, are given in Table 4.19.

Table 4.19 Calculated HOMO and LUMO energies and their differences for some tetrahedral isoelectronic species

	BH_4^-	CH_4	NH_4^+
Energy HOMO /kJ mol^{-1}	−1389	−1495	−1592
Energy of LUMO /kJ mol^{-1}	1061	473	241
HOMO–LUMO Gap /kJ mol^{-1}	2450	1968	1833

The HOMO is concentrated more on the central atom as the electronegativity is increased and the LUMO is concentrated more on the hydrogen atoms as can be seen from the data given in Table 4.20.

Table 4.20 Contributions of the HOMO and LUMO on the central atoms and the hydrogen atoms of some tetrahedral isoelectronic species

	BH_4^-	CH_4	NH_4^+
% on central atom in HOMO	37	48	58
% on H's in HOMO	63	52	42
% on central atom in LUMO	63	52	42
% on H's in LUMO	37	48	58

For a fuller discussion see D. M. P. Mingos and J. C. Hawes, *Struct. Bond.*, 1985, **63**, 1.

Since carbon and hydrogen have almost equal electronegativities the HOMO and LUMO are almost equally distributed, whereas for other ions the HOMO and LUMO reflect the electronegativity order: N > H > B. The concentration of the orbitals in the HOMO influences greatly the atoms which are better donor sites if the molecule acts as a Lewis base, and the concentration of orbitals in the LUMO influences the point of attack of nucleophiles on the molecule and consequently the relative distributions described above are significant for analysing chemical reactivity trends within the framework of molecular orbital theory.

More generally in XY_n ($n > 4$) molecules the orbitals on Y form a set of linear combinations which may be derived using standard group theoretical techniques. If the atoms Y are located around the atom X, such that they lie on a sphere and cover that sphere efficiently, then four of the linear combinations have symmetry properties and nodal characteristics identical to those of the s and p orbitals on the central atom. The remaining ($n - 4$) linear combinations have symmetry properties which match those of the d orbitals on X.

For example, in octahedral XY_6 the linear combinations of the orbitals of the six Y groups span the symmetry representations $a_{1g}(s)$, $t_{1u}(p)$, and $e_g(d_{z^2}, d_{x^2-y^2})$ (see Fig. 4.27).

The relevant molecular orbital diagram which results when these linear combinations combine with the atomic orbitals of the central atom to form bonding and antibonding combinations are illustrated in Figs 4.28 and 4.29. If the peripheral atoms are more electronegative than the central atom the orbitals of the former are displaced downwards relative to those of the central atom (see Fig. 4.28). If the peripheral atoms are less electronegative then they are displaced upwards relative to those of the central atom (see Fig. 4.29).

The resultant molecular orbital diagrams for these alternative bonding situations have many similarities because in both case the orbitals which have symmetry matching properties combine to form four stable bonding molecular orbitals and four very unstable antibonding molecular orbitals. In both situations the octet electronic configuration can be clearly identified with the occupation of the a_{1g} and t_{1u} molecular orbitals. These of course match in symmetry terms and nodal properties the filled valence orbitals of a noble gas.

The e_g linear combinations in Figs 4.28 and 4.29 do not have a symmetry match because only s and p orbitals have been utilized on the central atom and therefore they are non-bonding and completely localized on the Y atoms. In those molecules with six electron pairs (Fig. 4.28) the e_g combination is non-bonding and localized on Y, but the orbitals are occupied by 4 electrons. The e_g combination remains empty in these molecules with four electron pairs (Fig. 4.29).

The important difference between the two classes of molecule lies in the relative energies of the non-bonding e_g orbitals. In the six electron pair case the e_g orbitals are occupied and therefore this bonding mode is favoured if the Y atoms have high electronegativities. In the case where the e_g orbitals are unoccupied they need to be removed from the bonding region as much as possible, i.e. the Y atoms must be electropositive. This simple molecular orbital interaction diagram therefore provides a basis for rationalizing why SF_6 falls into one class and $[C(AuPPh_3)_6]^{2+}$ into the other.

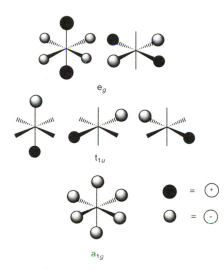

e_g

t_{1u}

a_{1g}

Fig. 4.27 Linear combinations of donor orbitals of the Y groups in octahedral XY_6

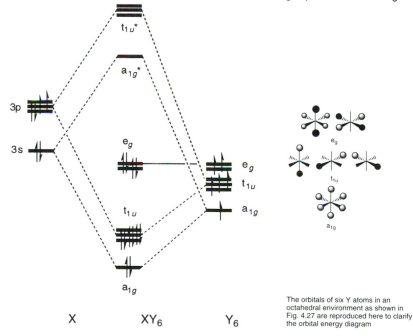

The orbitals of six Y atoms in an octahedral environment as shown in Fig. 4.27 are reproduced here to clarify the orbital energy diagram

Fig. 4.28 Molecular orbital diagram for XY_6 where Y is a third row atom which is more electronegative than X

The complementarity of the orbitals in Figures 4.28 and 4.29 is also noteworthy. The bonding orbitals have symmetries which match the s and p orbitals of the central atom and the non-bonding orbitals have symmetries which match the d orbitals of the central atom. This partitioning allows the bonding electron density to be concentrated in molecular orbitals which are derived from the most stable atomic orbitals and the non-bonding density on the peripheral atoms in linear combinations of atomic orbitals which have nodal properties, which match those of the atomic d valence orbitals of the central atom. The complementary principle is also applicable to the higher coordination number polyhedra given in Table 4.21.

Since the e_g combinations are noded between the Y atoms then the occurrence of overlap between the Y orbitals raises the energy of the e_g orbitals and makes them more inaccessible. In the hypervalent gold compounds the overlap between the gold 6s orbitals is sufficiently large to contribute such a destabilizing effect.

Table 4.21 The structures of the coordination polyhedra XY_n for various values of n

Value of n	XY_n structures
5	Trigonal bipyramid
6	Octahedron
7	Pentagonal bipyramid
	Capped octahedron
	Capped trigonal prism
8	Dodecahedron
	Square-antiprism
9	Tricapped trigonal prism

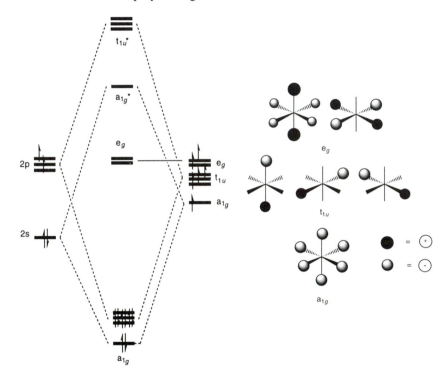

Fig. 4.29 Molecular orbital diagram for XY_6^{2+} where Y is less electronegative than X

The molecular orbital description developed above can provide, as in Fig. 4.28, a satisfactory description of the bonding in molecules such as SF_6 using just the s and p valence orbitals on the central bonding atom. However, it is also sufficiently flexible to permit the introduction of some interactions involving the 3d orbitals on the central atom.

The effect of such interactions are illustrated pictorially in Fig. 4.30. The 3d orbitals have two components of e_g symmetry (see Fig. 4.31) which match the e_g combinations on the ligands and this results in some stabilization of the latter. The orbital diagram has sufficient flexibility that this interaction can be drawn as rather weak—in a molecule where the 3d orbitals are high lying, e.g. SH_6, or stronger in a molecule such as SF_6 where the 3d orbitals are lower lying. In the isolated sulphur atom the 3d orbitals are more than 800 kJ mol^{-1} higher in energy than 3p and therefore too inaccessible to make a significant contribution to bonding. The energies of the 3d orbitals are known to be very sensitive to the residual charge on the atom. Therefore, in a molecule containing very electronegative groups such as fluorine then the 3d orbital energies are likely to be lower than in SH_6. As the 3d orbital energy is lowered then the interactions with the ligand orbitals are strengthened. Sophisticated molecular orbital calculations have suggested that in SF_6 the charge on sulfur is as large as +3.0e which would stabilize the 3d orbitals significantly.

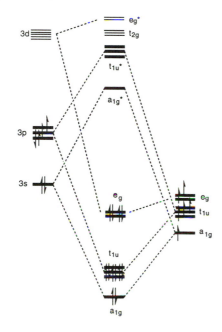

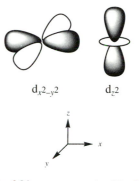

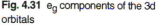

$d_{x^2-y^2}$ d_{z^2}

Fig. 4.31 e_g components of the 3d orbitals

Fig. 4.30 Introduction of 3d orbital interactions into the orbital interaction diagram for an octahedral molecule XY_6.

Electronic basis of the octet rule

The analysis developed above has important implications for providing a deeper understanding of the octet rule and its role in the structural chemistry of the p block elements. If used blindly the octet rule is no better than numerology—it gives one a satisfied feeling when it comes up with the right answer, but it does not give a deeper understanding of the fundamental problem and does not provide any advice about which direction to take when it fails to work.

An inert gas has a stable octet electron configuration because it has a complete set of occupied valence s and p orbitals, leading to a spherical electron distribution, which experiences a high effective nuclear charge. The high effective nuclear charge ensures that the valence electrons have a high ionization energy and contracts the orbitals preventing them from overlapping well with the orbitals of adjacent atoms. For the heavier inert gases, and most notably xenon, the valence shell is complete, but the valence orbitals are sufficiently far from the nucleus that they do not experience as large an attractive force. Consequently the ionization energy is smaller and the orbitals are sufficiently diffuse to overlap well with the orbitals of other atoms and this leads to the formation of a wide range of xenon compounds.

A molecule which conforms to the octet rule also has a set of four bonding and non-bonding molecular orbitals which have nodal properties which emulate those of the 'inert gas' p orbitals. The stabilization of these molecular orbitals by strong bond formation ensures that the ionization energies of the molecular orbitals is high and therefore the electrons occupying them are unavailable for additional bond formation. The Effective Atomic Number Rule begins to break down when the strength of the covalent bonding becomes weaker and where there is a large electronegativity difference between the peripheral atoms and the central atom.

4.7 Isoelectronic relationships in catenated and polyhedral molecules

Ring and three-connected molecules

The number of homonuclear element–element bonds in catenated molecules may generally be derived from the octet rule. In an element–element bond both atoms are sharing the same electron pair in order to achieve octets individually and the total valence electron count reflects this sharing. For example, Si_2H_6 has a total of 14 valence electrons which is two less than double the octet for each silicon atom, because the shared electron pair is not double counted. More generally for a molecule with N valence electrons and n skeletal atoms the number of localized two-centre two-electron element–element bonds, x, is:

For a review of Group 14 polyhedral molecules, see S. Nagare, *Acc. Chem. Res.*, 1995, **28**, 469.

$$x = \frac{8n - N}{2}$$

Examples of the application of this formula in ring and polyhedral molecules are given in Fig. 4.32.

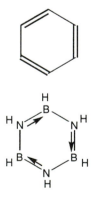

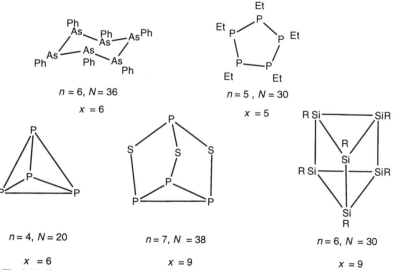

Fig 4.32 Examples of polyhedral and ring compounds with x element–element bonds

It follows that ring compounds are associated with $6n$ valence electrons and chain compounds with $6n + 2$ valence electrons, examples of such molecules and ions are given in Table 4.22.

Table 4.22 Examples of ring and chain molecules. In calculating the valence electron count organic radicals and halogens are considered as one electron donors

Ring compounds	No. of valence electrons	Chain compounds	No. of valence electrons
S_4	24	S_3^{2-}	20
S_6	36	S_4^{2-}	26
S_8	48	S_2Cl_2	26
$[PEt]_5$	30		
$[AsPh]_6$	36		
$[SnPh_2]_6$	36		
$[R_2SiO]_4$	48		
$[Cl_2PN]_3$	36		
$[F_2PN]_4$	48		

If the ring molecule has double bonds then this is also reflected in the total valence electron count. For example, benzene and borazine have a total of 30 valence electrons:

$$\text{Number of element–element bonds} = [8 \times 6 - 30] \div 2 = 9$$
$$\text{(i.e. 6 sigma bonds and 3 } \pi\text{-bonds)}$$

Three-connected polyhedral molecules, by definition form three localized bonds and therefore are associated with $5n$ valence electrons. Typical three-connected polyhedral skeletal geometries are illustrated in the margin. Specific examples of three-connected polyhedral molecules are given in Table 4.23.

Table 4.23 Examples of three-connected polyhedral molecules

Geometry	Example	No. of valence electrons
Tetrahedral	P_4	20
	C_4R_4	20
	Si_4^{4-}	20
Trigonal prismatic	C_6R_6	30
	Si_6R_6	30
Cubic	C_8H_8	40
	Si_8R_8	40
	$Al_4N_4R_8$	40
Cuneane	C_8H_8	40
Dodecahedral	$C_{20}H_{20}$	100

Molecules which have between $5n$ and $6n$ valence electrons have structures which are intermediate between three-connected polyhedra ($5n$) and ring

Tetrahedron

Trigonal prism

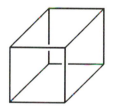

Cube

Cuneane-wedge

Three-connected polyhedra

compounds (6*n*). In general one bond of the three-connected polyhedron is broken for each of the electron pairs in excess of 5*n*. Examples of such molecules are illustrated in Fig. 4.33.

Four-connected polyhedral molecules

The main group fragments B–H, C–H, and N–H are only capable of forming three localized two-centre two-electron bonds because they only have three sp³ hybrid orbitals pointing towards the centre of the polyhedron.

The resulting three-connected polyhedral molecules utilize these orbitals and thereby satisfy the octet rule. If there are four adjacent atoms it is no longer possible for the E–H fragment to form two-centre two-electron bonds and therefore either a delocalized system of bonding has to occur, or the octet rule is violated. The former option occurs and for molecules which have four-connected polyhedral skeletons the skeletal bonding is best described within a completely delocalized molecular orbital framework.

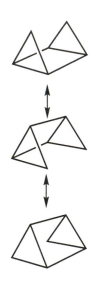

The above resonance forms contribute to an average structure with longer bonds between the two triangular ends of the ion as shown below:

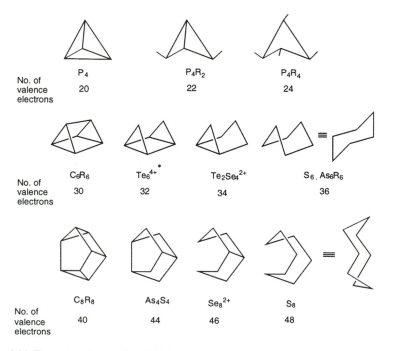

Fig. 4.33 Examples of molecules which have structures intermediate between three-connected and ring structures. *See the diagrams in the margin.

Specifically for octahedral $B_6H_6^{2-}$ the delocalized molecular orbitals may be derived by partitioning the three out-pointing hybrid orbitals into radial and tangential components as shown in Fig. 4.34.

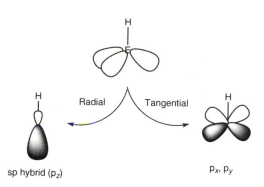

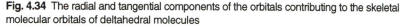

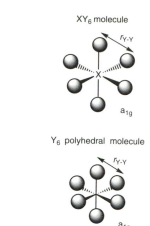

XY$_6$ molecule

Y$_6$ polyhedral molecule

sp hybrid (p$_z$)

Radial

Tangential

p$_x$, p$_y$

Fig. 4.34 The radial and tangential components of the orbitals contributing to the skeletal molecular orbitals of deltahedral molecules

In the XY$_6$ molecule the X–Y bonding distance causes the Y–Y distances to be longer than the sum of the covalent radii, whereas in Y$_6$ the Y atoms are in bonding contact

In symmetry terms the radial sp hybrids in octahedral $B_6H_6^{2-}$ transform as a_{1g}, e_g, and t_{1u}, i.e. identically to the Y orbitals in XY$_6$ in Fig. 4.27. There is an important difference, however, between the resultant spectrum of molecular orbitals in the two classes of compound. In $B_6H_6^{2-}$ the boron atoms are sufficiently close together that the sp hybrid orbitals on adjacent centres overlap strongly. Therefore, the a_{1g} linear combination is strongly bonding because all the overlaps are in-phase, the e_g combination is strongly antibonding because the next neighbour overlaps are out-of-phase and t_{1u} is non-bonding because there are no contributions from the orbitals on adjacent atoms. The orbital diagram for $B_6H_6^{2-}$ is shown in Fig. 4.35.

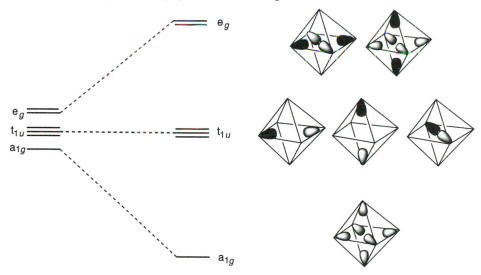

Fig 4.35 Radial bonding molecular orbitals in octahedral $B_6H_6^{2-}$

The tangential p$_x$, p$_y$ orbitals transform in the octahedral point group as t_{1u}, t_{2g}, t_{1g}, t_{2u} and the spectrum of molecular orbitals which results when the orbital overlaps are taken into account is shown in Fig. 4.36. The manifold of orbitals is symmetrical about the non-bonding level with six orbitals strongly bonding and six strongly anti-bonding.

If the radial and tangential sub-components of the molecular orbital diagrams are combined then it is evident that the bonding in $B_6H_6^{2-}$ is characterized by one very stable radial molecular orbital of a_{1g} symmetry and six bonding tangential molecular orbitals of t_{1u} and t_{2g} symmetry. This makes a total of seven ($n + 1$) bonding skeletal molecular orbitals. The mode of analysis can be extended to other four-connected polyhedral molecules and the conclusions follow a similar pattern, i.e. they all have one very stable radial bonding molecular orbital and n tangential bonding molecular orbitals. Some other examples, of four-connected polyhedral skeletons are illustrated in Fig. 4.37.

In a previous section it was noted that two-connected ring molecules and three-connected polyhedral molecules are characterized by a total of $4n$ and $5n$ valence electrons respectively. If the electrons involved in E–H bonding are included the four-connected molecules described above are characterized by a total of $4n + 2$ valence electrons, i.e. $2n$ are associated with the terminal E–H bonds and $2n + 2$ electrons occupy the radial and tangential bonding molecular orbitals. The fact that these molecules do not fall into the extrapolated electron count of $4n$, is a reflection of the fact that the delocalized skeletal molecular orbitals no longer reflect the number of connections between the atoms.

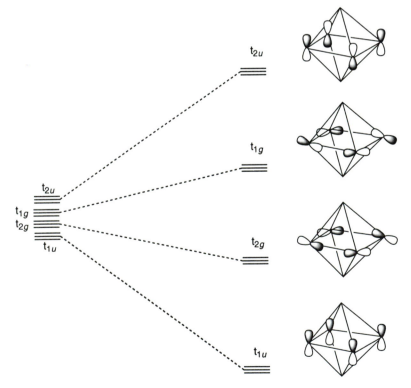

Fig 4.36 Tangential molecular orbitals in octahedral $B_6H_6^{2-}$

Octahedron   Square antiprism   Cuboctahedron

Fig. 4.37 Examples of four-connected polyhedral skeletons

Deltahedral molecules

In addition to the four-connected molecules described above, the p block elements form a series of molecules which adopt polyhedra which have exclusively triangular faces. These polyhedra are described as *deltahedra* (Gk, Δ represents a triangle) and they are illustrated in Fig. 4.38.

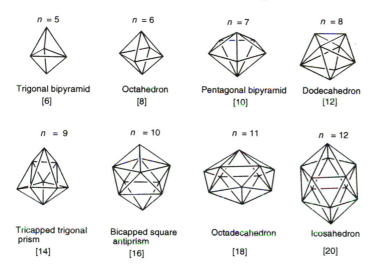

Fig. 4.38 Deltahedral skeletons. The numbers in square brackets refer to the number of triangular faces.

The vertices of the polyhedra, when projected on a spherical surface, represents the most efficient way of covering the spherical surface and they also have the maximum number of connectivities between the vertices for any spherical polyhedra The latter property makes them ideal skeletons for polyhedra, where the atoms are electron deficient, e.g. B–H and Al–H.

For reviews see, N. N. Greenwood, *Chem. Soc. Rev.*, 1992, **21**, 49 (boranes); D. G. Tuck, *Chem. Soc. Rev.*, 1993, **22**, 269 (low oxidation state indium); and H. Schrockel *et al.*, *Angew. Chem. Int. Ed.*, 1996, **35**, 129 (low oxidation state aluminium and gallium).

> A polyhedral collection of electron deficient atomic centres can optimize its total bonding energy by sharing the limited amount of electron density with as many nearest neighbours as possible.

The deltahedral molecules E_nH_n illustrated in Fig. 4.38 have manifolds of molecular orbitals which are closely related to those described above for the four-connected polyhedral molecules. Specifically, they all have $(n + 1)$

skeletal bonding molecular orbitals—1 radial and n tangential. Such molecules are characterized by a total of $4n + 2$ valence electrons—$2n + 2$ occupying the skeletal m.o.s and $2n$ occupying the terminal E–H bonds. Examples of deltahedral molecules which satisfy this electronic requirement are summarized in Table 4.24. The most important examples of this class of molecule are the borane anions $B_nH_n^{2-}$ ($n = 6$–12) and the isoelectronic carboranes $C_2B_nH_{n+2}$ ($n = 3$–10). There are also examples of 'naked' deltahedral clusters which have no terminal E–H bonds and where E–H is replaced by a lone pair.

Table 4.24 Examples of deltahedral molecules

Geometry	Example	No. of valence electrons
Trigonal bipyramidal	$B_5H_5^{2-}$	22
	$C_2B_3H_5$	22
	Sn_5^{2-}, Pb_5^{2-}, Tl_5^{7-}	22
Octahedron	$B_6H_6^{2-}$	26
	$C_2B_4H_6$	26
	Tl_6^{8-}	26
Pentagonal bipyramid	$B_7H_7^{2-}$	30
	$C_2B_5H_7$	30
Dodecahedron	$B_8H_8^{2-}$	34
	$C_2B_6H_8$	34
Tricapped trigonal prism	$B_9H_9^{2-}$	38
	$C_2B_7H_9$	38
Bicapped square-antiprism	$B_{10}H_{10}^{2-}$	42
	$C_2B_8H_{10}$	42
Octadecahedron	$B_{11}H_{11}^{2-}$	46
	$C_2B_9H_{11}$	46
Icosahedron	$B_{12}H_{12}^{2-}$	50
	$C_2B_{10}H_{12}$	50
	$Al_{12}R_{12}^{2-}$	50

The deltahedral molecules summarized in Table 4.24 are described as *closo-* because they have a complete skeleton which can be projected on a spherical shell. In addition there are other series of related polyhedral molecules which have less complete shells. The *nido-*deltahedra have one of the vertices, usually the most highly connected, removed from the *closo-*structures. Examples of these *nido-*deltahedra are illustrated in Fig. 4.39 (the *nido-* description comes from the nest-like appearance of the resultant polyhedral shell, Latin *nidus*—nest). A second series of partial polyhedral skeletons may be derived by removing two vertices from the parent deltahedra. For borane polyhedral molecules two adjacent vertices are removed—the first one used to generate the *nido* plus an adjacent one, but there are also examples of non-adjacent vertices being removed. Fig. 4.39 provides some examples of such polyhedral skeletons, which are described as *arachno* from their resemblance to patterns observed in spiders webs (Greek *arachno* —spider).

The structural pattern relating *closo*, *nido*, and *arachno* main group polyhedral molecules which is illustrated in Fig. 4.39 also has an electronic basis. A series of *closo*, *nido*, and *arachno*-molecules which are based on the same parent also have the same number of skeletal bonding electron pairs.

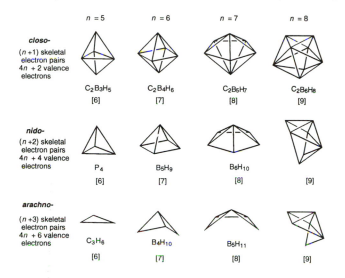

Fig. 4.39 Examples of *closo*-, *nido*-, and *arachno*–polyhedral skeletons. The numbers in square brackets refer to the number of skeletal bonding electron pairs.

Specifically the ions, $B_6H_6^{2-}$, $B_5H_5^{4-}$, and $B_4H_4^{6-}$ all have geometries derived from the octahedron. The *arachno*–borane $B_4H_4^{6-}$ has two alternative geometries based either on the loss of adjacent or opposite vertices. Each is associated with 7 skeletal bonding electron pairs. The borane anions $B_5H_5^{4-}$ and $B_4H_4^{6-}$ are too basic to be isolated as independent species, but are well documented in their protonated forms B_5H_{10} and B_4H_{10}, which retain the same skeletons on protonation. B_4H_{10} has a geometry based on the loss of two adjacent vertices while the Se_4^{2+} ion, which isoelectronic, has a square structure derived from the octahedron by the loss of two opposite vertices.

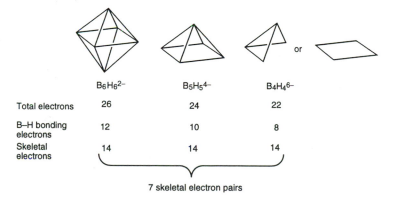

Fig 4.40 Geometric and electronic relationships on *closo*-, *nido*-, and *arachno*–borane

Therefore, the polyhedral molecules illustrated in Fig. 4.40 obey a structural paradigm which closely resembles that developed earlier in the chapter for molecular compounds. The important structural similarity is emphasized in Fig. 4.41 which illustrates how the polyhedral molecules with 7 skeletal electron pairs retain the basic octahedral geometry and molecular compounds with six bonding electron pairs leads to a series of octahedral molecules where the ligands are successively replaced by lone pairs.

For a more detailed discussion of these bonding principles see D. M. P. Mingos and D. J. Wales, *Introduction to Cluster Chemistry*, Prentice-Hall, Englewood Cliffs, New Jersey, 1990, and C. E. Housecroft, *Cluster Molecules of the p-Block Elements*, OUP, Oxford, 1994.

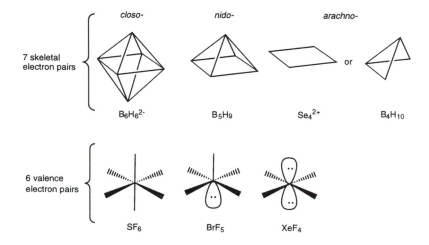

Fig. 4.41 Structural relationships between polyhedral cage molecules and VSEPR molecules

4.8 Zintl isoelectronic relationships

Many binary compounds are formed between the Group 1 and 2 elements and Groups 13–15, e.g. Na_3As, $CaIn_2$. The Zintl concept utilizes isoelectronic relationships in order to provide some insight into the solid state structures adopted by such compounds. The electronegativity difference between the atoms is sufficiently large in these compounds that, to a first approximation, an ionic representation is reasonable, e.g. Na_3As may be viewed as $3Na^+As^{3-}$. The isoelectronic relationship is then used as a basis for rationalizing the manner in which the Group 13–15 elements aggregate in the solid state. For example, NaSi is formulated as Na^+Si^- and the isoelectronic relationship between Si^- and P suggests that the silicon atoms in NaSi might form Si_4^{4-} tetrahedra analogous to the P_4 tetrahedra in elemental white phosphorus. This geometry is indeed the one observed in the crystal structure. The compound $BaSi_2 \equiv 2Ba^{2+}Si_4^{4-}$ also contains Si_4 tetrahedra.

Using similar reasoning, it could be argued that the phosphorus anions in $LiP \equiv Li^+P^-$ are isoelectronic with S and therefore LiP should have P_8^{8-} crowns similar to the S_8 units observed in monoclinic sulfur. Actually, the phosphorus atoms form helical spirals similar to those observed in elemental selenium and tellurium. Therefore, a strict isoelectronic relationship is not observed, but the Zintl concept does indicate the relevant group valency characteristics.

A further example is provided by $CaSi_2$ in which the $[Si_n]^{n-}$ component forms a layer structure resembling that observed in elemental arsenic.

In general for M_xA_y the total number of electrons in the formula unit may be used to estimate the type of ring or polyhedral molecule found in the structure for the A atoms. If it does not form element–element bonds the total valence electron count is 8, but as the number of electrons is reduced from 8 it can be used as an indicator of the number of A–A bonds which are present in the anionic component. Table 4.25 illustrates that when the number of valence electrons per atom is seven, dimeric A–A molecules are formed, when it is six, ring compounds are observed, and when it is five, three-connected polyhedral molecules result.

Diamond-like lattices are observed for the Tl^-, In^-, and Ga^- anions in the following: NaTl, LiGa, LiIn, and $BaTl_2$. The isoelectronic relationships between the anions and the Group 14 atoms, C, Si, Ge, and Sn form the basis of these structural similarities.

Many of the anionic species observed in Zintl phases in the solid state are not observed as stable solution species, because their large negative charges make them very nucleophilic and therefore sensitive to the slightest traces of moisture. Their synthesis in the solid state, which involves mixing and heating under dry conditions the appropriate ratios of the elements, is ideal for creating the anhydrous environment necessary for isolating such reactive species.

Table 4.25 Examples of the application of Zintl relationships.

Formula	Total number of valence electrons	Structure of anion	Comments
Na_3P	8	P^{3-}	
Na_2S	8	S^{2-}	
$\{Li_2S_2\}$	7	S_2^{2-}	Iso-structural
$Ni^{II}P$	7	P_2^{4-}	with Cl_2
$Fe^{II}S_2$	7	S_2^{2-}	
LiAs	6	As_4^{4-} ring	Rings or
$\{InP_3\}_{1/3}$	6	P_6^{6-} ring	infinite chains
$\{CaSi\}$	6	Zig-zag or helical chains	cf. S and Se
$\{K_4Ge_4\}_{1/4}$	5	Ge_4^{4-} tetrahedra	Isostructural with P_4
$\{BaSi_2\}$	5	Si_4^{4-} tetrahedra	
$\{CaC_2\}$	5	C_2^{2-} pairs	Isostructural with N_2
$\{CaSi_2\}$	5	Corrugated layers	Isostructural with α-As
NaTl	4	Diamond structure	
$\{SrGa_2\}$	4	Diamond structure	

For a more detailed discussion of Zintl compounds see U. Müller, *Inorganic Structural Chemistry*, John Wiley and Sons, Chichester, 1993.

4.9 Isoelectronic relationships in infinite solids

The isoelectronic relationships which have been discussed above may be extended to solid inorganic compounds which adopt infinite structures. Table 4.26 compares the properties of such an isoelectronic series.

In these isoelectronic series the coordination numbers of the ions in the structures decrease as the electronegativity difference decreases. For both series the ions have octahedral coordination geometries at the beginning of the series where the bonding is predominantly ionic, but the structure is based on the diamond structure at the end of the series with tetrahedral geometries about each atom. The zinc blende and wurtzite structures are closely related to the diamond structure since in both cases the geometries about each ion are tetrahedral. This change can be related to a combination of geometric and electronic effects. The size of the cation decreases across the series because of the increased charge, i.e. Na^+, Mg^{2+}, Al^{3+}, and the anion size increases, i.e. Cl^-, S^{2-}, P^{3-} therefore the ratio of the cation to anion radii decreases across the series and this favours the formation of lower coordination number structures in the solid state, even on a simple hard sphere model. Secondly, as the electronegativity difference decreases the bonding becomes more covalent and the octet rule is more likely to be applicable. The diamond, wurtzite, and zinc blende structures enable the atoms to form four tetrahedral sp^3 hybridized localized two-centre two-electron bonds and thereby conform to the octet rule.

Table 4.26 Isoelectronic compounds which have infinite structures

Compound	Structure	Inter-nuclear distance/pm	m.p./°C	Enthalpy of formation /kJ mol^{-1}
NaCl	NaCl	281	801	−411
MgS	NaCl	260	>2025	−346
AlP	Zinc blende	235	1050	−167
Si	Diamond	235	1410	0
LiF	NaCl	201	848	−616
BeO	Distorted wurtzite	190	2507	−608
BN	Wurtzite/zinc blende	157	Sublimes ~3000	−254
C	Diamond	154	> 3550	0

For a more detailed discussion see P. A. Madden and M. Wilson, *Chem. Soc. Rev*, 1996, 25, 339.

Table 4.27 summarizes the common types of infinite M_2X, MX, and MX_2 structures and provides specific examples of the compounds which adopt these structures. The following infinite structures are commonly observed for MX compounds: caesium chloride (8:8 cubic coordination), sodium chloride (6:6 octahedral coordination), and wurtzite (4:4 tetrahedral coordination). For MX_2 compounds the corresponding infinite structures are fluorite (8 cubic: 4 tetrahedral), rutile (6 octahedral: 3 trigonal), and crystabolite (4 tetrahedral: 2 angular). In all of these structures the cations and anions have nearest neighbours which are ions with opposite charges. The lattice energies in such compounds may be estimated using the Born–Landé equation given on page 178. For a given combination of cation and anion, z^+z^- and r^++r^- may be considered to be constant and consequently the magnitude of $\Delta_{latt}H^{\ominus}$ is influenced primarily by the Madelung constant, M, which provides a geometric summation of the attractive and repulsive forms for successive layers of ions surrounding the central ion.

Table 4.27 Summary of common infinite structural types

Caesium chloride, $CsCl$	**Body-centred cubic alloys**
CsX ($X = Cl, Br, I$)	$NaTl, AgLi, CuZn, CuPd$
TlX ($X = Cl, Br$)	
NH_4Cl	

Sodium chloride, $NaCl$
MO ($M = Mg, Ca, Sr, Ba, Cd; V, Mn, Co, Ni; Nb, Mo, Ta, Zr;$
                                     La, Eu, Ce, Yb, Pu)
MS ($M = Mg, Ca, Sr, Ba, Cd, Pb; Mn, Zr; La, Eu, U, Th, Pu$)
MX ($X = Cl, Br, I; M = Li, Na, K$)
MN ($M = Sc, Ti, V, Cr, Zr, Hf, Nb, Re; La, Pu$)
AgX ($X = F, Cl, Br$)

Nickel arsenide, $NiAs$
MX ($X = S, Se; M = V, Cr, Fe, Ni, Co$)
MTe ($M = Cr, Mn, Fe, Pd, Pt, U$)
MX ($X = As, Sb; M = Mn, Fe, Ni$)
CrH

Zinc blende, sphalerite, ZnS	**Wurtzite,** ZnS
MS ($M = Be, Mn, Zn$)	MN ($M = B, Al, Ga, In$)
MX ($X = Se, Te; M = Zn, Cd, Hg$)	ZnX ($X = O, S, Se$)
MX ($X = P, As, Sb; M = Al, Ga, In$)	BeO
BX ($X = P, As$)	MnX ($X = S, Se$)
SiC	SiC

Fluorite, CaF_2	**Anti-fluorite,** M_2X
MF_2 ($M = Ca, Sr, Ba, Cd, Hg, Pb, Hf, Eu$)	M_2O ($M = Li, Na, K, Rb$)
MO_2 ($M = Zr, Bi, Ce, U$)	M_2S ($M = Li, Na, K$)
MH_2 ($M = Ti, Cr, La, Er, Gd$)	

Rutile, TiO_2
CrX_2 ($X = F, Cl$)
MF_2 ($M = V, Fe, Co, Ni, Cu, Zn, Mg$)
MO_2 ($M = V, Cr, Mn, Nb, Mo, Ru, Rh, Pd, Ta,$
$Re, W, Os, Ir, Pt; Ge, Sn, Pb$)
MgH_2

Cadmium chloride, $CdCl_2$	**Cadmium iodide,** CdI_2
MCl_2 ($M = Mg; Mn, Fe, Co, Ni, Cd$)	MCl_2 ($M = Ti, V$)
TaS_2	MX_2 ($X = Br, I; M = Mg;$
	Ti, V, Cr, Fe, Cd)
	MX_2 ($X = S, Se; M = V,$
	$Zr, Hf, Pt; Sn$)

Anti-cadmium chloride, $M(OH)_2$
($M = Mn, Fe, Co, Ni, Zn, Cd$), Cs_2O

The value of M decreases with coordination number and consequently compounds where M and X are separated by a large electronegativity difference prefer to adopt infinite structures with high coordination numbers. As the electronegativity difference between M and X decreases covalent bonding effects become increasingly important. The strength of covalent bonding depends on a range of factors including the extent of overlap between the orbitals of M and X, the electronegativity difference, and the coordination number. In general the mean bond enthalpies in simple molecules decrease as the coordination number increases (see page 240) and therefore if the same trends apply in infinite solids one would anticipate a preference for lower coordination numbers in compounds where M and X have similar electronegativities.

Infinite structures have all the layers of anions interspersed with layers of cations. Consequently each anion and cation has exclusively ions of the opposite charge as nearest neighbours. In lower dimensional 'layer' structures adjacent layers of anions may be identified which do not have cations interspersed between them. Therefore, the anions are no longer exclusively surrounded by cations—on one side they make short contacts to cations but on the opposite side the nearest neighbours are more distantly coordinated anions. The cadmium chloride and cadmium iodide lattices provide the most commonly encountered examples of 'layer' structures. In these structures the metal ions are octahedrally coordinated but the anions have trigonal pyramidal coordination geometries.

It has been suggested that polarization effects favour the adoption of layer structures in preference to infinite structures. The polarization of a polarizable anion by a small highly charged cation can only create an induced dipole in the former if it occupies a low symmetry site in the lattice. The polarization of the anions in a layer structure occurs in a manner such that the negative ends of the dipoles are dragged down into the layer of cations. This negative charge 'screens' the repulsions between positively charged cations which are more concentrated in specific layers. The positive end of the dipole which points towards anions if the adjacent anion layer reduces the anion–anion repulsions. In an idealized ionic structure the anions sit on sites of such high symmetry that no dipoles are introduced by polarization effects.

The effects of covalency and polarizability may be illustrated by the structural changes associated with the following series of compounds.

	$MgCl_2$	$CaCl_2$	$SrCl_2$	$BaCl_2$
Structure:	Layer	Rutile	Fluorite	Fluorite

Increasing ionic character

$\longrightarrow$

Decreasing polarizing power of cation

	CaF_2	$CaCl_2$	$CaBr_2$	CaI_2
Structure:	Fluorite	Rutile	Rutile	Layer

Decreasing ionic character

$\longrightarrow$

Increasing polarizability of anion

The following generalizations are evident from the data in Table 4.27.

1. The CsCl structure is only observed for large cations where the bonding is primarily ionic and in alloys.
2. The NaCl structure is by far the most common structural type for this stoichiometry and is adopted for a wide range of halides, oxides, and sulfides.
3. The NiAs structure is preferred for compounds which have significant covalent character and have the possibility of forming metal–metal bonds, i.e. the chalcogenides of the transition metals.
4. The zinc blende and wurtzite structures are adopted by compounds which have a high degree of covalent character and cations and anions.

Similar generalizations are applicable to the MX_2 compounds listed in Table 4.27.

4.10 Isostoichiometric relationships

Isostoichiometric series of compounds, i.e. compounds which have a common formula, but different numbers of valence electrons are useful for documenting the effect of adding electrons on the bond lengths and angles in a molecule.

At its simplest the dioxygen molecule and the associated cation (O_2^+) and anions (O_2^- and O_2^{2-}) provide an example of such an isostoichiometric series. The relevant internuclear distance and force constants are summarized in Table 4.28.

Table 4.28 Properties of dioxygen and related cations and anions

	O_2^+	O_2	O_2^-	O_2^{2-}
Total number of valence electrons	11	12	13	14
Ground state electronic configuration	$(\pi)^4(\pi^*)^1$	$(\pi)^4(\pi^*)^2$	$(\pi)^4(\pi^*)^3$	$(\pi)^4(\pi^*)^4$
d(O–O)/pm	112	121	126	149
v(O–O)/cm^{-1}	1865	1580	1097	833
k(O–O)/(N m^{-1}) force constant	1632	1180	620	383
Number of unpaired electrons	1	2	1	0
Bond order	2.5	2	1.5	1

Within a molecular orbital framework a generalised delocalized orbital scheme of energy levels is proposed on the basis of symmetry arguments and calculations. This takes into account the energies of the contributing atomic orbitals and the orbital overlaps. If the atoms forming a molecule have s and p valence orbitals, eight molecular orbitals result and they are illustrated in Fig. 4.42. The electrons are introduced into the energy level scheme using the *aufbau* principle. This involves filling the lowest energy levels first and ensuring that no two electrons share the same set of quantum numbers.

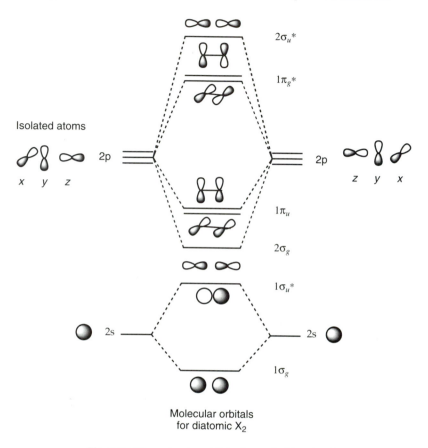

Isolated atoms

Molecular orbitals
for diatomic X_2

Fig. 4.42 The molecular orbital scheme for X_2 molecules

This simplified treatment ignores interactions between 2s and 2p orbitals on the different atoms which can alter the relative energies of $2\sigma_g$ and $1\pi_u$ in diatomics such as B_2, C_2, and N_2

Table 4.29 Some homonuclear diatomic molecules and their numbers of valence electrons

Molecule	No. of valence electrons
B_2	6
C_2	8
N_2	10
O_2	12
F_2	14
Ne_2	16

Of the species listed in Table 4.28, the ion O_2^+ has the fewest valence electrons (11) and results in the orbital occupations, $1\pi_u^4(1\pi_g*)^1$, and a formal bond order of 2.5 since there are 6 electrons occupying bonding σ and π molecular orbitals and only a single electron occupying the antibonding $1\pi_g*$ molecular orbital. In O_2, O_2^-, and O_2^{2-} the additional electrons occupy the $1\pi_g*$ orbitals and cause the bond order to fall progressively as shown in Table 4.28. This reduced bond order is reflected in longer O–O bond lengths, a lower v(O–O) stretching frequency, and a smaller force constant. All these parameters have been determined experimentally. It is also significant that the number of unpaired electrons in this series varies and this parameter may also be measured from the magnetic moments of the molecule and ions.

Similar comparisons may be made for the diatomic molecules B_2 to F_2 which have 6 to 16 valence electrons (see Table 4.29). The formal bond order correlates with force constants and bond enthalpies (see the plots in Fig 4.43). In this series the radii of the atoms are also changing, but these changes are smaller than those associated with changes in formal bond orders and therefore the orbital occupations are the dominant effect. The asymmetric appearance of the graphs is associated with these supplementary factors. It is noteworthy that the bond strengths reach a maximum for molecules with ten valence electrons where the formal bond order reaches its maximum of 3.0.

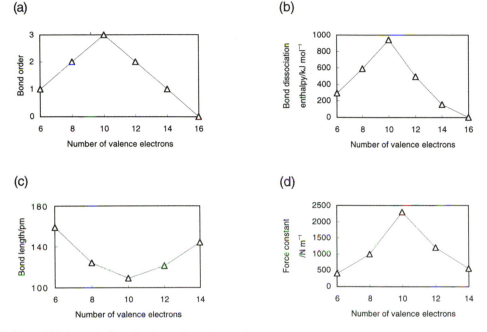

Fig. 4.43 Plots of (a) theoretical bond order, and experimentally observed (b) bond dissociation enthalpy, (c) bond length, and (d) force constants, for diatomic molecules against their numbers of valence electrons

The bond lengths and dissociation enthalpies of the related series of diatomics C_2, CN, CO, and CF show similar trends as shown by the data in Table 4.30.

Table 4.30 Bond distances and bond enthalpies for some diatomic molecules

	Bond distance/pm	Bond dissociation enthalpy /kJ mol^{-1}
C_2	124	603
CN	117	787
CO	113	1072
CF	127	548

In Table 4.31 the structural data for some triatomic molecules are summarized. In these series some molecules have been included which are not stable under normal conditions, but which may nonetheless be studied in the gas phase using spectroscopic techniques. The bond lengths and angles in such molecules are obtained very accurately using microwave spectroscopy.

Interestingly, in the first series of molecules the X–F bond length decreases in the order Be > B > C, but then rises again for N and O, i.e. the expected contraction in atomic radii is not reproduced. A similar trend is not apparent for the simple hydrides as shown by the data in Table 4.32.

Table 4.31 Data for some difluorides and dioxides of elements of the second row of the Periodic Table

	BeF_2	BF_2	CF_2	NF_2	OF_2
Number of valence electrons	16	17	18	19	20
D(X–F)/pm	150	149	130	135	141
F–X–F angle / °	180	117	105	103	103
	CO_2	NO_2	O_3		
Number of valence electrons	16	17	18		
D(X–F) /pm	116	119	127		
F–X–F angle / °	180	134	118		

Table 4.32 Bond distances and Σr_{max} values for some triatomic molecules

	BH_2	CH_2	NH_2	OH_2
d(X–H)/pm	118	111	102	96
Σr_{max}	109	93	82	73

The data in Table 4.32 suggest that the bond distances observed in the hydrides correspond approximately to the maximum in the $r_{max}(X) + r_{max}(H)$ for BH_2, i.e. reflect the criterion for maximum overlap, but depart from it increasingly for the more electronegative atoms. This suggests that the valence orbitals are contracting rapidly as one moves to the right and consequently the atoms cannot approach each other at sufficiently close distances to maximize their overlaps. Therefore the bonds become progressively weaker across the series.

The data presented above for the fluorides, oxides, and hydrides suggests that the discontinuity observed for the fluorides and oxides may have its origins in an electronic effect which has nothing to do with the contraction in atomic radii. The most commonly encountered explanation for the trends is based on the observation that the fluorides and oxides differ from the hydrides in having lone pairs on the outer atoms. It has been suggested that repulsions between the lone pairs and non-bonding electrons on the central atom may lengthen the bonds. This repulsive effect is particularly marked for atoms with small covalent radii, because the lone pairs on adjacent atoms are closer together. Therefore, the bond enthalpies of the second row elements N, O, and F are particularly affected. For atoms with larger radii the effect is hardly discernible.

For molecules with more than two atoms the shape of the molecule also changes as the number of valence electrons is increased. The columns in the structure matrix illustrated in Fig. 4.1 provide specific examples of such isostoichiometric series of molecules. The structural changes are also summarized in Table 4.33.

The structure matrix illustrates the way in which the shape of the molecule changes as a result of the incremental increase in the total number of valence electrons. The importance of attaining the octet configurations for the outer atoms and the number of residual lone pairs on the central atom has been emphasized in preceding sections.

Table 4.33 Shapes of main group molecules

	XY_n	EXY_n	E_2XY_n	E_3XY_n
Molecule	CO_2	O_3	OF_2	F_3^-
Geometry	Linear	Angular	Angular	Linear
No. of valence electrons	16	18	20	22
Molecule	BF_3	NF_3	ClF_3	
Geometry	Trigonal planar	Trigonal pyramidal	T-shaped	
No. of valence electrons	24	26	28	
Molecule	SiF_4	SF_4	ICl_4^-	
Geometry	Tetrahedral	Folded square	Square-planar	
No. of valence electrons	32	34	36	

The angular changes in intermediate radical molecules may also be traced as in the following series of nitrogen oxides:

	NO_2^+	NO_2	NO_2^-
O–N–O ($^\circ$)	180	134	115

In the first example there is no lone pair on the central nitrogen atom and in the last there is one. The radical NO_2 has a single electron occupying the hybrid orbital and therefore the O–N–O angle is intermediate.

Isostoichiometric relationships in more complex molecules

Isostoichiometric relationships may also be applied to more complex molecules and the following examples are illustrative rather than exhaustive.

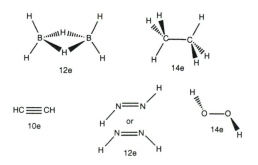

In general, increasing the number of valence electrons results in either a change in the multiple character of the bonds or a change from three-centre two-electron bonds by localized two-centre two-electron bonds. In turn these are replaced by three-centre four-electron bonds.

Table 4.34 summarizes the important isostoichiometric relationships for polyhedral molecules. Clearly as the number of electrons is increased the structure progressively opens up from deltahedral to three-connected to partial cages to rings.

Table 4.34 Summary of structures of polyhedral and ring compounds, X_n

$n = 6$	$n = 5$	$n = 4$	Structure type	Total no. of valence electrons
S_6^{2-}	S_5^{2-}	S_4^{2-}	Chain	$6n + 2$
$\{Cl_2PN\}_3$	$\{PPh\}_5$	P_4Ph_4	Ring	$6n$
			Intermediate	$5n + 2$
		P_4Ph_2		$5n + 4$
				$6n - 2$
	Not geometrically possible		Three-connected	$5n$
B_5H_{11}	B_4H_{10} $\equiv 5n+2$		arachno– deltahedron	$4n + 6$
B_6H_{10}	B_5H_9 $\equiv 5n$		nido– deltahedron	$4n + 4$
$B_6H_6^{2-}$	$C_2B_3H_7$		closo– deltahedron	$4n + 2$

Further reading

Chemical Bonding, M. Winter, OUP, Oxford, 1994.
Physical Inorganic Chemistry, S. F. A. Kettle, Spektrum, Oxford, 1996.
Chemical Structure and Bonding, H. B. Gray and R. L. DeKoch, Benjamin, Merlo Park, California, 1980.
An Introduction to Molecular Orbitals, Y. Jean, F. Volatron, and J. K. Burdett, OUP, Oxford, 1993.

5 Transition elements (d block), lanthanides, and actinides (f block elements)

5.1 Introduction

For the s and p block elements the commonly observed valencies and oxidation states are distinctive and characteristic of the Group. For example, sulfur, selenium, and tellurium form compounds with valencies of 2, 4, and 6, and phosphorus, arsenic, and antimony form compounds with oxidation states +3 and +5. Radical molecules with intermediate valencies are only observed either as reaction intermediates trapped in low temperature matrices or in sterically crowded molecules. Furthermore, when isostoichiometric series of molecules are compared it is clear that their structures change as a function of the total number of valence electrons. For example:

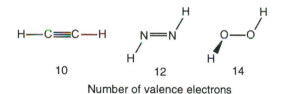

Number of valence electrons

In contrast the transition metals, the lanthanides, and the actinides exhibit many examples of isostoichiometric compounds which have closely related structures.

These elements caused considerable problems for Mendeleev when he was collating the available data in order to draw up the Periodic Table. His original Table, shown in Fig. 5.1, placed the transition metals, Ti, V, Cr, and Mn, under Si, P, S, and Cl in Groups IV–VII, primarily because both series of elements formed compounds with similar formulae in their highest oxidation states, e.g. $SiCl_4$, $TiCl_4$, ClO_4^-, and MnO_4^-. Iron, cobalt, and nickel were placed in Group VIII—a Group which was subsequently assigned to the noble gases when they were discovered some years later.

The quantum mechanical description of the atom, which has led to the Periodic Table as we currently know it, has provided a natural explanation for the occurrence of series of elements with partially filled d and f shells. For the s and p block elements the vertical comparisons are of paramount importance and the chemical differences across the series are generally quite dramatic. Indeed, the collective nouns for naming these elements reflect their vertical similarities, i.e. alkali metals, halogens, chalcogenides, etc. The transition metals, the lanthanides, and the actinides show many horizontal similarities and their names highlight these relationships.

Group	I — R_2O	II — RO	III — R_2O_3	IV RH_4 RO_2	V RH_3 R_2O_5	VI RH_2 RO_3	VII RH R_2O_7	VIII — RO_4
Series								
1	H 1							
2	Li 7	Be 9.4	B 11	C 12	N 14	O 16	F 19	
3	Na 23	Mg 24	Al 27.3	Si 28	P 31	S 32	Cl 35.5	
4	K 39	Ca 40	—44	Ti 48	V 51	Cr 52	Mn 55	Fe 56 Co 59
5	(Cu 63)	Zn 65	—68	—72	As 75	Se 78	Br 80	Ni 59 Cu 63
6	Rb 85	Sr 87	? Y 88	Zr 90	Nb 94	Mo 96	—100	Ru 104 Rh 104
7	(Ag 108)	Cd 112	In 113	Sn 118	Sb 122	Te 125	I 127	Pd 106 Ag 108
8	Cs 133	Ba 137	? Di 138	? Ce 140	—	—	—	— — —
9	(—)	—	—	—	—	—	—	—
10			? Er 178	?? La 180	Ta 182	W 184	—	Os 195 Ir 197
11	(Au 199)	Hg 200	Tl 204	Pb 207	Bi 208	—	—	Pt 198 Au 199
12	—	—	—	Th 231	—	U 240	—	

Fig. 5.1 Mendeleev's Periodic Table (1871 version)

acac

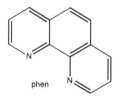

bipy

phen

The three transition series each contain ten elements and yet they are all metals. This behaviour is very different from that noted in the previous chapters where the s and p block elements undergo a transition from metallic to metalloid to non-metallic character. This raises the important question: why does the change in effective nuclear charge across the transition series (and also the lanthanide and actinide series) not lead to such a dramatic change in the properties of the elements?

The transition metals, the lanthanides, and the actinides form many series of isostoichiometric compounds which show similar structural properties. These similarities are observed for compounds which have infinite structures in the solid state and also molecular compounds. For example, the majority of the first row transition metals all form oxides MO which have structures derived from the sodium chloride lattice. In solution the metals form the aqua-ions $[M(OH_2)_6]^{x+}$ which have regular octahedral or slightly distorted octahedral geometries. In addition they form many series of complexes which have common stoichiometries and structures. Some examples of such compounds are given below (the formulae of the abbreviated ligands are shown in the margin).

$$[M(acac)_3] \qquad [M(CN)_6]^{4-} \qquad [M(CN)_6]^{3-}$$
$$[M(NH_3)_6]^{3+} \qquad [M(\eta\text{-}C_5H_5)_2] \qquad [M(\eta\text{-}C_6H_6)_2]$$
$$[M(bipy)_3]^{3+} \qquad [M(phen)_3]^{3+} \qquad MF_6$$

The occurrence of such series of isostoichiometric compounds which are also generally isostructural means that some of them have unpaired electrons. This raises the question: why are radical species so common for the transition series, the lanthanides, and the actinides, but uncommon for the s and p block elements?

The transition elements form compounds in a wide range of oxidation states, but unlike the p block elements they do not discriminate between even and odd oxidation states. For example, vanadium forms compounds in oxidation states -1 to $+5$, whereas phosphorus generally forms compounds in oxidation states $+3$ and $+5$. The even oxidation states of vanadium, when observed in mononuclear compounds, are paramagnetic. Therefore, the occurrence of such compounds also reflects the stability of radicals for the d and f block elements.

The transition metals have a unique ability to form stable compounds with unsaturated organic molecules such as CO, ethene, and ethyne. These molecules, which are not very polar and are not usually considered strong Lewis bases, nonetheless form very stable complexes with the transition metals.

All these properties of the transition series and those of the related lanthanides and actinides have their origins in the relative energies and nodal properties of their valence orbitals and their radial distribution functions. The latter influence the strengths of the covalent bonds formed between transition metals and ligands and the relative contribution of the s, p, d, and f orbitals to bonding. The aim of this chapter is to introduce the reader to those trends which are distinctive of the transition metals and to develop a coherent picture which accounts for their diverse chemical behaviour.

5.2 Trends associated with the isolated atoms

For successive atoms in the transition series one proton is added to the nucleus and one electron to the valence $(n + 1)$s and nd orbitals. The d orbitals lie relatively close to the nucleus and therefore the electrons entering these orbitals are effective at screening the other d electrons from the nucleus. Therefore, the effective nuclear charge increases across the series, but not as much as that for the atoms in the p subshells.

Table 5.1 Electronic configurations of the transition elements in their ground states

Sc	Ti	V	Cr	Mn	Fe	Co	Ni	Cu	Zn
$4s^23d^1$	$4s^23d^2$	$4s^23d^3$	$4s^13d^5$	$4s^23d^5$	$4s^23d^6$	$4s^23d^7$	$4s^23d^8$	$4s^13d^{10}$	$4s^23d^{10}$
Y	Zr	Nb	Mo	Tc	Ru	Rh	Pd	Ag	Cd
$5s^24d^1$	$5s^24d^2$	$5s^14d^4$	$5s^14d^5$	$5s^14d^6$	$5s^14d^7$	$5s^14d^8$	$4d^{10}$	$5s^14d^{10}$	$5s^24d^{10}$
La	Hf	Ta	W	Re	Os	Ir	Pt	Au	Hg
$6s^25d^1$	$6s^25d^2$	$6s^25d^3$	$6s^25d^4$	$6s^25d^5$	$6s^25d^6$	$6s^25d^7$	$6s^15d^9$	$6s^15d^{10}$	$6s^25d^{10}$

The increase in the effective nuclear charge, Z_{eff}, across the transition series has two interrelated consequences: the ionization energies of the atoms become larger and the sizes of the atoms decrease. The $(n + 1)$s and the nd orbitals of the transition elements have similar energies and therefore the metal atoms have ground state electronic configurations (see Table 5.1) which involve populating both the $(n +1)$ and nd shells. Fig. 5.2 illustrates schematically how the energies of the nd, $(n+1)$s, and $(n+1)$p orbitals vary across the series if electron–electron repulsion and exchange energy effects are ignored.

For a more detailed discussion of the relative energies of the 3d and 4s orbitals see M. Melrose and E. R. Surri, *J. Chem. Educ.*, 1996, **73**, 498.

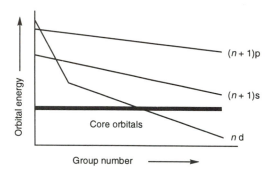

Fig. 5.2 A schematic illustration of the variation in orbital energies across the transition series

| ↑ | ↑ | ↑ |
Exchange energy = 3K

| ↑↓ | ↑ | |
Exchange energy = K

Table 5.2 Exchange energies for d^n configurations

M(g) configuration	Exchange energy
d^1	0
d^2	K
d^3	3K
d^4	6K
d^5	10K
d^6	10K
d^7	11K
d^8	13K
d^9	16K
d^{10}	20K

Although the nd subshell is stabilized relative to $(n+1)$s across the series the $(n+1)$s orbitals remain populated because this configuration reduces the total electron–electron repulsion energies. For the later transition metals the placing of electrons in the less stable s orbital is not predicted by the *aufbau* principle, which favours placing all the electrons in the 3d shell, once this orbital set becomes more stable than the 4s orbital. The 3d orbital set is more contracted than the 4s orbital and the electrons populating it experience more electron–electron repulsion. This is exacerbated because the d orbitals have to accommodate a total of 10 electrons rather than 2 associated with the s shell. The $(n+1)$s^xnd^{y} ($x = 1$ or 2, $y = 1$–10) configurations are energetically favoured because the electron repulsion experienced by two electrons—one in a d orbital and one in an s orbital—is smaller than that for a pair of electrons in the d orbital set. It is apparent from Table 5.1 that Cr and Cu differ from the other first row transition metals in having s^{1}d^n ($n = 5$ or 10) configurations rather than s^{2}d^{n-1}. This arises primarily from the favourable exchange energies associated with half-filled shells. The electron–electron repulsion between a pair of electrons with parallel spins is less than that for a pair of electrons with anti-parallel spins and the energy difference is equal to the exchange energy, K (see the full discussion in Chapter 1).

The exchange energies reach maxima for d^5 and d^{10} and therefore the s^{1}d^x ($x = 5$ or 10) are favoured relative to s^{2}d^{x-1} for Cr and Cu (see Table 5.2). When an electron is ionized from a neutral transition metal atom the nd orbital in the resulting M$^+$ ion is stabilized more than $(n+1)$s and therefore the ionized atoms have the more regular electronic configurations; $nd^x(n+1)s^0(n+1)p^0$ ($x = 1$–10). Since in most compounds of the transition elements there is a residual partial positive charge on the metal it is convenient when defining the d electron count of a metal to assume that all the electrons are accommodated in the d shell. So the effective d electron count in Cr^0, Cr^{II}, and Cr^{IV} compounds are d^6, d^4, and d^2 respectively although the neutral atom in the gas phase has the electronic configuration 3d^{5}4s^1.

Fig. 5.2 also suggests that at the end of the transition series the d orbitals enter into the core region and become unavailable as valence orbitals. This occurs after Group 11 of the Periodic Table, beyond which the d orbitals no longer behave as valence orbitals.

Fig. 5.3 illustrates the radial distribution functions for the 3d, 4s, and 4p orbitals of a cobalt atom.

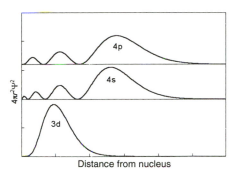

Fig. 5.3 Radial distribution functions for the valence orbitals of cobalt. The 4s and 4p functions are offset along the radial distribution function axis to aid clarity

The contracted nature of the 3d orbitals is clearly evident. In terms of the maximum probability in the radial distribution functions discussed in Chapter 1 the relative values are $3d(r_{max}) = 40$ pm; $4s(r_{max}) = 145$ pm which may be contrasted with the metallic radius of cobalt which is 125 pm. The bonding implications of the semi-core-like behaviour of the 3d orbitals is discussed in some detail below. In contrast the 4s and 4p orbitals are very diffuse and overlap more effectively with the orbitals of the ligands around the metal.

Variations in ionization energies

The valence d orbitals for the transition metals are much more core-like than the s and p valence orbitals for the representative elements. The relatively small change in effective nuclear charge across the transition series is sufficient to contract the sizes of the metal atoms and leads to an overall increase in ionization energies, but the variation is smaller than that for a group of p block elements. The first and second ionization energies of the first row transition metals are illustrated in Fig. 5.4. The first ionization process is associated with the loss of a 4s electron and the resulting M^+ ions all have d^n electronic configurations, i.e. the processes $s^2d^n \rightarrow d^{n+1}$ (Ti, Mn, Co, Ni, and Zn), $s^2d^3 \rightarrow d^4$ (V), and $s^1d^n \rightarrow d^n$ (Cr and Cu). This variation in electronic configurations makes the detailed analysis of the first ionization energies in Fig. 5.4 problematical. The second ionization processes, $d^n \rightarrow d^{n-1}$, consist of initial and final states which differ only in the number of electrons in the d shell and so are more amenable to a simple interpretation.

As with the p block elements the difference in exchange energies between $M^{n+}(g)$ and $M^{(n+1)+}(g)$ plays a significant role in influencing the ionization energies. The difference in exchange energies for $M^{n+}(g)$ and $M^{(n+1)+}(g)$ for d^1 to d^{10} configurations are shown in Table 5.3.

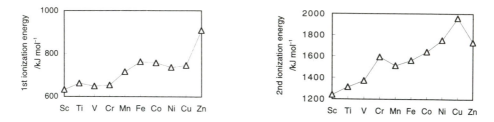

Fig. 5.4 First and second ionization energies of the first row transition elements, Sc–Zn

Table 5.3 Changes in exchange energy for the ionization process $M^{n+}(g) \rightarrow M^{(n+1)+}(g)$

$M^{n+}(g)$ configuration	Change in exchange energy
d^1	0
d^2	K
d^3	$2K$
d^4	$3K$
d^5	$4K$
d^6	0
d^7	K
d^8	$2K$
d^9	$3K$
d^{10}	$4K$

The difference in exchange energies reaches maxima at d^5 and d^{10}. It follows that the ionization energies for d^5 and d^{10} configurations are increased because of the associated loss of exchange energies and those for d^1 and d^6 atoms, which involve no loss in exchange energies, are reduced. The effect of these differences of exchange energy on the ionization energies is illustrated schematically in Fig. 5.5. The resulting 'humped' appearance of the ionization energies for the $M^+(g)$ ions, which all have d^n configurations and illustrated in Fig. 5.4, may be interpreted in terms of these exchange energy differences. The relatively high ionization energies for Cr^+ (d^5) and Cu^+ (d^{10}) are particularly noteworthy.

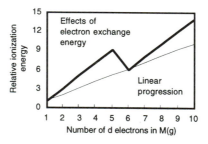

Fig. 5.5 A diagram showing the effect of exchange energy variations on an otherwise general increase in ionization energy as the number of d electrons changes

Electronegativities

Table 5.4 summarizes the electronegativities of the transition metals. As anticipated from their positions in the Periodic Table the calculated values for the transition metals lie between those of the s and p block elements, c.f. Ca 1.00 and Ga 1.81.

Given that the series consists of ten elements this variation is remarkably small and reflects the relative efficiency of the d electrons in shielding the nuclear charge. The electronegativity increase across the series arises primarily from the increasing ionization energies. A direct consequence of this variation in electronegativities is that the bonding in an isostoichiometric series of compounds becomes slightly more covalent across a particular row of transition metals.

Table 5.4 Allred–Rochow electronegativities of the transition metals

3rd row element (1st transition series)	χ	4th row element (2nd transition series)	χ	5th row element (3rd transition series)	χ
Sc	1.20	Y	1.11	La	1.08
Ti	1.32	Zr	1.22	Hf	1.23
V	1.45	Nb	1.22	Ta	1.33
Cr	1.56	Mo	1.30	W	1.40
Mn	1.60	Tc	1.36	Re	1.46
Fe	1.64	Ru	1.42	Os	1.52
Co	1.70	Rh	1.45	Ir	1.55
Ni	1.75	Pd	1.35	Pt	1.44
Cu	1.75	Ag	1.42	Au	1.42
Zn	1.66	Cd	1.46	Hg	1.44

The data in Table 5.4 also suggest that the electronegativities vary little for the elements in the same column. Furthermore, the electronegativities have the following relative values:

1st transition series > 3rd transition series > 2nd transition series

This trend has its origins in the interpolation of the lanthanides between the second and third transition series which results in a higher effective nuclear charge for the elements of the third series. The elements of the third transition series have radii which are very similar to those for the second series—indeed this effect is sometimes described as a *lanthanide contraction*. It has also been suggested that the valence orbitals of the third series of transition elements are influenced by relativistic effects and this phenomenon is discussed on page 26.

Summary
The relatively small variations in electronegativities and ionization energies across the three transition series lead to much smaller variations in their properties than those observed for the p block elements described in Chapter 3. Furthermore, the contracted (semi-core-like) behaviour of the d orbitals leads to the formation of many isostoichiometric compounds. The variations in ionization energies and electronegativities are even smaller for the lanthanides and actinides and consequently they show smaller variations in their properties across the series.

5.3 Physical properties of the metals

Melting points, boiling points and enthalpies of atomization

The data shown in Table 5.5 for the fourth row elements, which include the first transition series, suggest that in general the melting points, boiling points, and enthalpies of atomization of these elements are closely correlated.

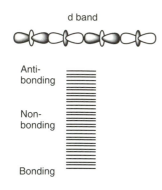

d band

Anti-
bonding

Non-
bonding

Bonding

Fig. 5.6 A diagram representing a d band of orbitals

The $\Delta_{at}H^{\ominus}$ of these elements reflect the strength of the metal–metal bonds in their metallic structures. The s and d block elements (K to Zn) show an overall increase in $\Delta_{at}H^{\ominus}$ up to V and then there is a marked decrease to zinc. The $\Delta_{at}H^{\ominus}$ values for Cr and Mn are anomalously low for reasons associated with the $4s^13d^5$ and $4s^23d^5$ configurations of these atoms.

Table 5.5 Melting and boiling points and standard enthalpies of atomization for the third row elements

	K	Ca	Sc	Ti	V	Cr	Mn	Fe	Co
m.p./ $^{\circ}$C	64	845	1539	1675	1900	1890	1244	1535	1495
b.p./ $^{\circ}$C	774	1487	2727	3260	3400	2480	2097	3000	2900
$\Delta_{at}H^{\ominus}$ / kJ mol^{-1}	90	177	390	469	502	397	284	406	439

	Ni	Cu	Zn	Ga	Ge	As	Se	Br
m.p./ $^{\circ}$C	1453	1083	419	30	937	817	217	−7
b.p./ $^{\circ}$C	2732	2595	907	2403	2830	subl.	685	59
$\Delta_{at}H^{\ominus}$ / kJ mol^{-1}	427	341	130	277	376	287	207	111

d band partially occupied (1–2 d electrons per metal atom)—weak metal–metal bonding

d band half occupied (4–6 delectrons per metal atom)—strong metal-metal bonding

d band fully occupied (10 d electrons per metal atom)—weak metal-metal bonding

Fig. 5.7 Diagrams illustrating the extent of d band filling to the metal–metal bond strength

In general, for a metal which has a close-packed solid state structure the maximum binding energies are associated with an electronic configuration where the valence shells are half-filled. The band structure of a metal corresponds to an infinite set of delocalized molecular orbitals which extend throughout the structure. Half of these closely spaced orbitals are bonding and half are antibonding. The most stable of these molecular orbitals have no antibonding nodes between orbitals on adjacent atoms and the least stable have nodes between each adjacent pair of orbitals. In the middle of the bands the number of bonding and antibonding relationships are approximately equal and the orbitals are non-bonding. These features of the band structure of a metal are shown schematically in Fig. 5.6.

If the d band of molecular orbitals are filled in an *aufbau* fashion the $\Delta_{at}H^{\ominus}$ should increase from a low initial value for d^0 to a maximum at d^5 (the half-filled shell), and a minimum for the completely filled shell (d^{10}) as indicated by the diagrams in Fig. 5.7. The s orbitals also generate a band which gives maximum binding energies for s^1 configurations. For the transition elements the maximum binding energies (and $\Delta_{at}H^{\ominus}$ values) should be achieved for the $4s^13d^5$ configuration and minimum for the $4s^23d^{10}$ configuration.

Although zinc does have the smallest $\Delta_{at}H^{\ominus}$ the maximum value is actually achieved for vanadium which has five valence electrons. For the transition elements the 3d electrons have particularly high electron–electron repulsion energies and consequently the idealized *aufbau* filling is not achieved. Some of the molecular orbitals which are metal–metal bonding are only occupied by single electrons and this forces the additional electrons to occupy more antibonding metal–metal orbitals. Although the occupation of these orbitals reduces the metal–metal bonding their occupation is favoured because the electron–electron repulsion effects are less than if the electrons were forced to occupy a bonding orbital which already contains a single electron.

Such elements therefore have lower than anticipated $\Delta_{at}H^{\ominus}$ values. These *aufbau* and non-*aufbau* fillings of metal orbital bands are illustrated in the diagrams shown in Fig. 5.8.

The large exchange energies associated with the $4s^1 3d^5$ (Cr) and $4s^2 3d^5$ (Mn) configurations lead these elements to have non-*aufbau* configurations when the metallic band structures are populated and therefore they have the anomalously low $\Delta_{at}H^{\ominus}$ values shown in Fig. 5.9.

Aufbau filling of d band with the spins pairing up.

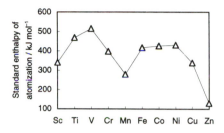

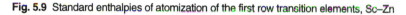

Fig. 5.9 Standard enthalpies of atomization of the first row transition elements, Sc–Zn

Non-*aufbau* filling of a d band. Some of the spins remain parallel leading to magnetic properties and occupation of higher energy and more anti-bonding levels in the band. This results in a weakening of the metal–metal bonding.

Fig. 5.8 *Aufbau* and non-*aufbau* fillings of metal d band orbitals

Fig. 5.10 illustrates the $\Delta_{at}H^{\ominus}$ values for the second and third row transition elements respectively. For these elements the larger sizes of the 4d and 5d orbitals (as compared with those of the 3d orbitals) reduces the extent of electron–electron repulsion and the filling of the band structures occurs in a more *aufbau* fashion. The $\Delta_{at}H^{\ominus}$ values consequently follow the theoretical curve more closely. For the second row the maximum occurs at Nb (5 valence electrons) and W (6 valence electrons) for the third row and only a small dip in the graph is observed for the next element.

It is also noteworthy that the enthalpies of atomization follow the order: 3rd row > 2nd row > 1st row. These differences reflect the relative overlaps between the metal d valence orbitals, i.e. 5d–5d > 4d–4d > 3d–3d. Down a column the nd orbitals become less core-like and overlap more effectively with the d orbitals of adjacent atoms. Specifically the r_{max} (nd) increase as follows: Co, 40 pm; Rh, 68 pm; Ir, 77 pm. This leads to stronger metal–metal bonding and higher $\Delta_{at}H^{\ominus}$ values.

(a)   (b)

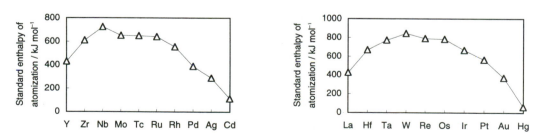

Fig. 5.10 Standard enthalpies of atomization of (a) the second (Y–Cd) and (b) the third (La–Hg) row transition elements

Theoretically the filling of the d and s metallic band structures should lead to a single parabolic curve with a maximum binding energy for the half-filled shell. The large exchange energies associated with the metal ions in the middle of the first transition series leads to a double humped appearance. This almost disappears for the second and third row elements where the binding energies are greater because the $nd-nd$ overlaps are larger and the interelectronic repulsion energies are smaller.

Metallic radii and densities

Across a row of the Periodic Table the sizes of the atoms generally decrease because the effective nuclear charge of the atoms increase. The data given in Table 5.6 confirm this trend for the pre-transition elements and the transition elements, with the exception of manganese and copper which have larger radii than anticipated.

As noted above the electron–electron repulsion energies for the d electrons of the first row transition elements are large and therefore filling of the band structure representing the delocalized molecular orbitals in a metal does not always follow a simple *aufbau* procedure. The $4s^2 3d^5$ electronic configuration of manganese and the associated *exchange energy* cause the population of more antibonding sections of the band structure and longer metal–metal bonds result.

The decreasing radii and increasing relative atomic masses lead to an increase in densities across the series. The large variations in densities across the series has important technological implications. For example, the fact that titanium has a density about 60% of that of iron has not been lost on aeroplane manufacturers.

For a more detailed discussion see D. Pettifor, *Bonding and Structure of Molecules and Solids*, Clarendon Press, Oxford, 1995.

Table 5.6 Metallic radii and densities of the elements K–Se of the fourth row of the Periodic Table

	K	Ca	Sc	Ti	V	Cr	Mn	Fe
Metallic radius/ pm	235	197	164	147	135	130	135	126
Density / kg m^{-3}	860	1540	3000	4500	6100	7200	7440	7860
	Co	Ni	Cu	Zn	Ga	Ge	As	Se
Metallic radius / pm	125	125	128	137	141	137	139	140
Density / kg m^{-3}	8860	8900	8920	7130	5910	5320	5730	4790

Structures

The structures of elements of Groups 3–12 are summarized in Table 5.7 and it is apparent that elements in many of the Groups share a common structural type.

Given that the energy differences between cubic close-packed, hexagonal close-packed and body-centred cubic structures for a given element are usually quite small it is not too surprising that for some of the columns a common structure is not observed. The first-row transition metals commonly have different structures because, for these metal atoms, the interelectronic repulsion effects are large and contribute significantly to the total energy.

Table 5.7 Structures of elements of Groups 3–12

Group(s)	Elemental structures
3	Sc, Y hexagonal close packed; La non-standard close packed
4	All hexagonal close packed
5 and 6	All body-centred cubic
7	Mn unique structure, Tc, Re hexagonal close packed
8	Fe body-centred cubic; Ru, Os hexagonal
9	Co hexagonal close packed, Rh, Ir cubic close packed
10 and 11	All cubic close packed
12	Zn, Cd distorted hexagonal close packed, Hg distorted cubic close packed

Electrical resistivities

Table 5.8 contains the values of the electrical resistivities (ρ) of the transition elements. There is no clear correlation across the transition series. The resistivities decrease from Groups 3 to 6 and then there is a marked increase at Group 7 followed by a decrease from Groups 8 and 10 for the first row and 8 and 9 for the second and third rows. A significant increase is then observed for the Group 12 elements.

Table 5.8 The electrical resistivities (ρ) of the transition elements at 298 K (the units are $\Omega\,m/10^{-8}$)

1st row element	ρ	2nd row element	ρ	3rd row element	ρ
Sc	61	Y	57	La	57
Ti	42	Zr	40	Hf	35.1
V	24.8	Nb	12.5	Ta	12.45
Cr	12.7	Mo	5.2	W	5.65
Mn	185	Tc	22.6 (373 K)	Re	19.3
Fe	9.71	Ru	7.6	Os	8.12
Co	6.24	Rh	4.51	Ir	5.3
Ni	6.84	Pd	10.8	Pt	10.6
Cu	1.673	Ag	1.59	Au	2.35
Zn	5.961	Cd	6.83	Hg	94.1

5.4 Chemical properties of the metals

Standard electrode potentials

The variations in standard reduction potentials given in Table 5.9 underline the decreasing reducing abilities of the transition elements across the series, i.e. the standard reduction potentials become less negative across the series.

The decrease in the reducing abilities of the metals across the transition series has its origins primarily in the increasing ionization energies of the atoms. These are not compensated for by the more favourable hydration enthalpies which are associated with the reduction in radii for the M^{2+} ions. The relevant thermochemical cycle showing the factors which influence the reduction of the M^{2+}(aq) to the solid metal is given in Fig. 5.11.

The anomalies at Mn and Cu for $E^\ominus$ (M^{2+}(aq)/M) may also be related to the larger sum of the first two ionization energies, $I_1 + I_2$, for these metals (see Fig 5.4).

Table 5.9 Standard reduction potentials of the first row transition elements

	$E^\ominus$ [M^{2+}(aq)/M] / V	$E^\ominus$ [M^{3+}(aq)/M] / V
Ca	−2.87	
Sc		−2.1
Ti		−1.2
V	−1.2	−0.86
Cr	−0.91	−0.74
Mn	−1.18	−0.28
Fe	−0.44	−0.04
Co	−0.28	0.4
Ni	−0.25	
Cu	0.34	
Zn	−0.76	

The higher ($I_1 + I_2 + I_3$) ionization enthalpies for the later transition metals are not compensated for by the more favourable enthalpies of hydration for the M^{3+}(g) ions and consequently the M^{3+}(aq) ions are not observed for Ni, Cu, and Zn

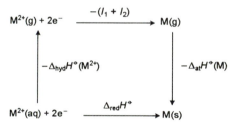

Fig. 5.11 A thermochemical cycle for the standard enthalpy change for the reduction of M^{2+}(aq) to the solid metal, M(s)

The standard electrode potentials discussed above have a direct bearing on the reactions of the metals with acid. However, it must be emphasized that the values of $E^\ominus$ refer only to the thermodynamic aspects of the reaction and do not provide any information on the kinetic factors which may affect the rate of the reaction of the metal with acid.

Titanium and vanadium dissolve in HCl(aq) giving [Ti(OH$_2$)$_6$]$^{3+}$ and [V(OH$_2$)$_6$]$^{3+}$ (and some chloroaquo ions); Cr, Mn, Fe, Co, and Ni give [M(OH$_2$)$_6$]$^{2+}$ ions. Oxidizing acids such as concentrated HNO$_3$ have a passivating effect (e.g. with Cr and Fe) due to the formation of a layer of impermeable oxide on the surface of the metal. Copper only dissolves in oxidizing acids, which is consistent with its positive $E^\ominus$ value.

The earlier elements in the series are also very reactive towards O$_2$ particularly if they are in a finely powdered form. Of course the reaction of iron with oxygen in the presence of water ('rusting') is of particular importance. At the other end of the series Co is oxidized by air only above 300°C forming Co$_3$O$_4$ and CoO. Copper oxidizes in air at red heat to give CuO and if the temperature is raised Cu$_2$O forms. In moist atmospheres copper becomes slowly covered with a green coating of basic carbonate.

Oxidation state variations

Relative stabilities of oxidation states

The standard electrode potentials $E^{\ominus}(M^{3+}/M^{2+})$ for the first transition series elements are summarized in Table 5.10.

Table 5.10 Standard reduction potentials for the first transition series. The numbers in brackets refer to theoretically calculated values

	$E^{\ominus}[M^{3+}/M^{2+}]$ / V		$E^{\ominus}[M^{3+}/M^{2+}]$ / V
Sc	(−2.6)	Fe	0.77
Ti	(−1.1)	Co	1.93
V	−0.26	Ni	(4.2)
Cr	−0.41	Cu	(4.6)
Mn	1.60	Zn	(7.0)

Broadly speaking these data indicate that for the earlier transition metals the $M^{2+}(aq)$ ion is reducing, and for the later transition metals the $M^{3+}(aq)$ ion is more strongly oxidizing. Nevertheless the trend across the series is not regular, $Fe^{3+}(aq)$ is less strongly oxidizing than a simple extrapolation would have suggested and $Mn^{3+}(aq)$ is more strongly oxidizing. The Hess's law thermodynamic cycle for the redox process, shown in Fig. 5.12, suggests that the dominant term is the third ionization energy of the $M^{2+}(g)$ ions and the anomalies in $E^{\ominus}$ may be related directly to the non-regular behaviour of I_3, the third ionization energy of the metal atom.

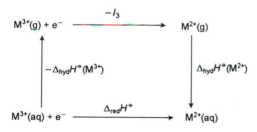

Fig. 5.12 A thermodynamic cycle for the enthalpy change for the reduction of $M^{3+}(aq)$ to $M^{2+}(aq)$

The I_3 for $Mn^{2+}(g)$ is larger than expected because the loss of an electron from this ion with a half-filled d^5 shell represents a significant loss of exchange energy. In contrast $Fe^{2+}(g)$ has a slightly lower than anticipated I_3 because it has a d^6 configuration and formation of $Fe^{3+}(g)$ (d^5) results in no loss of exchange energy (see Fig. 5.13 and the discussion on pages 276–278).

The conclusion from this analysis of $E^{\ominus}(M^{3+}(aq)/M^{2+}(aq))$ data that the higher oxidation state ion becomes progressively less stable from left to right across the transition series is quite a general one. The observation that the Group oxidation states 8, 9, 10, and 11 are not observed for Fe, Co, Ni, and Cu is one extreme consequence of this. The absence of many +4, +5, and +6 oxidation state compounds for Co, Ni, and Cu are further manifestations of the increasing ionization energies of the atoms across the series. Indeed, the Group oxidation states are only observed up to manganese for the first row, ruthenium for the second row, and osmium for the third row metals. The data also suggests that irregularities in the electrode potentials across the series may often be related to differences in the ionization potentials of the isolated metal atom or ion.

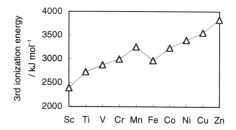

Fig. 5.13 The third ionization energies (I_3) of the first row transition elements, Sc–Zn

5.5 General bonding considerations

Introduction

In the aqua-complexes and the halides the transition metals form isostoichiometric compounds and the d orbitals behave as if they are semi-core-like, i.e. they are not exerting a strong stereochemical or stoichiometric preference. The transition metals also form the following isostoichiometric complexes: $[M(acac)_3]$, $[M(CN)_6]^{4-}$, $[M(CN)_6]^{3-}$, $[M(NH_3)_6]^{3+}$, $[M(bipy)_3]^{3+}$, $[M(phen)_3]^{3+}$. It is significant that the majority of examples have oxygen, nitrogen, or chlorine atoms bonded to the transition metals and the metals in intermediate oxidation states—usually +2 and +3.

Some ligands are, however, able to come close enough to the metal to produce good overlap with the 3d orbitals. For example, hydrogen is able to approach the metal sufficiently closely to overlap well with the metal d orbitals and form a series of molecular complexes, which resemble those of the p block elements. The strong participation of the metal d orbitals makes these complexes conform to the Effective Atomic Number (EAN) rule and therefore their formulae and structures change across the Periodic Table. Examples of such compounds are shown below.

The hydride ligand, because of its small size, is able to approach the metal nucleus sufficiently closely to overlap effectively with the metal nd valence orbitals

$[ReH_9]^{2-}$	$WH_6(PR_3)_3$	$[WH_5(PR_3)_4]^+$
$ReH_7(PR_3)$	$RuH_6(PR_3)_2$	$IrH_5(PR_3)_3$ $WH_4(PR_3)_4$
$OsH_4(PR_3)_3$	$ReH_3(PR_3)_4$	
$[RuH_6]^{4-}$	$[OsH_3(PR_3)_3]^-$	$IrH_3(PR_3)_3$

It is noteworthy than many of these complexes also have bulky phosphine ligands attached to the metal. These phosphine ligands have two roles—they make the compounds soluble in non-reactive organic solvents and they reduce the rates of decomposition reactions by placing the metal hydrides in a sterically protected environment.

Ligands may also involve the d orbitals more fully by forming multiple bonds. The effect is in some respects self-sustaining because as the multiple bond character is increased then the ligand approaches the metal closely which in turn increases the extent of overlap between the two orbitals (see the diagram in the margin). Therefore, such ligands continue to form shorter and shorter bonds to the metal until the electronic repulsion effects involving their core electrons balances the increasingly strong covalent bonding effects.

The contraction in bond lengths associated with the changes in formal bond orders may be gauged by the tungsten compound shown in the margin which has been formulated with single, double, and triple bonds between tungsten and carbon within one molecule. It is noteworthy that the single and triple bond lengths differ by nearly 50 pm.

Multiple bonding

Low oxidation states

In low oxidation state compounds of the transition elements, multiple bonding is achieved by transferring electron density from filled metal orbitals of π symmetry to empty π^* orbitals on the ligands. The most commonly encountered ligand of this type is CO. This ligand forms σ-bonds to a transition metal by donating electron density from the lone pair orbital on carbon to a suitable acceptor orbital on the metal as indicated by the orbital diagram shown in Fig. 5.14.

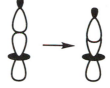

The effect of shortening the metal–ligand distance on the overlap between a ligand orbital and a metal d orbital

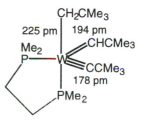

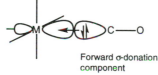

Forward σ-donation component

Fig. 5.14 A representation of the σ-donation of electron density from the CO ligand to the metal, M

Such σ-bonding is supplemented by back donation from filled d-orbitals on the metal to empty π^*-orbitals of the ligand both of which have π-symmetry with respect to the M–C bond as indicated by the orbital diagram shown in Fig. 5.15.

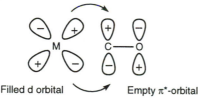

Filled d orbital   Empty π^*-orbital

Back-donation π^*-component

Fig. 5.15 A representation of the π-donation of electron density from a filled metal orbital to the anti-bonding π^* orbital of the CO ligand

The complementary σ and π components ensure that the metal remains electroneutral. In classical bonding terms the resultant changes in multiple bond character are represented by the resonance forms shown in the margin.

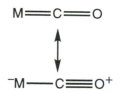

In contrast to octahedral SF_6 discussed on pages 246 and 259 the 3d orbitals are more stable than the s and p valence orbitals and therefore participate more in the bonding.

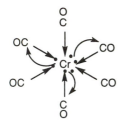

The back-donation effects may also be represented by the orbital box diagram of Fig. 5.16, which shows clearly the implications for the EAN rule. Such representations emphasize that the EAN rule is associated with the full utilization of the nd, $(n + 1)$s, and $(n + 1)$p valence orbitals either in electron acceptance or back-donation.

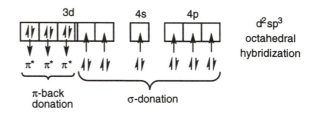

Fig. 5.16 An orbital box diagram showing how the EAN rule operates when back donation occurs

Table 5.11 M–C and C–O stretching frequencies for some metal–carbonyl species

	ν (M–C) /cm^{-1}	ν (C–O) /cm^{-1}
$[Mn(CO)_6]^+$	416	2101
$[Cr(CO)_6]$	441	1981
$[Mn(CO)_6]^-$	460	1859

In situations where the CO is unable to enter into π bonding interactions the CO stretching frequency is much higher, e.g. Me$_3$AlCO (2185 cm^{-1}), HCO$^+$ (2184 cm^{-1}), (η-C$_5$Me$_5$)$_2$Ca(CO) (2158 cm^{-1}), *Angew. Chem. Int. Ed.*, 1995, **34**, 791.

The back donation occurs from filled metal d orbitals to empty CO π^* orbitals. Therefore as the degree of back donation is increased the multiple bonding between metal and carbon is developed at the expense of the C–O multiple bond. These effects may be confirmed experimentally by studying the bond lengths, force constants, or stretching frequencies in isoelectronic carbonyl complexes.

For example, in the series of carbonyl compounds given in Table 5.11 the ν(M–C) and ν(C–O) stretching frequencies reflect the way in which the M–C bond strengthens and the C–O bond weakens as the negative charge on the complex increases.

As the negative charge is increased the dominant effect involves the back donation component which increases and strengthens the M–C bond at the expense of the CO bond. Other ligands which are capable of forming multiple bonds of this type include NO, CNR, CS, CR$_2$ (carbenes), and CR (carbynes).

Ethyne and ethene also stabilize low oxidation states by utilizing forward and back donation components and the relevant orbital interactions are illustrated in Fig. 5.17. In these complexes the ligand is not utilizing a lone pair as the forward donation component, but a filled π-orbital. Therefore, such ligands bond in a sideways-on or π-fashion.

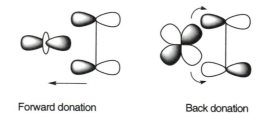

Forward donation   Back donation

Fig. 5.17 Representations of forward and back donation of electron density when ethyne or ethene act as ligands

In valence bond terms these bonding effects are represented by the following canonical forms:

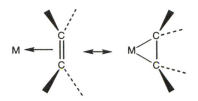

As the electron density on the metal is increased the back donation component is enhanced and the double bond character of the ethene is reduced. This is reflected in a longer (and weaker C–C) bond and a bending back of the substituents on the carbon atoms. Ligands such as PF_3 are also able to stabilize low oxidation states. The forward donation component from the phosphorus lone pair orbital is complemented by back donation from the filled metal d orbitals to the ligand σ^* (P–F) antibonding molecular orbitals, as shown in Fig. 5.18. The σ^* (P–F) antibonding orbitals have the same symmetries as the empty phosphorus 3d orbitals and therefore the back donation may also involve some transfer of electron density into these orbitals.

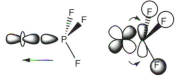

Fig. 5.18 A representation of the forward and back donation of electron density when PF_3 acts as a ligand

The forward and back donation effects discussed above may also occur through the σ and σ^*-orbitals of a molecule such as dihydrogen H_2. The symmetry aspects of the problem are identical although the precise symmetry designations of the orbitals in the free ligand are different. The appropriate orbital interactions for the forward and back donation of electron density are shown in Fig. 5.19. A specific example of such a dihydrogen complex is illustrated in the margin.

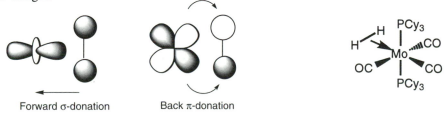

Forward σ-donation   Back π-donation

Fig. 5.19 Representations of the forward and back donation of electron density when dihydrogen acts as a ligand

The forward and back donation effects result in a weakening of the H–H bond and, in the limit, this results in the dissociation of the molecule and the formation of a di(hydrido) complex of the metal, whose formal oxidation state is two higher than that in the original complex as shown in Fig. 5.20.

Fig. 5.20 General structures indicating the change from a dihydrogen complex to a dihydrido-complex

> Low oxidation state complexes of the transition metal are stabilized by ligands which are able to accept electron density from filled metal d orbitals into low lying antibonding orbitals of π-symmetry. The relative π-acid strengths follows the order:
>
> $$NO > CO > RNC > PF_3 > PCl_3 > P(OR)_3 > PR_3 > RCN.$$

High oxidation states

In high oxidation state transition metal complexes multiple bonding is achieved by the donation of electron density from filled ligand orbitals to empty d orbitals with π-symmetries on the metal. The oxo (O^{2-}) and nitrido ligands (N^{3-}) are particularly effective in forming multiple bonds of this type and the relevant orbital interactions are shown in Fig. 5.21.

The multiple bonding in such complexes is represented by the following valence bond diagrams.

$$M\equiv N \qquad M\!\!\stackrel{\longleftarrow}{=}\!\!O \qquad M\!\!\stackrel{\longleftarrow}{=}\!\!NR$$

Each of these ligands is able to utilize both of its p_π orbitals in forming multiple bonds and therefore the extreme covalent bond representations emphasize the triple bonded nature of the interaction. Such ligands are able to overlap efficiently with the contracted 3d orbitals of the metal and this leads to strong multiple bonds. These multiple bonding effects are particularly important for metals in high oxidation states which have high charge/size ratios and are therefore capable of attracting electron density from the ligands. The resultant transfer of electron density leads to approximate electroneutrality at the metal centre.

The relative π-donor abilities of these ligands reflects the formal negative charge on the ligand, i.e. $N^{3-} > O^{2-} \gg F^-$ and $P^{3-} > S^{2-} \gg Cl^-$ and the π-donor abilities of the halogens is so small that in general it is ignored. Recently the range of π-donor ligands of this type has been extended to P, Se, and Te and it has been suggested that the degree of multiple bond character decreases in the order: $O \gg S > Se > Te$.

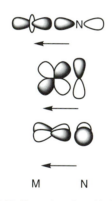

Fig. 5.21 Examples of p orbitals of nitrogen overlapping with the d_{z^2}, d_{xz}, and d_{yz} orbitals of a metal, M, leading to triple bond formation

See for example, L. G. Leipoldt *et al.*, *Adv. Inorg. Chem.*, 1994, **40**, 241 and D. E. Wigley *et al.*, *Prog. Inorg. Chem.*, 1994, **42**, 239.

The data in Table 5.12 are in agreement with this general trend.

Table 5.12 The Zr–E bond distances for the compounds $Cp_2Zr(py)E$ (structures shown in the margin)

E	Zr–E(pm)	% Decrease relative to sum of covalent radii
O	180	10
S	233	7
Se	248	7
Te	273	5

Data taken from G. Parkin *et al.*, *J. Amer. Chem. Soc.*, 1994, **116**, 606.

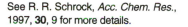

The strength and chemical importance of this type of multiple bond is reinforced by the following remarkable reaction which at room temperature leads to the scission of the N_2 triple bond which has a dissociation enthalpy of 964 kJ mol^{-1}.

See R. R. Schrock, *Acc. Chem. Res.*, 1997, **30**, 9 for more details.

The structures shown below also indicate how the multiple bonds may be redistributed between the ligand and the metal.

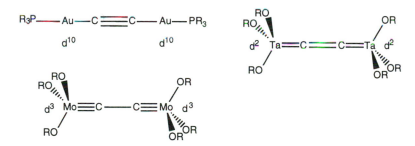

The d-electron counts indicated refer to the number of d-electrons associated with the metal centres in their formal oxidation states.

π-Donor ligands do not necessarily have to be single atom donors and there are examples of ligands which can donate in a σ and a π-manner which mirrors that found in π-acceptor olefin complexes. The peroxo-ligand O_2^{2-} provides an example of such a ligand, the forward donation from its σ-type (O–O π bonding) and π-type (O–O π^* antibonding) orbitals are shown in Fig. 5.22.

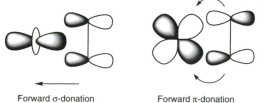

Forward σ-donation   Forward π-donation

Fig. 5.22 The forward σ-type and π-donations from the π bonding and π* antibonding orbitals of the peroxide ion

Consequences of multiple bonding

Complexes which have ligands which form strong multiple bonds have the metal nd, $(n + 1)$s, and $(n + 1)$p valence orbitals heavily involved in covalent bond formation and therefore they behave like molecular compounds of the post-transition metals.

For π-acceptor complexes, ML_n, the π-back donation effects lead to a significant energy break in the spectrum of molecular orbitals above the $(9 - n)$d orbitals which are involved in back donation, i.e. the EAN rule applies.
For π-donor complexes the corresponding energy gap occurs below the d orbital manifold, i.e. d^0 electronic configurations are favoured.

1. The strong covalent interactions introduced by strong π-donors and π-acceptors lead to molecules with characteristic closed shell electronic configuration because the formation of strong σ- and π-bonds leads to large energy separations between bonding and antibonding molecular orbitals.

2. Since the complexes achieve electroneutrality at the metal, the metals are not strong Lewis acid centres and the resultant compounds are generally molecular and do not form infinite structures.

3. The low oxidation state complexes with π-acceptor ligands generally conform to the effective atomic number (EAN) rule and have a total of 18 electrons in the coordination shell environment. Examples of such complexes are:

$Cr(CO)_6$	$Fe(CO)_5$	$Ni(CO)_4$
	$Mn(CO)_4(NO)$	$Co(CO)_3(NO)$
		$Fe(CO)_2(NO)_2$
$Cr(PF_3)_6$	$Fe(PF_3)_5$	$Pt(PF_3)_4$

It is important to note that the adherence to the Effective Atomic Number Rule results in changes in the formulae of the compounds across the series in a manner reminiscent of that observed for post-transition elements, e.g. CF_4, NF_3, OF_2, and HF, to which the octet rule applies.

$V(CO)_6$ represents an interesting exception—it remains a monomeric radical rather than forming a dimeric metal–metal bonded compound. Possibly steric effects prevent the formation of a V–V bond in the putative dimer.

4. In general such complexes form metal–metal bonded compounds rather than forming paramagnetic radicals. For example, the 17 electron molecules $Co(CO)_4$ and $Mn(CO)_5$ form the dimeric metal–metal bonded compounds, shown below, which obey the EAN rule rather than remaining as radicals.

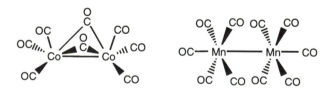

This behaviour is also reminiscent of that of main group compounds. For example, SnR_3 and PR_2 form dimers which obey the octet rule rather than remaining as radicals:

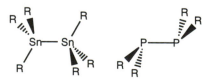

For both classes of compound the dimerization process becomes increasingly unfavourable if large organic groups are introduced into the molecules and kinetically stabilized radicals may be isolated using such strategies.

5. Complexes with strong π-donor ligands generally are associated with d^0 electronic configurations at the metal centre. The presence of low d-electron counts maximizes the possibilities for donating electron density from filled p_π ligand orbitals to empty d orbitals. Some specific examples of such complexes are illustrated below:

Metal–metal bond formation is only significant when the metal ion has a formal oxidation state of +3 or lower. Presumably in higher oxidation states the d orbitals are too contracted to give good overlaps. The metal–metal interactions become stronger down a column of elements,
e.g. W–W > Mo–Mo >> Cr–Cr

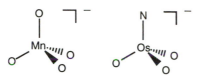

Such compounds are again reminiscent of the oxo-anions of the post-transition elements ClO_4^-, SO_4^{2-}, PO_4^{3-}, and SiO_4^{4-}.

6. It is possible to form compounds with d^1, d^2, d^3, and d^4 configurations with π-donor ligands if the electrons can be accommodated in orbitals which are non-bonding and the metals are in a high oxidation state. If the metals are in their lower oxidation states, the compounds either remain as radicals or form metal-metal bonded molecules. The range of tetrahedral $[MO_4]^{n-}$ anions and their d-electronic configurations are summarized in Table 5.13.

Table 5.13 Examples of tetrahedral oxo-complexes and their colours

d^0	CrO_4^{2-} yellow	MnO_4^- purple		
d^1	CrO_4^{3-} blue	MnO_4^{2-} green		
d^2	CrO_4^{4-} black	MnO_4^{3-} blue	FeO_4^{2-} red	
d^3		MnO_4^{4-} black		
d^4			FeO_4^{4-} black	CoO_4^{3-} blue-black

The compounds with the highest d^n configurations have only been observed in the solid state as Ba or Sr salts, because they react with the solvents which are used to dissolve them. For example, $[Fe^{VI}O_4]^{2-}$ has a d^2 configuration and its magnetic properties are consistent with the presence of two unpaired-electrons and it shows no tendency to dimerize. In contrast $Mo(OR)_3$ forms a dimeric compound with a triple metal–metal bond (shown in the margin).

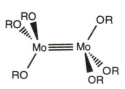

Ligands which are weak σ-donors and do not have π-orbitals which interact strongly with the metal d orbitals have a small energy gap within the d-manifold and consequently do not show well defined closed shell electronic configurations.

Compounds which do not have ligands which can overlap strongly with the metal d orbitals by σ- or π-bonding, e.g. Cl, F, Br, OH_2, and NH_3, have the following properties.

1. They form series of isostoichiometric compounds which do not conform to the EAN rule. Examples of such compounds are listed in Tables 5.14 - 5.16.

Table 5.14 The dihalides of the first transition series (stable indicated by √, × indicates that the compound is unstable or unknown)

	Sc	Ti	V	Cr	Mn	Fe	Co	Ni	Cu	Zn
MF_2	×	×	√	√	√	√	√	√	√	√
MCl_2	×	√	√	√	√	√	√	√	√	√
MBr_2	×	√	√	√	√	√	√	√	√	√
MI_2	×	√	√	√	√	√	√	√	√	√

Table 5.15 The trihalides of the first transition series(stable indicated by √, × indicates that the compound is unstable or unknown)

	Sc	Ti	V	Cr	Mn	Fe	Co	Ni	Cu	Zn
MF_3	√	√	√	√	√	√	√	√	×	×
MCl_3	√	√	√	√	*	√	×	×	×	×
MBr_3	√	√	√	√	×	√	×	×	×	×
MI_3	×	√	√	√	×	×	×	×	×	×

* decomposes at −40°C

Table 5.16 Chlorides of the second and third transition series

Second transition series						
$ZrCl_2$		$MoCl_2$				$PdCl_2$
$ZrCl_3$	$NbCl_3$	$MoCl_3$		$RuCl_3$	$RhCl_3$	
$ZrCl_4$	$NbCl_4$	$MoCl_4$	$TcCl_4$	$RuCl_4$		$PdCl_4$
	$NbCl_5$	$MoCl_5$				
		$MoCl_6$				

Third transition series							
						AuCl	
	$TaCl_2$	WCl_2		$OsCl_2$		$PtCl_2$	$HgCl_2$
$HfCl_3$	$TaCl_3$	WCl_3	$ReCl_3$	$OsCl_3$	$IrCl_3$		
$HfCl_4$	$TaCl_4$	WCl_4	$ReCl_4$	$OsCl_4$	$IrCl_4$	$AuCl_3$	
	$TaCl_5$	WCl_5	$ReCl_5$	$OsCl_5$		$PtCl_4$	
		WCl_6	$ReCl_6$				

2. Since these compounds have the d-electrons occupying very contracted and almost core-like orbitals then many of the compounds are paramagnetic. For example, the magnetic properties of some first row transition metal compounds are listed in Table 5.17.

It is noteworthy that for d^4–d^7 ions in these octahedral complexes two alternative magnetic moments (μ_{eff}) are observed, depending on the ligands in the complex. This aspect is discussed in more detail later in this chapter.

Table 5.17 Magnetic properties of some octahedral complexes

d-electron configuration	Example	μ_{eff}/μ_B
d^1	K_3TiF_6	1.70
d^2	K_3VF_6	2.79
d^3	CrF_3	3.85
d^4	K_3MnF_6	4.95
d^4	$K_3Mn(CN)_6$	3.2
d^5	Na_3FeF_6	5.85
d^5	$K_3Fe(CN)_6$	2.4
d^6	K_3CoF_6	5.53
d^6	$K_4Fe(CN)_6$	0
d^7	CoI_2	5.03
d^7	$[Co(diars)_3](ClO_4)_2$	1.92
d^8	$(NH_4)_2Ni(SO_4)_2.6H_2O$	3.3
d^9	$(NH_4)_2Cu(SO_4)_2.6H_2O$	1.9

3. The absence of strong metal–ligand interactions leaves the metal ion with a significant residual positive charge and partially vacant d orbitals on the metal which are not involved in π-bonding with π-donor ligands. This has two implications. In the solid state many of the simple oxides and halides form infinite structures. Some examples are shown below.

CaO	TiO	VO	MnO	FeO	CoO	NiO	ZnO
$\leftarrow$			sodium chloride			$\rightarrow$	wurtzite

Many halides are Lewis acids in solution and readily expand their coordination numbers to reduce the residual positive charge on the metal. Examples are:

$$TiCl_4 + 2NR_3 \rightarrow TiCl_4(NR_3)_2$$

$$CrCl_3 + 3NH_3 \rightarrow CrCl_3(NH_3)_3$$

4. It is clear that unlike the metal carbonyl complexes the stoichiometries of the resultant compounds are not controlled by the EAN rule, their stoichiometries are controlled primarily by the donor properties of the ligands and their steric requirements.

5. The presence of a contracted d shell in such complexes results in related complexes with the metals displaying a range of oxidation states and d-electron configurations, e.g. $[Fe(CN)_6]^{4-}$, $[Fe(CN)_6]^{3-}$.

Table 5.18 Oxidation states of the first row transition elements in aqueous environments

Element	Well documented oxidation states
Sc	3
Tc	2 3 4
V	2 3 4 5
Cr	2 3 4 5 6
Mn	2 3 4 5 6 7
Fe	2 3 4 5 6
Co	2 3 4 5
Ni	2 3 4
Cu	1 2 3

As has been noted previously the post-transition elements form compounds in a range of oxidation states, e.g. Sn^{II}, Sn^{IV}, Tl^{I}, Tl^{III}, but they generally differ by units of 2 unless metal–metal bonds are formed.

The transition metals not only show a wide range of oxidation states, but the even and odd oxidation states are both well represented as shown in Table 5.18. Tables 5.19 and 5.20 provide specific examples of compounds in the highest and lowest oxidation states, respectively, of the specific elements.

Table 5.19 Examples of fluoride compounds in the highest oxidation state. Compounds shown in italics correspond to those predicted by the Group Number.

ScF_3	TiF_4	VF_5	CrF_6	MnF_4	FeF_3	CoF_3	NiF_2	CuF_2	ZnF_2
YF_3	ZrF_4	NbF_5	MoF_6	TcF_6	RuF_6	RhF_6	PdF_4	AgF_2	CdF_2
LaF_3	HfF_4	TaF_5	WF_6	ReF_7	OsF_7	IrF_6	PtF_6	AuF_5	HgF_2

The following general points are worth emphasizing.

1. The Group oxidation state is not achieved beyond Group 7 (Mn) for the first row and Group 8 for the second and third row (Ru and Os). In a previous section it was noted that the +3 oxidation state becomes progressively more oxidizing across the first transition series, primarily because of the increasing third ionization energies of the metal atoms. More generally, it becomes progressively more difficult to regain the energy required to form the metal atom in the group oxidation state by chemical bond formation.

Table 5.20 Examples of compounds of the elements of the first transition series in their lower oxidation states

Ti	V	Cr	Mn	Fe	Co	Ni
$Ti(CO)_2(PF)_3(dmpe)$	$[V(CO)_6]^-$	$[Cr(CO)_5]^{2-}$ $Cr(NO)_4$	$[Mn(CO)_5]^-$ $Mn(NO)_3(CO)$ $Mn(phthalocyanin)_2$	$[Fe(CO)_4]^{2-}$ $Fe(NO)_2(NO_2)$	$[Co(CO)_4]^-$ $Co(CO)_3(NO)$	$Ni(CO)_4$ $[Ni_2(CO)_6]^{2-}$
Zr $Zr(bipy)_3$	Nb $[Nb(CO)_5]^{3-}$ $[Nb(CO)_6]^-$	Mo $[Mo(CO)_5]^{2-}$	Tc $[Tc(CO)_5]^-$	Ru $[Ru(CO)_4]^{2-}$ $[Ru(diphos)_2]^{2-}$	Rh $[Rh(CO)_4]^-$	Pd $Pd(PPh_3)_3$
Hf $(arene)Hf(PMe_3)$	Ta $[Ta(CO)_5]^{3-}$ $[Ta(CO)_6]^-$	W $W(CO)_6$	Re $[Re(CO)_5]^-$	Os $Os(CO)_5$	Ir $Ir(CO)_3(PPh_3)^-$ $Ir[(NO)_2(PPh_3)_2]^+$	Pt $Pt(PPh_3)_3$

2. High oxidation state compounds are generally associated with oxygen and fluorine ligands. The low dissociation energy of fluorine (F_2) and the strong bonds formed between fluorine and transition metals makes it particularly favourable for forming compounds in high oxidation states. Furthermore, the fluoride ligand is the most difficult halide anion to oxidize and therefore can survive in the highly oxidizing environment of the metal in its highest oxidation state. Oxygen and nitrogen are able to stabilize metals in high oxidation state because of their ability to form strong bonds with significant multiple bond character.

There are some examples of high oxidation state compounds with ligands which do not form such strong bonds with transition metals, e.g. H and CH_3, and yet provide examples of compounds in high oxidation states, e.g. $[ReH_9]^{2-}$ and $W(CH_3)_6$. It is likely that these compounds are thermodynamically unstable but kinetically inert. Their synthesis in non-aqueous and non-protic solvents and the high activation energy for their decomposition makes their isolation possible.

Even in aqueous solutions the pH influences the ability to isolate compounds in high oxidation states. Redox couples with H^+ on the left-hand side of the equation become less oxidizing as the pH increases and they are therefore less able to oxidize the solvent. Therefore, basic and oxidizing solutions (e.g. NaOCl) are frequently used to generate oxo-compounds, e.g. FeO_4^{2-}, in high oxidation states.

3.   The low oxidation state examples are stabilized by ligands which are strong π-acceptors, e.g. CO and NO^+. The partial multiple bond character of the metal–ligand bond and the transfer of electron density from metal to ligand provides an important stabilizing interaction.

4.   The complexes in intermediate oxidation states are found with ligands such as NH_3, OH_2, Cl, Br, etc. which do not overlap strongly with the d orbitals of the transition metals. The resultant d orbitals therefore remain essentially core like and are able to accommodate unpaired-electrons in orbitals which are thermodynamically non-bonding and kinetically inert because they are well shielded from the chemical environment. Also they do not overlap sufficiently well with the d orbitals of neighbouring atoms to form diamagnetic metal–metal bonded compounds.

Formal d-electron count, oxidation state, and total valence electron count

From the discussion above it is apparent that the formal metal oxidation state of the metal in a transition metal complex and the total number of valence electrons which surround the metal are the important parameters for classifying metal complexes. Therefore, the following section describes how these terms are defined and applied to a range of coordination and organometallic complexes.

Total valence electron count calculation

1.   Assume for convenience that the metal atom is neutral and contributes all the valence electrons in the nd and $(n+1)s$ shells for bonding as shown for the elements in Table 5.21, i.e. the Periodic Group number.

Table 5.21 Number of valence electrons associated with the transition metal atoms

	Sc La Ac	Ti Zr Hf	V Nb Ta	Cr Mo W	Mn Tc Re	Fe Ru Os	Co Rh Ir	Ni Pd Pt	Cu Ag Au
$x+y$ in $(n+1)d^x s^y$	3	4	5	6	7	8	9	10	11

2.   The ligands contribute the following numbers of valence electrons when bonded only to a single metal atom:

1-electron donors: H   Cl   CN   SR NR_2 CH_3
                       Br   OR PR_2 C_6H_5
                       I

2-electron donors: CO PR_3 SR_2 CNR N_2 $=CR_2$ C_2R_4
                            NH_3 OR_2
                            py

3-electron donors: NO (linear)   CR   N   NNR (linear)   allyl
                       η-C_3H_5 cyclopropenyl η-C_3R_3 P

hapto-nomenclature

The Greek letter η prefixing a ligand formula indicates that the ligand is π-bonded to the metal. The presence of a superscript, e.g. η^3, η^4, indicates that 3 or 4 of the atoms of the ligand are coordinated in a π manner. Where all the donor atoms of the ligand are coordinated to the metal, the superscript may be omitted

For an authoritative discussion of nomenclature see *Nomenclature of inorganic chemistry: Recommendations* (1990) Ed. G. J. Leigh, official IUPAC publication.

4-electron donors: $\eta\text{-}C_4R_4$ $\eta\text{-}C_4H_6$ ⫤O ⫤NR
trimethylenemethane $C(CH_2)_3$
5-electron donors: $\eta\text{-}C_5R_5$ $\eta\text{-}C_5H_7$

3. Some ligands have ambiguous electron counts depending on the mode of coordination and the strength of π-bonding. The ligands NO and N_2R in linear modes with the metal centre are 3e donors, but when the coordination is non-linear they act as 1e donors as shown in Fig. 5.23.

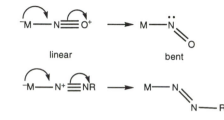

linear   bent

Fig. 5.23 The donor properties of NO and N_2R dependent upon the coordination mode

RN and O groups can act as 4e or 2e donors as indicated by the diagrams below.

4e donors   2e donors

The cyclopentadienyl ligand can be a fully $\eta^5\text{-}C_5H_5$ 5-electron donor or in a 'slipped' structure as a $\eta^3\text{-}C_5H_5$ 3-electron donor as shown by the diagrams below.

$\eta^5\text{-}C_5H_5$ 5-electron donor $\eta^3\text{-}C_5H_5$ 3-electron donor
'slipped' structure

In compounds having bridging ligands and/or metal–metal bonds the following additional considerations apply.

C_2H_2 (ethyne) can act as a 4e donor by using both its filled π-orbitals or as a 2e donor by using only one of them as indicated by the diagrams in the margin.

Ethyne acting as a 4e donor

Ethyne acting as a 2e donor

4. Ligands coordinated to one metal centre may utilize lone pairs to coordinate to other metals as shown in Fig. 5.24.

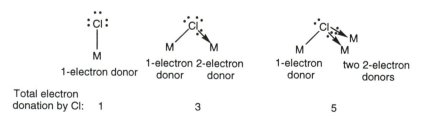

1-electron donor   1-electron 2-electron   1-electron   two 2-electron
donor   donor   donor   donors

Total electron
donation by Cl: 1   3   5

Fig. 5.24 Diagrams showing how a chlorine atom can act as a 1e, 3e, or 5e donor

5. Bridging ligands with no additional lone pairs, e.g. CO, CNR, CR$_2$, are assumed to contribute 1 electron to each metal as indicated in the margin for CO.

6. Metal–metal bonds involve sharing of the metal's electrons, but as far as calculating the total number of valence electrons is concerned the electrons involved in the metal–metal bond from the adjacent metal atom are added to the valence electrons around the first metal centre. Examples are shown in Fig. 5.25.

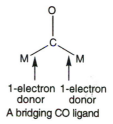

1-electron 1-electron
  donor   donor
A bridging CO ligand

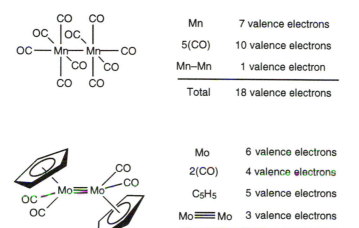

Mn	7 valence electrons
5(CO)	10 valence electrons
Mn–Mn	1 valence electron
Total	18 valence electrons

Mo	6 valence electrons
2(CO)	4 valence electrons
C$_5$H$_5$	5 valence electrons
Mo≡Mo	3 valence electrons
Total	18 valence electrons

Fig. 5.25 Electron counting in Mn$_2$(CO)$_{10}$ and Cp$_2$Mo$_2$(CO)$_4$

d-Electron count and formal oxidation state

The d-electron count of a metal in a complex is the number of electrons associated with the metal in its formal oxidation state. The formal oxidation state of the metal is established as follows.

1. The monatomic ligands are assumed to have the charge which results in their octet being completed, e.g.

$$F^-, O^{2-}, N^{3-} \text{ and } Cl^-, S^{2-}, P^{3-}$$

2. Neutral ligands make no contribution to the charge, e.g.

$$CO, NH_3, C_2R_4, PR_3, CNR \text{ are typical neutral ligands}$$

3. Ligands derived from acids or hydrides bear the charge of the completely deprotonated conjugate base, e.g.

$$NO_3^-, SO_4^{2-}, CO_3^{2-}, PO_4^{3-}$$

$$PR_2^-, SR^- \text{ as the conjugate bases of } PR_2H \text{ and } SRH$$

$$CH_3^-, CH_2Ph^- \text{ as the conjugate bases of } CH_4 \text{ and } H_3CPh$$

4. Cyclic conjugated ligands are assumed to have the charge which is associated with the Hückel $4n + 2$ electrons contributing to their aromatic ring system (where n is the number of rings), e.g. $\eta\text{-}C_5H_5$ as the anion, $C_5H_5^-$, $\eta\text{-}C_6H_6$ as the neutral molecule, C_6H_6, and $\eta\text{-}C_7H_7$ as the cation, $C_7H_7^+$, are all examples of single ring systems with $4n + 2 = 6$ electrons.

Some ambiguities remain, however. For example, ligands with $4n$ electrons e.g. $\eta\text{-}C_4H_4$ and $\eta\text{-}C_8H_8$ may achieve the $4n + 2$ rule criterion either by losing or gaining two electrons resulting in $C_4H_4^{2+}$ or $C_4H_4^{2-}$ and $C_8H_8^{2+}$ or $C_8H_8^{2-}$.

EAN rule

1. The 18 electron (EAN) rule works well only for the d-block metals. It is not applicable to the f-block metals.

2. It works best for compounds with the metals in low oxidation states.

3. Ligands which are good σ-donors and π-acceptors utilize all the valence orbitals in bonding and consequently their compounds obey the rule.

4. Complexes which contain a combination of good σ-donor ligands and π-acceptor ligands conform to the rule, e.g. $Cr(NH_3)_3(CO)_3$ and $Cr(\eta\text{-}C_6H_6)(CO)_3$.

5. The compounds which obey the rule are also kinetically inert towards substitution reactions.

6. Exceptions to the rule occur at the two ends of each of the transition series where the nd, $(n + 1)s$ and $(n + 1)p$ valence orbitals are less well matched in energy.

5. Linear NO is formally considered as NO^+ and bent NO as NO^-. Once the sum of the charges on the ligands has been established the metal formal oxidation state is that which, when added to the total charges associated with the ligands, leads to the observed total charge on the complex. For example, all the complexes shown in Fig. 5.26 have a formal cobalt oxidation state of $+3$ (Co^{III}).

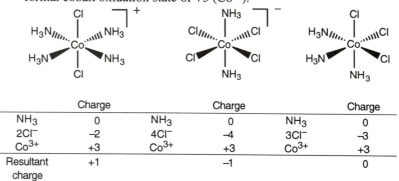

	Charge		Charge		Charge
NH_3	0	NH_3	0	NH_3	0
$2Cl^-$	-2	$4Cl^-$	-4	$3Cl^-$	-3
Co^{3+}	$+3$	Co^{3+}	$+3$	Co^{3+}	$+3$
Resultant charge	$+1$		-1		0

Fig. 5.26 Examples of calculations of resultant charges of some Co^{III} complexes

6. Metal–metal bonds do not contribute to the formal oxidation state of the metal as indicated by the examples shown in Fig. 5.27.

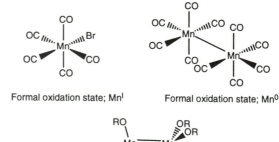

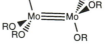

Fig. 5.27 Examples of oxidation state calculations emphasizing the importance of ignoring metal–metal bonds in the calculations

The EAN rule is widely applicable to compounds of transition metals, particularly when they are bonded to π-acid and organometallic ligands. The following section provides some indications of the range of compounds to which the rules may be applied. All the compounds illustrated conform to the EAN rule. For each compound the formal metal oxidation state and the associated d-electron count are also indicated. In Figs. 5.28–5.33 the compounds have been organized to illustrate the following general points.

1. Combinations of rings, CO, NO, H, and PR_3 ligands may be brought together in any combination that leads to a total of 18 valence electrons as shown in Fig. 5.28.

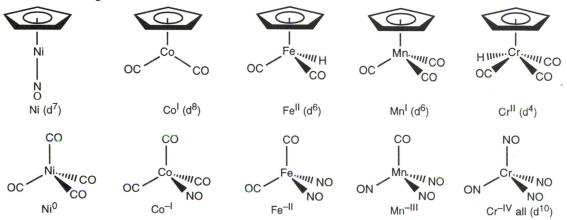

Fig. 5.28 Examples of compounds in which there are totals of 18 valence electrons. The d configurations of the participating metals are indicated

2. The carbocyclic π-bonded ligands may be replaced by hetero-substituted rings, e.g. PC_4H_4, RBC_5H_6, and $(RB)_2C_3H_3^-$ as shown by the examples shown in Fig. 5.29.

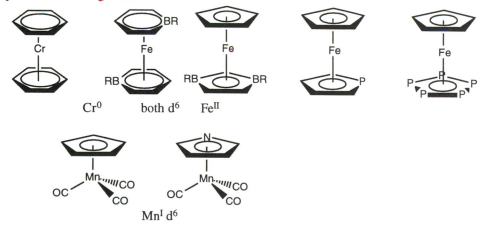

Fig. 5.29 Examples of complexes with hetero-substituted ring ligands

3. Sandwich compounds may have different combinations of π-bonded ligands above and below the metal as shown by the examples of Fig. 5.30.

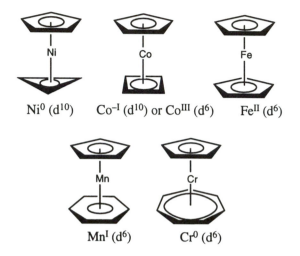

$Ni^0 (d^{10})$ $Co^{-I} (d^{10})$ or $Co^{III} (d^6)$ $Fe^{II} (d^6)$

$Mn^I (d^6)$ $Cr^0 (d^6)$

The above 19 and 20 electron structures provide interesting exceptions to the EAN rule. The carbocyclic ring ligands are good σ-donors and poor π-acceptors and therefore in the absence of other good π-accepting ligands their sandwich compounds do not always conform to the EAN rule.

Fig. 5.30 Examples of sandwich compounds with different π-bonded rings above and below the metal centre

4. Carbocyclic ligands may be replaced with open dienes or other π-systems which contribute the same number of π-electrons. The following compounds illustrate some alternative η^4 4-electron π-donors.

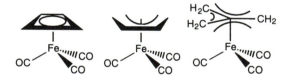

5. The sandwich structures may be transformed into bent structures as in the following examples.

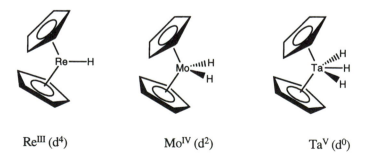

$Re^{III} (d^4)$ $Mo^{IV} (d^2)$ $Ta^V (d^0)$

The hydrides may in some cases be replaced by alternative one-electron donors such as H, CH_3, or Cl.

6. The rule is also widely applicable to compounds containing single and multiple bonds as shown by the examples of Fig. 5.31.

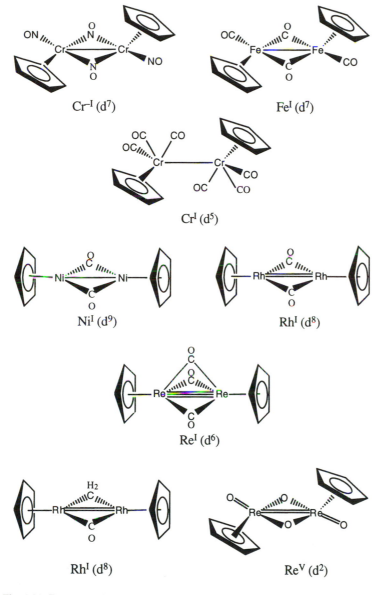

Fig. 5.31 Examples of compounds containing single and multiple metal–metal bonds

7. If the carbocyclic ligand has more carbon atoms than can be accommodated by the EAN rule, only those atoms which result in an 18 electron count bond to the metal. Some examples of cyclooctatetraene, C_8H_8, complexes which illustrate this principle are shown in Fig. 5.32.

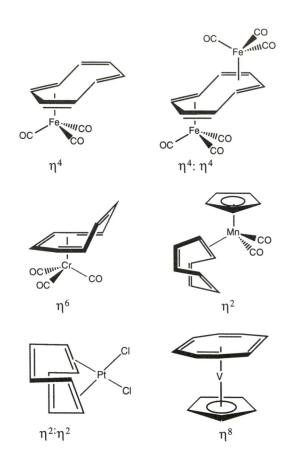

Note that in these compounds the ligand is folded in such a way that it is no longer effectively conjugated with the carbon atoms coordinated to the metal

Fig. 5.32 Examples of cyclooctatetraene complexes

See R. Poli, *Chem. Rev.*, 1996, **96**, 2135 for a discussion of organometallics which do not conform to the EAN rule.

8. The EAN rule is also useful for accounting for the reactions observed for organometallic compounds. A sequence of reactions for $[Fe_2(\eta\text{-}C_5H_5)_2(CO)_4]$ in which all the products conform to the EAN rule is illustrated in Fig. 5.33.

Summary
The EAN rule is hugely important in much transition metal chemistry and especially those compounds with π-acceptor ligands and with metal-metal bonds. The compounds are thermodynamically stable because they are fully utilizing their orbitals in the σ-bonding framework and in back donation to suitable low lying orbitals on the ligands. In addition the resulting large energy gap between the highest occupied molecular orbital and the lowest unoccupied molecular orbital makes these compounds also kinetically inert.

It should be emphasized that although a pair of compounds might be isoelectronic and conform to the EAN rule their chemical properties could be very different. The alternative ligand combinations could result in very different residual charges on the metal which in turn greatly influences the acid–base character and redox properties of the metal. This aspect is discussed in more detail in Section 5.6 (p. 346).

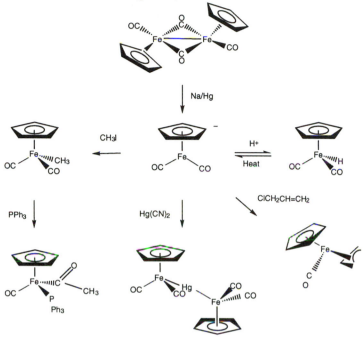

Fig. 5.33 A reaction sequence in which all the products conform to the EAN rule

To illustrate the interplay of oxidation state, d-electron count, and total electron count it is instructive to consider the range of compounds observed for a specific metal, the example chosen being molybdenum.

Molybdenum compounds

Table 5.22 summarizes the important classes of molybdenum compounds as a function of the total valence electron count and the d-electron count. The most noticeable feature of such a plot is the absence of entries in the top left diagonal area. The compounds which would occupy this area would have low oxidation states and valence electron counts of 10–16.

These are absent because the π-acid ligands which are necessary to stabilize the low oxidation states generally show a strong preference to form compounds which conform to the EAN rule. Table 5.22 only documents compounds which are sufficiently thermodynamically stable and kinetically inert to be stored in a bottle or container albeit sometimes under an atmosphere of an inert gas such as dinitrogen. Compounds such as $Mo(CO)_5$, $Mo(CO)_4$, and $Mo(CO)_3$ would be entries suitable for inclusion in these parts of the Table, but they have only been observed and spectroscopically characterized in inert gas matrices at low temperatures and do not have an independent existence as compounds which could be stored in containers in gram quantities.

Table 5.22 Summary of total valence electron counts and d-electron counts for molybdenum compounds. The entries in italics have ambiguous total electron counts

	Total number of valence electrons				
d^n	10	11	12	13	14
d^6					
d^5					
d^4					
d^3			*$Mo_2(OR)_6$* *$Mo_2(NR_2)_6$*		
d^2	*$Mo_2(NMe_2)_4$* $Mo(SBu^t)_4$				$MoCl_6^{2-}$ $MoCl_4(PR_3)_2$ $MoCl_4(OR_2)_2$
d^1		MoF_5 (g)		Mo_2Cl_{10} $[MoF_6]^-$ $MoOCl_3(OR_2)$	
d^0			$MoCl_6$ MoF_6 $Mo(OR)_6$ *MoO_2Cl_2 (g)* MoO_4^{2-}		$[MoOF_4]_4$ $[MoOCl_5]^-$

M. L. H. Green, *J. Organometal. Chem.*, 1995, **500**, 127, has provided a detailed account of the application of valency/EAN count matrices for interpreting the properties of d-block organometallic compounds.

The empty orbitals in these molecules make them strong Lewis acids and therefore highly reactive. The following areas of the plot are of particular interest.

1. Top right; low oxidation state compounds with π-acid ligands which conform to the EAN rule.
2. Bottom right; high oxidation state compounds which conform to the EAN rule. These are generally stabilized by good σ-donor ligands such as H and PR$_3$.
3. Bottom left; high oxidation state complexes with π-donor ligands which have d^0, d^1, and d^2 configurations. Such complexes have rather ambiguous total electron counts.

The bond lengths in oxo, nitrido, and imido-complexes suggest that the bonds which they make to transition metals have considerable multiple bond character and this should be reflected in their valence bond canonical forms, as in the right-hand side of Fig. 5.34. In contrast F, Cl, Br, I, and SR show little multiple bond character and therefore their filled π-orbitals do not make a major contribution to the bonding.

Total number of valence electrons				Oxidation state
15	16	17	18 (EAN rule)	
			$Mo(CO)_6$ $Mo(PR_3)_6$ $Mo(CNR)_6$ $Mo(CO)_3(trien)$	Low oxidation state (0)
			$[Mo(\eta\text{-}C_5H_5)(CO)_3]_2$	1
	$Mo(dppe)_2I_2$ $Mo(CN)_6^{4-}$ $[Mo_2Cl_8]^{4-}$		$Mo(CO)_3(PR_3)_2X_2$ $Mo(CNR)_7^{2+}$ $Mo(CN)_7^{5-}$	2
$Mo(CN)_6^{3-}$ $MoCl_3py_3$ $MoCl_3(thf)_3$ $MoCl_3(PR_3)_3$		$Mo(diars)(CO)_3Br^+$ $Mo(CN)_7^{4-}$	$[Mo_2Cl_9]^{3-}$	Intermediate oxidation state (3)
	$MoCl_4(PR_3)_3$		$[Mo(CN)_8]^{4-}$ $[MoF_8]^{4-}$ $MoH_4(PR_3)_4$	4
$MoO(NCS)_5^{2-}$ $[MoOCl_4(NCMe)]$		$[MoF_8]^{3-}$ $[Mo(CN)_8]^{3-}$		5
	$[MoOCl_4(MeCN)_2]$		$MoH_6(PR_3)_3$	High oxidation state (6)

For example, $Mo(NMe_2)_4$ may be described as a 10 or 18 electron compound depending on the contribution to π-bonds made from the lone pairs on NMe_2 as indicated in Fig. 5.34.

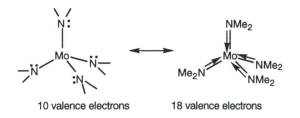

10 valence electrons   18 valence electrons

Fig. 5.34 Alternative descriptions of the compound $Mo(NMe_2)_4$

In complexes where the ligands are effective π-donors the number of filled π orbitals on the ligands frequently exceeds the number of available empty orbitals on the metal. An upper limit to the total number of valence electrons which may be donated from the ligands is set by the availability of orbitals on the central atom. Therefore, the EAN rule is achieved in these complexes by donation from the ligands to the metal. For example, MoO_4^{2-} may be formulated with electron counts between 8 and 18 as shown in Fig. 5.35. The one shown with a total electron count of 24 is not permitted because it would require the participation of twelve valence orbitals on the metal and only nine are available.

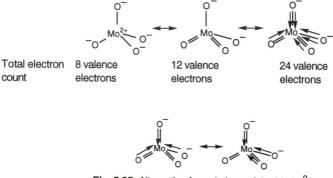

| Total electron count | 8 valence electrons | 12 valence electrons | 24 valence electrons |

Fig. 5.35 Alternative formulations of the MoO_4^{2-} ion

Therefore, in complexes with strong π-donor ligands, which have low d-electron counts, the EAN rule may be achieved by donation from the π−orbitals of the ligands. Therefore, such compounds are not strong Lewis acids and may be isolated as stable entities. Because of the ambiguities associated with the effectiveness of π-donation for such ligands the plot in Table 5.22 has been based on the assumption that an oxo-ligand is a 2-electron donor, OR is a 3-electron donor, and NMe_2 is a four-electron donor. However, the entries have been placed in italics to emphasize the possible ambiguities.

4. Middle diagonal area; medium oxidation state complexes with 13–16 valence electrons. Ligands which are neither good π-acceptors or σ-donors predominate in this region, e.g. Cl, NR_3, OR_2, SR_2, etc.

Finally, the approximately chess board-like appearance of the plot is striking and arises from the following unfavourable combinations which lead to empty boxes: even total electron counts with odd d-electron counts and odd total electron counts with even d-electron counts. These combinations do not occur unless the compound forms a metal-metal bond, e.g. $[Mo(CO)_3(\eta\text{-}C_5H_5)]_2$ in its monomeric form is a 17 electron complex, but with a metal–metal bond in the dimer it comes into the EAN column at the extreme right. Similarly, $W_2Cl_9^{3-}$ achieves an 18 electron configuration with a triple metal-metal bond. Since the driving force for forming such compounds is the attainment of an EAN configuration they do not contribute to other similar positions with 16, 14, etc. total electron configurations. Similar plots may be constructed for other transition elements and they represent a convenient way of bringing together much structural information for one element.

Summary

The EAN rule is a manifestation of a more important unifying principle for transition metal complexes, which states that in stable and inert complexes the nd, $(n + 1)s$, and $(n + 1)p$ valence orbitals of a transition metals are fully utilized in strong bonding interactions. In complexes with π-acceptor ligands this is achieved by donation from the ligand orbitals to empty valence orbitals on the metal and back donation from filled d orbitals on the metal to empty π^* orbitals on the ligands. The total valence electron count of the d-electrons and the electrons donated by the ligands leads to the EAN rule. In complexes with π-donor ligands the full utilization of metal orbitals is achieved by donation from the ligands' σ and π-orbitals, but since the number of filled π orbitals on the ligands may exceed the number of empty orbitals on the metal the total valence electron count in the molecule appears to exceed the EAN rule. In these circumstances it is helpful to emphasize the d-electron count of the metal in its formal oxidation state.

Those ligands which are not capable of interacting strongly with the d orbitals cause only a small splitting of the d manifold and give rise to series of isostoichiometric complexes, which generally do not conform to the EAN rule.

Coordination numbers and geometries

Introduction

The transition elements form an overwhelming variety of neutral, cationic, and anionic coordination compounds with coordination numbers of 2–9 and for a particular coordination number some alternative geometries are observed. The purpose of this section is to introduce some sense of order into this wealth of structural data. Tables 5.23–5.26 give some indication of the complexity of the problem by summarizing some of the coordination geometries of the first row transition elements.

From the outset it is important to note that the most important coordination number and geometry for first-row transition metals is octahedral. With the exception of copper all the metals form some regular octahedral complexes in some of their oxidation states. Copper(II) does form many 6-coordinate complexes but they are not regular octahedra. In general they display tetragonally distorted structures of the type shown in the margin.

Regular
octahedron

Tetragonally distorted
octahedron

4-coordinate
tetrahedron

5-coordinate
trigonal bipyramid

6-coordinate
octahedron

The octahedral geometry is also observed commonly in infinite solids, e.g. MO (M = V, Mn, Fe, Co, and Ni) have sodium chloride (6:6) structures, MO_2(M = Ti, V, Cr, and Mn) have rutile (6:3) structures. The second most commonly observed coordination polyhedron is the tetrahedron. It is observed in tetrahalide complexes, e.g. $[FeCl_4]^-$, $[CoCl_4]^{2-}$, $TiCl_4$, and VCl_4; low oxidation state carbonyls and nitrosyls, e.g. $Ni(CO_4)$, $Co(CO)_3(NO)$, $Fe(CO)_2(NO)_2$, and high oxidation state nitrido- and oxocomplexes, e.g. $[MnO_4]^-$, $[CrO_4]^{2-}$, and $[OsO_3N]^-$.

Table 5.23 Coordination numbers and geometries of compounds of Ti and V

Ti	V
8-dodecahedral in $TiCl_4(diars)_2$	8-dodecahedral in $VCl_4(diars)_2$
7-pentagonal bipyramidal in $TiCl(S_2CNMe_2)_3$	7-pentagonal bipyramidal in $[V(CN)_7]^{2-}$
6-octahedral in TiO_2 (infinite)	6-octahedral in $VCl_4(py)_2$
6-octahedral in $TiCl_4(PEt_3)_2$	6-octahedral in $V(CO)_6$
$TiCl_3(NCCH_3)_3$	6-trigonal prismatic in $V(S_2C_2R_2)_3$
	5-square pyramidal in $VO(acac)_2$
5-trigonal bipyramidal in $[TiCl_5]^-$	5-trigonal bipyramidal in VF_5
4-tetrahedral in $TiCl_4$	4-tetrahedral in $[VO_4]^{3-}$
3-trigonal planar in $Ti(N(SiMe_3)_2)_3$	3-trigonal planar in $V(N(SiMe_3)_2)_3$

Generalizations

The earlier transition metals are more likely to provide examples of complexes with coordination numbers greater than six, particularly if the ligands are small. Table 5.23 gives examples of 7- and 8-coordinate transition metal complexes of Ti and V. Some high coordinate complexes of the later transition metals may be isolated if the ligands have small bite angles, e.g. NO_3^- in $[Co(NO_3)_4]^{2-}$. There are few examples of 9-coordinate complexes of the first row transition elements, but the geometry is more commonly observed for the second and third row transition metals. The decreasing size of the metal ions across the transition series is the major reason for this trend. For M^{3+} the ionic radius contracts from 76 pm for Ti^{3+} to 62 pm for Ni^{3+}.

Table 5.24 Coordination numbers and geometries of compounds of Cr and Mn

Cr	Mn
8-dodecahedral in $CrH_4(dmpe)$	8-dodecahedral in $[Mn(NO_3)_4]^-$
7-capped trigonal prismatic in $Cr(CO)_2(diars)_2X$	7- in $MnH_3(dmpe)_2$ (geometry not defined because H atoms are not located by X-rays)
6-octahedral in $Cr(CO)_6$	
6-octahedral in $[Cr(OH_2)_6]^{3+}$	6-octahedral in $[Mn(OH_2)_6]^{2+}$
5-square pyramidal in $[CrOCl_4]^-$	5-square pyramidal in $[MnCl_5]^{2-}$
5-trigonal bipyramidal in $CrCl_3(NMe_3)_2$	5-trigonal bipyramidal in $MnI_3(PMe_3)_2$
4-tetrahedral in $[CrO_4]^{2-}$	4-tetrahedral in $[MnO_4]^-$
4-tetrahedral in $[CrO_4]^{3-}$	4-tetrahedral in $[MnO_4]^{2-}$
4-tetrahedral in $Cr(NEt_3)_4$	
3-trigonal planar in $Cr(N(SiMe_3)_3)_3$	

The observed coordination geometries for 9, 8, and 7-coordination are those which minimize the electrostatic repulsion between the ligands, i.e. those polyhedra which on average place the ligands as far apart as possible on the surface of a sphere. The resultant polyhedron is not necessarily the most highly symmetric. For example, for 8-coordination the high-symmetry cube is rarely observed in coordination compounds. This geometry gives rise to greater ligand–ligand repulsions than the dodecahedron and the square antiprism, which have lower symmetries.

Table 5.25 Coordination numbers and geometries of compounds of Fe and Co

Fe	Co
8-dodecahedral in $[Fe(NO_3)_4]^-$	8-dodecahedral in $[Co(NO_3)_4]^{2-}$
7-pentagonal bipyramidal in $[Fe(edta)H_2O]^-$	
6-octahedral in $[Fe(OH_2)_6]^{3+}$	6-octahedral in $[Co(NH_3)_6]^{3+}$
6-octahedral in $[Fe(CN)_6]^{4-}$	6-octahedral in $CoCl_2$ infinite
5-trigonal bipyramidal in $Fe(CO)_5$	5-trigonal bipyramidal in $[Co(NCMe)_5]^+$
5-square pyramidal in $FeCl(S_2CNR_2)_2$	5-square pyramidal in $[Co(NCPh)_5]^+$
4-tetrahedral in $[FeO_4]^{2-}$	4-tetrahedral in $[CoCl_4]^{2-}$
4-tetrahedral in $[FeCl_4]^-$	4-tetrahedral in $[Co(CO)_4]^-$
4-tetrahedral in $Fe(NO)_2(CO)_2$	
	4-square planar in $[Co(CN)_4]^{2-}$
3-trigonal planar in $Fe(N(SiMe_3)_2)_3$	3-trigonal planar in $Co(N(SiMe_3)_2)(OBu^t)_2$
	2-linear in $Co(N(SiMe_3)_2)_2$

Table 5.26 Coordination numbers and geometries of compounds of the elements Ni and Cu

Ni	Cu
6-octahedral $[Ni(NH_3)_6]^{2+}$	6-octahedral – tetragonal distorted in $[Cu(OH_2)_6]^{2+}$
5-square pyramidal/trigonal bipyramidal in $[Ni(CN)_5]^{3-}$	5-trigonal bipyramidal in $[CuCl_5]^{3-}$
4-tetrahedral in $[NiCl_4]^{2-}$	4-flattened tetrahedral (D_{2d}) in $[CuCl_4]^{2-}$
4-tetrahedral in $Ni(CO)_4$	4-tetrahedral in $[Cu(CN)_4]^{3-}$
4-square planar in $[Ni(CN)_4]^{2-}$	4-square planar in $[Cu(py)_4]^{2+}$
3-trigonal planar in $[Ni(NPh_2)_3]^-$	3-trigonal planar in $[CuCl_3]^-$
	2-linear in $[CuCl_2]^-$

The minimum repulsion polyhedra for 7- to 9-coordination are illustrated in Fig. 5.36. The energy differences between the alternative geometries for 5, 7, 8, and 9 coordination are generally small and this leads to two important observations.

The observed solid state structure need not be one of the idealized geometries illustrated in Fig. 5.36, but may adopt an intermediate structure. Indeed, the energy differences between the structures may be so small that different geometries are observed according to the counter ion used to crystallize the coordination complex. Secondly, the compounds are generally 'fluxional' in solution. They do not maintain the structure observed in the solid state, because the transfer of energy to the molecule by collisions in solution is sufficient to alter the structure of the complex, e.g. from pentagonal bipyramid to capped octahedron. Therefore, these molecules are highly flexible in solution and attempts to observe them using for example NMR spectroscopy results in the observation of an average structure at room temperature.

A statistical analysis of the coordination geometries of the transition metals has been reported, D. Venkataraman *et al.*, *J. Chem. Educ.*, 1997, **74**, 915.

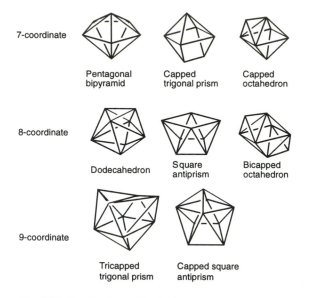

Fig. 5.36 Coordination polyhedra for 7, 8, and 9-coordination

Since the preferred coordination number and geometry for transition metals is 6-octahedral, in order to achieve a higher coordination number it is necessary to try to build in the following factors.

1. The ligand should be small and capable of interacting strongly with the metal ion via strong electrostatic interactions or covalent bonds. The ligands F^-, CN^-, and H^- have such properties.

2. If the ligand is bidentate with a small bite angle it is particularly effective at stabilizing complexes with high coordination numbers. These ligands may be neutral or anionic, e.g. NO_3^- and SO_4^{2-}, diars and dmpe.

3. A polydentate ligand which has 6–9 donation sites can wrap around the metal ion and thereby achieve a high coordination number. Low coordination numbers are achieved by using large sterically crowded ligands. For example, PPh_3 and $N(SiMe_3)_2$ are particularly effective at stabilizing coordination numbers of 3 and 4.

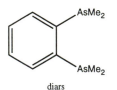

diars

dmpe

d-Orbital splittings in transition metal complexes

Molecular orbital analysis

In Section 4.6 it is noted that the ligands at the vertices of the spherical coordination polyhedra (ML_m) illustrated in Fig. 5.36 have linear combinations of orbitals which match the s, p, and $(m - 4)$d orbitals on the central atom. Fig. 5.37 illustrates how these linear combinations interact with the valence orbitals of a transition metal. For a transition metal the relative orbital stabilities are $nd > (n+1)s \gg (n+1)p$. The overlaps between the ligand linear combinations and the $(n+1)s$ and $(n+1)p$ metal orbitals are large and therefore resulting in bonding and antibonding molecular orbitals which are separated by large energy gaps. The nd orbitals overlap less effectively with the symmetry matching linear combinations of ligand orbitals and give rise to a smaller energy separation between bonding and antibonding molecular orbitals. The resultant frontier molecular orbitals for a transition metal complex are localized mainly on the metal and are shown in Fig. 5.37. In complexes which conform to the Effective Atomic Number rule (EAN rule) there is a complementarity between the ligand combinations and the non-bonding and weakly antibonding orbitals localized on the metal. Taken together they make a complete set of functions which emulate those of an inert gas. Specifically a spherical ML_m complex has $m\sigma$ bonding molecular orbitals and $(9 - m)$ non-bonding and weakly antibonding orbitals localized on the central atom which have the appropriate symmetry properties to form π-bonds with the ligands.

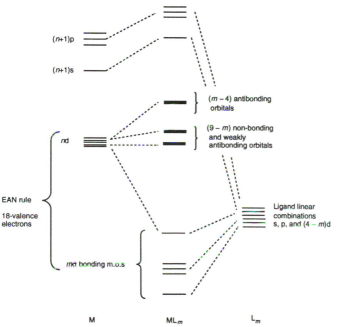

For a more detailed discussion of these complementary orbital interactions see D. M. P. Mingos and J. C. Hawes, *Struct. Bond.*, 1985, **63**, 1, and for a more detailed analysis of the molecular orbital treatment see Y. Jean, F. Volatron, and J. K. Burdett, *An Introduction to Molecular Orbitals*, OUP, Oxford, 1993.

Fig. 5.37 Generalized molecular orbital diagram for transition metal complexes

Fig. 5.38 shows a more specific molecular orbital diagram for octahedral metal complexes. In an octahedral complex the metal $(n+1)$s and $(n+1)$p orbitals form a set four strongly bonding and antibonding molecular orbitals by overlapping with the symmetry matching ligand linear combinations (a_{1g} and t_{1u}). The metal $d_{x^2-y^2}$ and d_{z^2} orbitals overlap less effectively with the e_g combination of ligand orbitals to give bonding e_g and antibonding e_g* molecular orbitals. The former are localized mainly on the ligand and the latter mainly on the metal. The metal d_{xz}, d_{yz}, and d_{xy} orbitals form a non-bonding triply degenerate set of t_{2g} symmetry. An octahedral complex which conforms to the EAN rule has the electronic configuration: $(a_{1g})^2(t_{1u})^6(e_g)^4(t_{2g})^6$. This configuration is stabilized if the non-bonding t_{2g} orbitals are stabilized. If the ligands are π-acceptor ligands then this may be achieved by the back donation effects described previously.

In general the ligand donor orbitals are associated with atoms with a higher electronegativity than the metals and consequently the ligand linear combinations are more stable than the metal d orbitals (as shown in Figs. 5.37–5.39). Therefore, these molecular orbital diagrams imply that a ML_m complex has m molecular orbitals which lie below the d orbitals and are localized predominantly on the ligands and the number of electrons which populate the d manifold correspond precisely to those associated with the metal in its formal oxidation state.

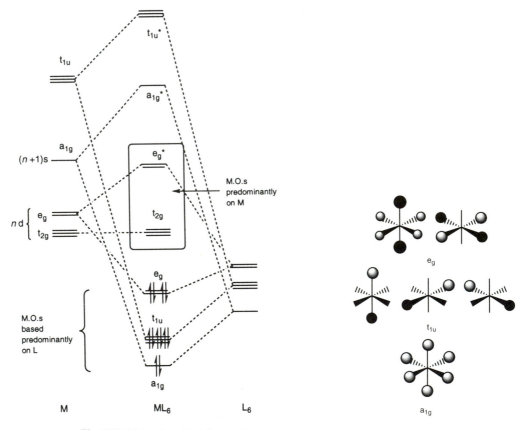

Fig. 5.38 Molecular orbital diagram for an octahedral complex

For example, octahedral $[Cr(NH_3)_6]^{3+}$ has 12 electrons occupying the $(a_{1g})^2(t_{1u})^6(e_g)^4$ set of orbitals and three electrons in the lower t_{2g} component of the d manifold. This corresponds to the d^3 electron configuration for Cr^{III} in this complex. Therefore, the molecular orbital diagram defines the closed-shell requirements for a complex to conform to the EAN rule and connects the number of electrons populating the d manifold to the d-electron count of the metal ion in its formal oxidation state.

Fig. 5.39 shows a molecular orbital diagram for a tetrahedral metal complex. The orbital complementarity is: a_1 (s), t_2 (p), e ($d_{x^2-y^2}$ and d_{z^2}), and t_2 (d_{xy}, d_{xz}, d_{yz}). These orbitals, if populated, taken together generate the complete set of valence orbitals of a noble gas atom.

In a tetrahedral complex the ligand orbitals transform as a_1 and t_2. The a_1 ligand combination forms a strongly bonding and antibonding pair of molecular orbitals—a_1 and $a_1{}^*$ when it overlaps with the metal $(n+1)s$ orbital. The t_2 ligand combination has the correct symmetry to overlap with both the (d_{xz}, d_{yz}, d_{xy}) and (p_x, p_y, p_z) sets of metal orbitals. The greater overlap with the metal $(n+1)p$ orbitals results in strongly bonding t_2 and strongly antibonding $2t_2{}^*$ molecular orbitals and the smaller overlap with the metal nd orbitals yields a weakly antibonding $1t_2{}^*$ set of molecular orbitals localized mainly on the metals and forming part of the d manifold. The metal $d_{x^2-y^2}$ and d_{z^2} orbitals, which have e symmetry, remain non-bonding.

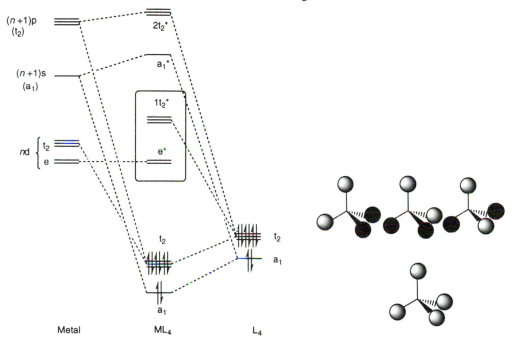

Fig. 5.39 Molecular orbital diagram for a tetrahedral complex

A tetrahedral complex which satisfies the EAN rule, e.g. $Ni(CO)_4$, is associated with the following filled molecular orbitals $(a_1)^2(t_2)^6(e)^4(1t_2{}^*)^6$ and corresponds to the filling of all the bonding and non-bonding orbitals and the set of weakly antibonding $1t_2{}^*$ molecular orbitals. The way in which the antibonding character of $1t_2{}^*$ is reduced by d–p orbital mixings is discussed further on page 320.

It has been noted above that the nd valence orbitals of transition metals do not overlap strongly with ligand orbitals unless the internuclear distances are very short. Consequently the energy gap between the non-bonding and antibonding d orbitals is relatively small unless π-acceptor ligands are present which are capable of overlapping strongly with the d orbitals. Therefore, the frontier orbitals of transition metal complexes which are indicated by the boxes in Figs. 5.38 and 5.39 are localized mainly on the metal. For complexes with small d orbital splittings, isostoichiometric and isostructural compounds are observed. The chemical and structural consequences of these orbital splittings are particularly important in octahedral complexes and are discussed in more detail in a subsequent section. The d orbital splitting diagrams for other coordination polyhedra are illustrated on pages 318–322.

Valence bond analysis

An alternative description of the bonding in transition metal complexes may be derived using the hybridization schemes developed from valence bond theory. A transition metal atom has nd, $(n + 1)s$, and $(n + 1)p$ valence orbitals and the metal–ligand bonds in a ML_m complex may be described in terms of $d^x sp^3$ hybrid orbitals ($x = m - 4$). The relevant hybrids in ML_9–ML_6 complexes are summarized in Table 5.27.

See also R. B. King, *J. Chem. Educ.*, 1996, **73**, 993 and D. M. P. Mingos and Lin Zhenyang, *Struct. Bond.*, 1990, **72**, 73 for a more detailed discussion of hybridization schemes.

Table 5.27 Hybridization schemes for high coordination numbers

Coordination number	Structure	Hybridization	Non-bonding orbitals
9	Tricapped trigonal prism (TTP)	$d^5 sp^3$	none
8	Dodecahedron (DD)	$d^4 sp^3$	$d_{x^2-y^2}$
8	Square antiprism (SA)	$d^4 sp^3$	d_{z^2}
8	Bicapped trigonal prism (BTP)	$d^4 sp^3$	d_{z^2}
7	Pentagonal bipyramid (PB)	$d^3 sp^3$	(d_{xz}, d_{yz})
7	Capped trigonal prism (CTP)	$d^3 sp^3$	d_{yz},
7	Capped octahedron (CO)	$d^3 sp^3$	$d_{x^2-y^2}, d_{xy}$
6	Octahedron (OCT)	$sp^3 d^2$	(d_{xz}, d_{yz}, d_{xy})
6	Trigonal Prism (TP)	$sp^3 d^2$	$d_{z^2}\ (d_{xy}, d_{x^2-y^2})$

Degenerate orbitals are enclosed in brackets

Table 5.28 Examples of 9-coordinate complexes

Complex	Structure
$[Ho(H_2O)_9](EtSO_4)_3$	TTP
$[Pr(H_2O)_9](BrO_3)_3$	TTP
$WH_6(MePh)_3$	TTP
$[ReH_9]^{2-}$	TTP
$[TcH_9]^{2-}$	TTP
$Tl(NO_3)_3(H_2O)_3$	CSAP

In a nine-coordinate transition metal complex all valence orbitals are used to generate $d^5 sp^3$ hybrids which may be utilized in forming metal–ligand bonds as shown in Fig. 5.40 and, therefore, the EAN rule will be obeyed only for those metal ions with d^0 configurations. Examples of such complexes are summarized in Table 5.28. Such complexes usually have tricapped trigonal prismatic geometries (TTP) although at times capped square antiprismatic (CSAP) geometries are observed.

For lower coordination numbers not all of the d orbitals are used in generating the appropriate hybrids which point directly at the ligands and therefore some of the d orbitals remain non-bonding.

The non-bonding d orbitals and the orbitals used in metal–ligand σ-orbitals make a complementary set which when taken together represent an electronic configuration which corresponds to that of a 'pseudo' inert gas. For example, in an ML_8 dodecahedral (DD) complex the metal d_{xy}, d_{xz}, d_{yz}, and d_{z^2} orbitals are hybridized with the s and p orbitals to form a set of localized d^4sp^3 hybrid orbitals which point towards the ligands. The $d_{x^2-y^2}$ orbital is non-bonding and is not involved in σ-bonding as shown in Fig. 5.41.

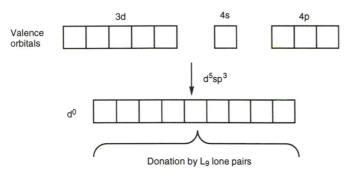

Fig. 5.40 The formation of d^5sp^3 hybrid orbitals in ML_9 complexes

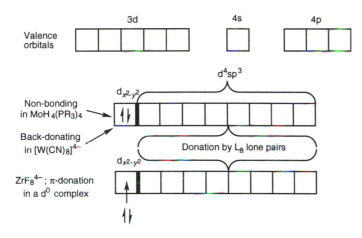

Fig. 5.41 The formation of d^4sp^3 hybrid orbitals in ML_8 complexes

In a complex which obeys the Effective Atomic Number rule the d^4sp^3 hybrids will be occupied by electron pairs donated by the ligands (8 electron pairs) and the ninth electron pair occupies the $d_{x^2-y^2}$ metal-based orbital. Since this orbital is non-bonding it can be depopulated without disturbing significantly the total bonding energy. Therefore, it is not uncommon to find examples of d^0, d^1, and d^2 dodecahedral ML_8 complexes, and some relevant examples are given in Table 5.29. Although the $d_{x^2-y^2}$ orbital is non-bonding with respect to the σ-framework it is capable of entering into π-interactions with the ligands. The orbital interactions illustrated in Fig. 5.42 suggest that π-acceptor ligands favour a d^2 configuration and π-donor ligands a d^0 configuration.

The square antiprismatic (SA) and bicapped trigonal prismatic (BTP) geometries are also based on d^4sp^3 hybrids for σ-bonding but use the d_{xy}, d_{xz}, d_{yz}, and $d_{x^2-y^2}$ orbitals leaving d_{z^2} non-bonding. Consequently these

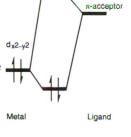

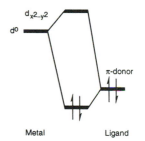

Fig. 5.42 d-Orbital interactions with π-acceptor and π-donor ligand orbitals

geometries are also commonly observed for d^0, d^1, and d^2 configurations. See Table 5.29 for some specific examples.

The relatively small overlaps between the d orbitals and the ligand orbitals means that the antibonding metal–ligand molecular orbitals derived from the metal d orbitals are lower lying than those associated with the s and p orbitals, Therefore the d orbitals split into a set of non-bonding and weakly antibonding molecular orbitals. These d based orbital energies are illustrated in the margins on pages 318–322.

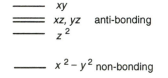

xy
xz, yz   anti-bonding
z^2

$x^2 - y^2$ non-bonding

Dodecahedron

Table 5.29 Examples of 8-coordinate complexes (see Table 5.27 for structural abbreviations)

Complex	d^n	Structure
$La(acac)_3.3H_2O$	0	BTP
$[ZrF_8]^{4-}$	0	BTP
$K_3Cr(O_2)_4$	1	DD
$NbCl_4(diars)_2$	1	DD
$H_4Mo(PPh_3)_4$	2	DD
$H_4Mo(PPhMe_2)_4$	2	DD
$H_4W(CN)_8.6H_2O$	2	SAP
$K_4W(CN)_8.2H_2O$	2	DD

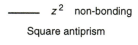

xz, yz
xy, $x^2 - y^2$

z^2 non-bonding

Square antiprism

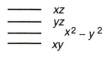

xz
yz
$x^2 - y^2$
xy

z^2 non-bonding

Bicapped trigonal prism

In seven coordinate pentagonal bipyramidal (PB), capped octahedral (CO), and capped trigonal prismatic (CTP) complexes a similar pattern emerges with d^3sp^3 hybrids forming the σ-bonded framework and the complementary pair of d orbitals remaining non-bonding, i.e. d_{xz} and d_{yz} in pentagonal bipyramidal; d_{yz} and $d_{x^2-y^2}$ in capped trigonal prismatic and d_{xy} and $d_{x^2-y^2}$ in the monocapped octahedral case. Examples of such complexes with $d^0 \rightarrow d^4$ configurations are commonplace and some typical examples are given in Table 5.30. Once again π-acceptor ligands which stabilize the two d orbitals not involved in σ-bonding will in general favour the formation of complexes where the metal ion has a d^4 EAN configuration and π-donor ligands favour a d^0 configuration. The d orbital splitting diagrams for 7-coordinate complexes are shown in the margin on page 319.

Table 5.30 Examples of 7-coordinate complexes (See Table 5.27 for structural abbreviations)

Complex	d^n	Structure
$[NbF_7]^{2-}$	0	CTP
$[ZrF_7]^{3-}$	0	PB
$MoBr_4(PMe_2Ph)_3$	2	CO
$[V(CN)_7]^{4-}$	2	PB
$H_4Os(PMe_2Ph)_3$	4	PB
$[Mo(CN^tBu)_7]^{2+}$	4	CTP
$W(CO)_4Br_3$	4	CO

The non-bonding orbitals and orbital splittings for octahedral and trigonal prismatic complexes are also illustrated in the margin. The presence of π-acceptor ligands stabilizes these orbitals and when filled with 6 electrons an EAN configuration is achieved. In both cases there are 3 non-bonding orbitals. Those complexes which conform to the EAN rule invariably have octahedral geometries. The trigonal prismatic geometry is observed more commonly for d^0–d^2 complexes with sulfur ligands or in organometallic complexes such as WMe_6. Specific examples of octahedral and trigonal prismatic complexes are given in Table 5.31.

Table 5.31 Examples of 6-coordinate complexes
(See Table 5.27 for structural abbreviations)

Complex	d^n	Structure
$Mo(S_2C_2H_2)_3$	0	TP
$(Ph_4As)[Nb(S_2C_6H_4)_3]$	0	TP
$[TiF_6]^{2-}$	0	distorted OCT
$Re(S_2C_2Ph_2)_3$	1	TP
MoS_2	2	TP
OsF_6	2	OCT
$[Cr(NH_3)_6]^{3+}$	3	OCT
IrF_6	3	OCT
$Mo(CO)_2(S_2CNPr^i_2)_2$	4	distorted TP
PtF_6	4	OCT
$[Mn(H_2O)Cl_5]^{3-}$	5	OCT
$Cr(CO)_6$	6	OCT

For 5- and 4-coordinate complexes the corresponding hybrid orbitals and complementary non-bonding or weakly antibonding d orbitals are less clearly defined (see Table 5.32).

Table 5.32 Hybridization schemes in 4- and 5-coordinate complexes

	Hybridization	Non-bonding or weakly antibonding complementary orbitals
Trigonal bipyramid	sp^3d or spd^3	$(d_{xz}, d_{yz})(d_{x^2-y^2}, d_{xy})$
Square pyramid	sp^3d or spd^3	$(d_{xz}, d_{yz}) d_{xy}, d_{z^2}$
Tetrahedron	sp^3 or sd^3	$(d_{z^2}, d_{x^2-y^2}) (d_{xy}, d_{xz}, d_{yz})$

Some of the d orbitals are strictly non-bonding in the sense described above. Others point directly at the ligands and have the same symmetries as the higher energy p orbitals. For example, in a trigonal bipyramidal ML_5 complex the hybridization may equally well described as either d^3sp or dsp^3. The d_{z^2}, s, and p_z orbitals contribute equally to both hybridization possibilities, but the d_{xy} and $d_{x^2-y^2}$ orbitals have the same symmetry as the p_x and p_y orbitals and their relative contributions depends on their relative energies and their overlaps with the ligand orbitals. The d_{xz} and d_{yz} orbitals remain non-bonding by symmetry in both hybridization possibilities. The relevant 'orbital box diagrams' for the alternative hybridization schemes are illustrated in Fig. 5.43.

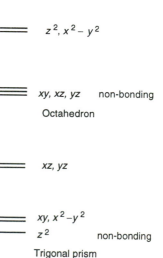

The dsp^3 hybridization is favoured for complexes with high d electron counts and d^{3}sp for low electron counts. In complexes where there is effective overlap between the metal 3d and ligand orbitals, e.g. in π–acid complexes, there will be significant mixing between the d and p orbitals and neither description is satisfactory. The relevant d–p orbital mixings are illustrated in Figs. 5.44 and 5.45. In these situations it is perhaps more realistic to describe the σ-bonding in terms of s(d/p)^{2}dp hybrids as shown in Fig. 5.43 and with two (d/p) hybrids remaining non-bonding and occupied by electron pairs from the metal.

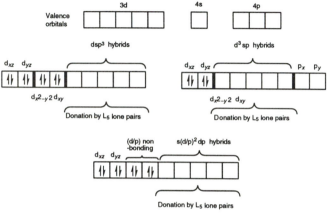

Fig. 5.43 Alternative hybridization schemes in trigonal bipyramidal complexes

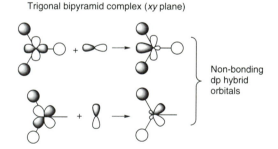

Fig. 5.44 d–p orbital mixings in a trigonal bipyramidal complex

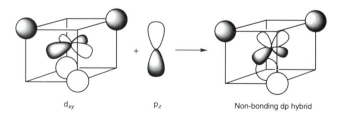

Fig. 5.45 The d$_{xy}$ and p$_z$ atomic orbital mixing is relevant to a tetrahedral complex. Similar hybrids are formed by the d$_{xz}$–p$_y$ and d$_{yz}$–p$_x$ combinations

A d^8 metal ion contributes just the right number of electrons to fill the non-bonding d_{xz} and d_{yz} orbitals and the two of the non-bonding (d/p) hybrids. The resulting complex conforms to the EAN rule since all nine valence orbitals are occupied by electron pairs originating either from the metal or the ligands. Examples of five-coordinate complexes are given in Table 5.33.

Table 5.33 Examples of 5-coordinate complexes

Complex	d^n	Structure
$[RuCl_2(PPh_3)_3]$	6	SP
$[Co(CNPh)_5]^+$	8	SP
$Fe(CO)_5$	8	TBP
$[Ni(CN)_5]^{3-*}$	8	SP/distorted TBP

* $[Cr(en)_3][Ni(CN)_5] \cdot 1.5 H_2O$ has two crystallographically distinct $[Ni(CN)_5]^{3-}$ ions

Similar orbital mixings and ambiguous hybridization possibilities are possible for tetrahedral complexes (see Table 5.32), because the p_x, p_y, and p_z orbitals have the same symmetry transformation properties as d_{xy}, d_{xz}, and d_{yz}. The relevant interaction diagrams are illustrated in Fig. 5.46. The d^3s hybridization scheme is more appropriate for complexes such as MnO_4^- with low d electron counts and sp^3 hybridization for complexes such as $Ni(CO)_4$ with high electron counts. When the d–p mixing is extensive then the hybrids are more accurately described as $s(d/p)^3$ and three non-bonding (d/p) hybrids are also generated by the orbital mixings illustrated in Fig. 5.45. A tetrahedral complex which conforms to the EAN rule requires a central metal atom or ion with a d^{10} configuration. These electrons occupy the non-bonding d_{z^2} and $d_{x^2-y^2}$ orbitals and the non-bonding (d/p) hybrids.

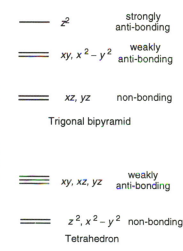

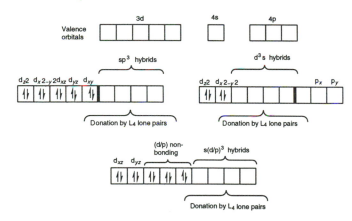

Fig. 5.46 Alternative hybridization schemes in tetrahedral complexes.

It is also important to note that the d orbital splittings in tetrahedral complexes shown in the margin are complementary to those in octahedral complexes. In the latter the $(d_{z^2}, d_{x^2-y^2})$ pair of orbitals lie above the degenerate (d_{xy}, d_{xz}, d_{yz}) set, whereas in the former the $(d_{z^2}, d_{x^2-y^2})$ pair lie below the (d_{xy}, d_{xz}, d_{yz}) set.

In the examples discussed above the presence of π-acceptor ligands which stabilize the complementary d orbitals lead to a preference for full occupation of these d orbitals and an 18 electron EAN rule configuration. For such complexes the necessary d–p orbital mixings are only achieved if the d–p promotion energy is not too large. At the end of the transition series and particularly for the second and third row elements the large d–p promotion energies lead to many complexes which have 16 or even 14 valence electrons and which do not obey the EAN rule. Square planar dsp^2, trigonal sp^2, and linear sp complexes are preferred in such situations because they reduce the p orbital contributions to the hybrids. Examples of such complexes are given in Table 5.34.

For a rare example of five-coordinate Au^{III} see the structure of $[AuCl_3(PMe_3)_2]$ reported by C. A.. McAuliffe *et al.*, *Angew. Chem. Int. Ed.*, 1996, **35**, 2342.

Table 5.34 Examples of square planar, trigonal, and linear transition metal complexes

Geometry	Total number of electrons	d^n	Examples
Square planar	16	d^8	$[PtCl_4]^{2-}$
			$[RhCl(CO)(PPh_3)_2]$
			$[AuCl_4]^-$
Trigonal	14	d^{10}	$AuCl(PPh_3)_2$
Linear	12	d^{10}	$[Ag(NH_3)_2]^+$

From right to left across the Periodic Table, for an isoelectronic series of compounds the d–p promotion energy of the metal becomes smaller and therefore the occurrence of 16 and 14 electron complexes diminishes. For example, for the isoelectronic d^8 ions Os^0, Ir^I, Pt^{II}, and Au^{III} the square planar ML_4 geometry is never observed in Os^0 compounds and the trigonal bipyramidal ML_5 complexes which conform to the EAN rule is preferred. Ir^I forms both square planar and trigonal bipyramidal complexes, Pt^{II} and Au^{III} rarely form trigonal bipyramidal complexes. Similarly, the preference for linear geometries in d^{10} ions decrease in the series: $Hg^{II} > Au^I > Pt^0$.

Square-planar 16 electron complexes are also observed for the earlier transition metals if the metal is associated with strong π-donor ligands, e.g. $[Os(NR)_2(PR_3)_2]$, $(Os(IV), d^4)$ and $W(OR)_4$ $(W(IV), d^2)$. The metal electronic configuration is reduced in these complexes to compensate for additional electrons donated by the ligands. Tetrahedral 16 electron complexes of the early transition metals occur with ligands which are less effective π-donors and are more sterically demanding, e.g. $Ti(\eta\text{-}C_5H_5)_2Cl_2$ (in this example the centroids of the cyclopentadienyl rings are viewed as two of the coordination sites of the tetrahedron).

The d orbital splitting diagram for a square planar complex is shown in Fig. 5.47. The d_{xz}, d_{yz}, and d_{xy} orbitals retain the non-bonding character associated with the octahedron, because they do not overlap with the ligand orbitals. The $d_{x^2-y^2}$ orbital overlaps with the ligands in the square plane in an identical manner to the octahedron, but the d_{z^2} orbital is stabilized because of the loss overlap with the two ligands removed from the z axis to convert the

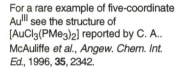

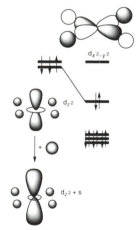

Fig. 5.47 d-Orbital splitting diagram for a square planar complex

octahedral complex into a square planar complex. Therefore, the square planar complex is associated with three non-bonding d orbitals and a weakly antibonding d_{z^2} orbital. The antibonding character of the latter is reduced further by mixing with the s orbital and consequently the square planar geometry is ideal for accommodating eight d electrons in a low spin configuration.

For d^8 complexes, the relative stabilities of square planar and tetrahedral geometries depends on a subtle mixture of steric and electronic effects. Steric effects favour the tetrahedral geometry which places the ligands farther apart, whereas electronic effects favour the square planar geometry. Therefore, as the splitting between the $d_{x^2-y^2}$ and the d_{z^2} orbitals becomes larger and the size of the central metal increases the square planar geometry becomes favoured. For example, in the nickel, palladium, and platinum group. Nickel forms both square planar and tetrahedral complexes, but palladium and platinum form exclusively square planar d^8 complexes. These metals have larger radii and therefore the steric repulsions between the ligands are reduced and also their d orbitals overlap more strongly with the ligands and consequently the separation between $d_{x^2-y^2}$ and the d_{z^2} increases.

Metals which show a preference for forming 16-electron complexes also form metal–metal bonded compounds, and the resulting compounds may be rationalized using the total valence electron count if it is recognized that each metal attains a 16 rather than 18 electron configuration. Examples of such complexes are illustrated in the margin.

An interesting possibility for isomerism can arise for those metals which lie at the 18/16 electron borderline. In the series of binuclear rhodium(I) compounds with phosphido-bridges shown in Fig. 5.48, the metals may come together with square planar geometries and not form a metal–metal bond because each metal atom attains a 16 electron configuration, or alternatively a double bond between the metal atoms is generated if the metals achieve an 18 electron configuration. Between these extremes is the intermediate case where one of the metals achieves an 18 electron configuration and the second one a 16 electron configuration resulting in a formal metal–metal bond order of one.

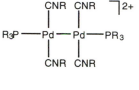

Pd^I – Pd^I

d^9 – d^9

32 valence electrons

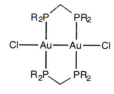

Au^{II} – Au^{II}

d^9 – d^9

32 valence electrons

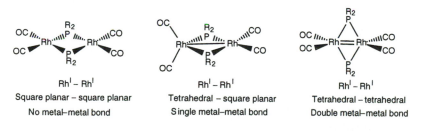

Rh^I – Rh^I	Rh^I – Rh^I	Rh^I – Rh^I
Square planar – square planar	Tetrahedral – square planar	Tetrahedral – tetrahedral
No metal–metal bond	Single metal–metal bond	Double metal–metal bond

Fig. 5.48 Some rhodium(I) complexes with phosphido-bridges showing the dependence of the metal-metal bond order upon the geometries of the two participating metal centres

The production of electrophilic metal cationic complexes of Ti, Zr, Hf, and Ni (see T. Marks *et al.*, *J. Am. Chem. Soc.*, 1996, **118**, 7900 and M. Bochmann, *J. Chem. Soc., Dalton Trans.*, 1996, 255) has proved to be very important in the development of a new generation of homogeneous Zeigler–Natta catalysts for olefin polymerization which have 14 and 16 valence electrons.

Those compounds which lie at the borderline between obeying and disobeying the EAN rule are of great importance, because the ability to switch from 18 to 16 electron configurations enables them to catalyse a

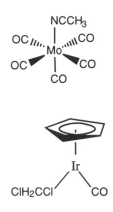

number of important transformations. A catalytic cycle requires a metal to coordinate the substrates of interest at the metal centre and then allows it to undergo reactions and transformations which convert it into the final product. Of course this would be fruitless unless the final product can be released from the complex to enable it to start at the beginning of the catalytic cycle.

The square planar complexes of the platinum metals are ideal for undertaking these transformations, because they are in a sufficiently low oxidation state to coordinate the substrates of interest such as CO and C_2H_4 and are sufficiently nucleophilic to oxidatively add reagents such as H_2 and MeI. Furthermore, the ability to go back and forth readily between 16 and 18 electron states enables them to facilitate migration reactions between the molecules on the metal centre.

Molecules which rigorously conform to the EAN rule are generally poor catalysts because they are substitutionally inert and cannot readily coordinate the substrates. Molecules such as $Mo(CO)_6$ can only generate 16 electron reactive intermediates by dissociation of CO, which can only be achieved by photolysis or by using high temperatures. Neither situation is conducive to be incorporated into a catalytic cycle which could operate at rapid rates at room temperatures. Sometimes, this obstacle is overcome by forming complexes which conform to the EAN rule, but which have labile and weakly held ligands coordinated to the metal. On dissociation of these ligands the reactive 16 electron complex is generated at room temperatures. Examples of some complexes with weakly held ligands are illustrated in the margin.

The weakly held CH_2Cl_2 molecule in the latter example readily dissociate and the resultant 16 electron complex is sufficiently reactive to oxidatively add C–H bonds from methane and cyclohexane.

Table 5.35 Summary of geometric preferences

Diamagnetic spin paired complexes	
d^{10}	Tetrahedral
d^8	Square planar, trigonal bipyramid, square pyramid
d^6	Octahedral
d^4	Pentagonal bipyramid, capped octahedron
d^2	Dodecahedron, square anti-prism
d^0	Tricapped trigonal prism
Paramagnetic high spin complexes	
d^9	Tetragonal, distorted octahedral,
	Flattened tetrahedron
d^7	Tetrahedral
d^5	High spin tetrahedral
d^3	Octahedral

Summary
The analysis developed above has underlined why the d-electron count of the central metal and the total number of valence electrons associated with the complex are the two most important variables in transition metal chemistry. In metal ML_m complexes which conform to the EAN rule m $d^{n-4}sp^3$ hybrids are formed which are used for the metal–ligand σ-bonds and the remaining complementary d orbitals are fully occupied. These orbitals are additionally stabilized by back π-donation effects.

In complexes with π-donor ligands the $d^{n-4}sp^3$ hybrids are also used for σ-bonding, but the complementary d orbitals remain completely empty or partially occupied. These vacancies allow some donation from the filled p_π orbitals on the ligands into the empty d_π orbitals. Such complexes do conform to the EAN rule if the ligands are good π-donors and the number of holes in the d_π set is taken into account in the electron summation. Therefore, it is necessary to specify the number of d electrons which are associated with the metal in its formal oxidation state in order to establish the number of available holes in the d shell. When the number of filled p_π orbitals exceeds the number of holes available, only the linear combinations which have a symmetry match with the empty d_π orbitals are relevant. The other linear combinations are redundant, and if counted, give the impression that the EAN rule is exceeded.

In complexes where the ligands are neither strong π-donors or acceptors, series of isostoichiometric compounds are commonly observed. In such compounds the energy separations between the components of the d manifold are not large and electrons can occupy both the non-bonding and antibonding components. Such complexes are frequently paramagnetic and their geometries are dictated primarily by electrostatic effects. However, the population of the antibonding components of the d orbitals may result in small distortions. These distortions are particularly evident when an incomplete set of the antibonding orbitals are occupied. For example, d^5 high spin complexes are invariably more symmetric than d^4, and d^{10} complexes are more symmetric than d^9. A summary of geometric preferences is given in Table 5.35.

Stereochemical inactivity of lone pairs

The lone pairs in p block molecules are generally stereochemically active and they have the same weight as ligands in deciding the observed geometry. For example, to a first approximation CH_4, NH_3, and OH_2 may be viewed as tetrahedral with either lone pairs or hydrogen atoms occupying the vertex positions.

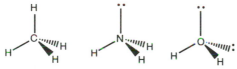

Transition metal complexes have up to five lone pairs and yet their stereochemical role is so minimal that their presence is frequently omitted from structural formulae. For example, $Cr(CO)_6$ is represented as:

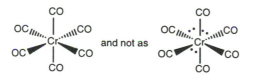

This difference is a direct consequence of the relative ordering of the valence orbitals and their nodal properties. A molecule such as NH_3 takes up a pyramidal rather than a planar geometry (see diagram in the margin) because the incorporation of some s character into the non-bonding p orbital has a stabilizing effect because the 2s orbital is more stable than the 2p.

In a transition metal complex the lone pair occupies a d orbital which is non-bonding, but since the relative stabilities of the orbitals is $nd > (n + 1)s \gg (n + 1)p$ there is no energy gain by the admixture of some s or p character. Therefore, in a transition metal complex with lone pairs the preferred geometry is one which places them in d orbitals which are noded in

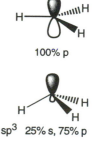

100% p

sp^3 25% s, 75% p

the metal–ligand directions and is therefore non-bonding. The d orbital splitting diagrams illustrated on pages 318–322 demonstrate how this is achieved in specific examples.

The illustration of lone pairs in molecular compounds of the p block elements also serves to emphasize their Lewis base properties. In general, transition metal complexes in medium to high oxidation states have ionization energies associated with the lone pairs which are so high that they do not function effectively as Lewis bases. Such complexes are more likely to behave as Lewis acids because of the empty d orbitals which they possess. However, low oxidation state complexes may function as strong Lewis bases, particularly if good electron-donating ligands are coordinated to the metal, as illustrated by the following examples.

$$Pt(PPh_3)_3 \ + \ H^+ \ \rightleftharpoons \ [PtH(PPh_3)_3]^+$$

$$IrCl(CO)(PPh_3)_2 \ + \ SO_2 \ \rightleftharpoons \ [Ir(SO_2)Cl(CO)(PPh_3)_2]$$

The availability of lone pairs in such complexes also encourages oxidative addition reactions with substrates including H_2, $HSiR_3$, MeI, Cl_2, Br_2, I_2, C–H bonds.

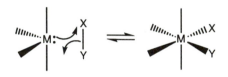

The presence of single electrons rather than electron pairs in transition metal complexes leads to radical species. Once again, in medium to high oxidation states the stability of the d orbitals and also their contracted nature which places them relatively close to the nucleus makes these radical species chemically unreactive compared to radicals of the p block elements. However, in low oxidation state compounds and particularly those of the second and third row transition metals these radicals become more reactive and lead to either radical abstraction or dimerization reactions. The latter leads to metal–metal bonded compounds.

See A. E. Merbach *et al.*, *Adv. Inorg. Chem.*, 1995, **42**, 2 for a more detailed discussion.

The non-bonding d electrons in transition metal complexes also exert a subtle and interesting role in the substitution reactions of metal complexes. High pressure n.m.r. studies have indicated that the mechanism of the exchange reaction:

$$[M(OH_2)_6]^{x+} \ + \ {}^*OH_2 \ \rightleftharpoons \ [M({}^*OH_2)(OH_2)_5]^{x+} \ + \ OH_2$$

$$(x = 2 \text{ or } 3)$$

changes across the series. On the left-hand side of the series where there are few lone pairs the reaction proceeds primarily by an associative mechanism. In these cases the incoming nucleophile (${}^*OH_2$) has empty non-bonding d orbitals which can accept the electron pair. The mechanism becomes progressively more dissociative towards the right of the series because the presence of electrons in the non-bonding d orbitals makes the electron donation from $^*OH_2$ unfavourable.

Complexes which conform to the EAN rule undergo nucleophilic substitution reactions *via* a dissociative mechanism which results in a 16 electron intermediate, e.g.

$$Ni(CO)_4 \rightleftharpoons Ni(CO)_3 + CO \quad \text{(Rate determining step)}$$

$$Ni(CO)_3 + PPh_3 \longrightarrow Ni(CO)_3(PPh_3)$$

An associative mechanism is disfavoured because of the repulsions between the lone pair on the ligand and the lone pairs on the central atom, and furthermore the resulting intermediate would have 20 valence electrons and therefore the additional electrons are required to occupy antibonding metal–ligand molecular orbitals.

If the ligands coordinated to the metal are capable of accepting electron pairs then an associative mechanism is observed. Ligands which have this property generally are capable of coordinating to the metal either as n or $(n - 2)$ electron donors. For example, NO is capable of acting either as a 3 electron donor (linear) or a 1 electron donor (bent) and the resultant five coordinate intermediate in the substitution reactions of $Co(CO)_3(NO)$ conforms to the EAN rule by switching the coordination mode of the NO ligand.

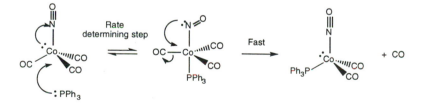

Other ligands which are capable of undergoing substitution *via* an associative pathway are cyclopentadienyl and benzene which can change their *hapto*-number by a ring slippage mechanism.

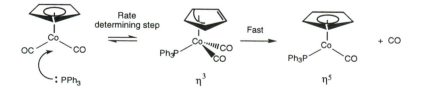

The starting complex, the final product, and the transition state complex thereby conform to the EAN rule.

Orbital splittings and stabilization energies in octahedral complexes

The orbital splittings described in the previous section are particularly useful for rationalizing the properties of series of isostoichiometric and isostructural compounds. The octahedral geometry is the most dominant in transition metal chemistry and therefore the d-orbital splittings associated with them are discussed in more detail in this section. The five d orbitals illustrated in Fig. 5.49 do not experience the same overlap with the ligands in an octahedral complex. The metal $d_{x^2-y^2}$ and d_{z^2} orbitals which point directly at the ligands overlap with them, but the d_{xz}, d_{yz}, and d_{xy} orbitals which have their lobes lying between the x, y, and z axes have zero overlaps.

S_σ is defined as the overlap integral for the overlap between the ligand σ-donating orbital and the metal d_{z^2} orbital when the ligand is located along the z-axis.

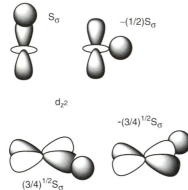

d_{z^2}

$d_{x^2-y^2}$

d_{xz}, d_{yz}, d_{xy}

Overlap = 0

Fig. 5.50 The relative overlaps (S) of ligand orbitals with metal d orbitals

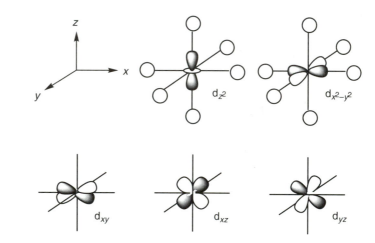

Fig. 5.49 The five d orbitals in an octahedral complex. Note that only d_{z^2} and $d_{x^2-y^2}$ overlap with the ligand orbitals.

The overlaps between the $d_{x^2-y^2}$ and the d_{z^2} orbitals and the ligand orbitals depend on the location of the ligand relative to the orbital. For example, if the ligand is located along the z axis the overlap with d_{z^2} is equal to S_σ. If the ligand is moved into the xy plane the overlap changes sign and is reduced in magnitude to $-\frac{1}{2} S_\sigma$. The overlap between a ligand located along the x or y axes and the $d_{x^2-y^2}$ orbital is equal to $\pm(\frac{3}{4})^{1/2}$ depending on whether the ligand and metal orbital have matching signs. The relative overlaps (S) and their signs are indicated in Fig. 5.50. According to the *angular overlap model* the total stabilization energy associated with a metal d orbital resulting from overlap between metal and ligand orbitals is proportional to the sum of the squares of the overlaps at each ligand location, βS^2 (where β is a constant for a specific ligand and depends on the difference in orbital energies of the ligand and metal orbitals). The relevant calculations of the stabilization energies for an octahedral complex are summarized in Fig. 5.51. Therefore the total stabilization energies associated with $d_{x^2-y^2}$ and d_{z^2} are $3\beta S_\sigma^2$. The corresponding d_{xz}, d_{yz}, and d_{xy} orbitals have an energy of 0, because they have zero overlaps with the ligand orbitals, i.e. they are non-bonding. This simple calculation therefore confirms that the d orbitals in an octahedral complex split into two degenerate sets $d_{x^2-y^2}$ and d_{z^2} (e_g) and d_{xz}, d_{yz}, and d_{xy} (t_{2g}).

For a detailed account of the angular overlap model see H. B. Gray and R. L. DeKock, *Chemical Structure and Bonding*, Benjamin/Cummings, Menlo Park, CA, 1980

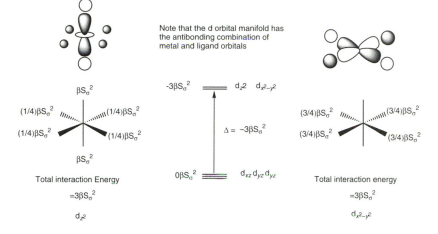

Note that the d orbital manifold has the antibonding combination of metal and ligand orbitals

Fig. 5.51 Overlaps between metal d_{z^2} and $d_{x^2-y^2}$ and ligand orbitals

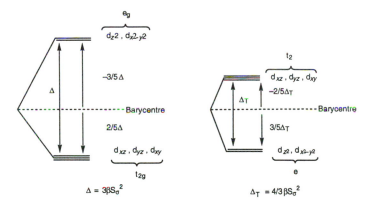

Fig. 5.52 Octahedral and tetrahedral splittings of d orbital energies

The orbitals shown in Fig. 5.51 correspond to those shown in the box of the more complete molecular orbital diagram shown in Fig. 5.38. The energy difference of $3\beta S_\sigma^2$ between these orbital sets may be represented by Δ. For a weighted zero point on the energy scale (known as a barycentre) the relative energies of the d orbitals are $+\frac{2}{5}\Delta(t_{2g})$ and $-\frac{3}{5}\Delta(e_g)$ as shown in Fig. 5.52. An identical orbital splitting diagram may be derived from the electrostatic *crystal field theory*. This results because both models incorporate the basic symmetry aspects of the problem.

In a tetrahedral complex the orbital energies of the d_{xz}, d_{yz}, and d_{xy} and the $d_{x^2-y^2}$ and d_{z^2} orbitals are reversed and the orbital splitting is $\frac{4}{9}\Delta$ for the octahedron. The relevant interaction diagrams and overlaps are illustrated in Fig. 5.53. These orbitals correspond to those shown in the box of the more complete molecular orbital diagram in Fig. 5.39.

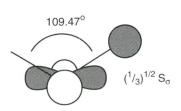

109.47°

$(1/3)^{1/2} S_\sigma$

Overlap between a ligand orbital and the metal d_{xy} orbital in a tetrahedral complex

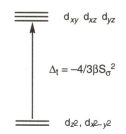

$$\Delta_t = -4/3\beta S_\sigma^2$$

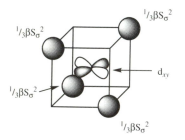

Fig. 5.53 Orbital interaction energies and energy diagram in a tetrahedral complex

d_{xz}, d_{yz}, d_{xy}

overlap = S_π

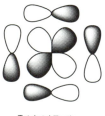

Total stabilization energy = $4S_\pi^2$

Fig. 5.54 π-Orbital overlaps and stabilization energies

The magnitude of the splitting energy, Δ, is described as the ligand field splitting energy and depends on the nature of the ligands, L. The following series has been established experimentally for octahedral complexes from electronic spectroscopic data:

$$I^- < Br^- < S^{2-} < \underline{S}CN^- < Cl^- < NO_3^- < F^- < C_2O_4^{2-} < H_2O < \underline{N}CS^-$$
$$< NH_3 < en < bipy < phen < NO_2^- < PPh_3 < \underline{C}N^- < \underline{C}O$$

Underlined atoms are donor atoms

This series cannot be interpreted simply in terms of electrostatic effects, particularly since such neutral ligands such as CO and NH_3 appear higher in the spectrochemical series than small anionic ligands such as F^-. The strength of covalent bonding in such complexes depends primarily on the *electronegativity* of the donor atom (the lower the electronegativity the more efficient the electron pair donation in the coordinate M–L bond), and the π-bonding character of the ligand. The latter is important because although the M–L σ–covalent bond occurs through the metal d_{z^2} and $d_{x^2-y^2}$ orbitals, the metal d_{xz}, d_{yz}, and d_{xy} orbitals have the correct symmetry properties to overlap with the π-orbitals of the ligands as shown in Fig. 5.54. The spectrochemical series therefore represents a superposition of the following ligand characteristics.

Electronegativity trends:

$$F < O < N < C$$
$$Cl < S < P < Si$$

As the electronegativity of the ligand decreases then the energy difference between the metal and ligand orbitals becomes smaller and the extent of covalent interaction between the orbitals increases, i.e. β becomes larger.

The effect of overlap between the ligand π-orbitals and the metal d_{xz}, d_{yz}, and d_{xy} orbitals depends on whether the ligands are π-donors or π-acceptors. π-donors have filled orbitals which lie below the d_{xz}, d_{yz}, and d_{xy} orbitals and consequently have the effect of reducing Δ, whereas π-acceptors have empty orbitals which lie above the d_{xz}, d_{yz}, and d_{xy} orbitals and have the effect of increasing Δ. The orbital interactions responsible for these effects are illustrated in Figs. 5.55 and 5.56 and typical π-donors and acceptors are summarized below.

π-donors	no π-character	π-acceptors
F^-, O^{2-}	NH_3, H^-	CO, CN^-
Cl^-, S^{2-}	en	phen, bipy

In the spectrochemical series ligands with large Δ values are good σ-donors and π-acceptors. Those with small Δ values are poor σ-donors and good π-donors.

Covalency effects also determine the role of the central metal atom in the spectrochemical series. Specifically, the extent of overlap between the metal d orbitals and the ligand orbitals increases down a Group of the Periodic Table—3d < 4d < 5d, because the d orbitals become less core-like and overlap more effectively. The increase in the metal oxidation state leads to more effective donation from the ligand to the metal d orbitals and therefore larger Δ values.

The following trends in Δ are therefore observed:

$$Fe^{3+} < Co^{3+} < Rh^{3+} < Ir^{3+} < Pt^{4+}$$

The increased covalency in the metal–ligand bonds also leads to an increase in Δ across the transition series:

$$Mn^{2+} < V^{2+} < Co^{2+} < Fe^{2+} < Ni^{2+}$$

For first row transition metals the energy gap, Δ, is comparable to the electron–electron repulsion energy and therefore the usual *aufbau* principle cannot be applied automatically. Specifically a metal complex with four d-electrons may take up *low spin* $(t_{2g})^4$ or *high spin* $(t_{2g})^3(e_g)^1$ configurations, with only the latter conforming to the *aufbau* principle. The relative *stabilization energies* (SE) are:

| Low spin | $(t_{2g})^4$ | $2/5 \Delta \times 4 =$ | $8/5\Delta$ |
| High spin | $(t_{2g})^3 e_g$ | $2/5\Delta \times 3 - 3/5\Delta =$ | $3/5\Delta$ |

The difference in stabilization energies is Δ because an electron is being transferred from e_g to t_{2g} and the former lies at Δ above the latter. For a d^5 complex the corresponding difference between high and low spin complexes is 2Δ because two electrons are relocated from e_g to t_{2g}.

| Low spin $(t_{2g})^5$ | $2/5\Delta \times 5 = 10/5\Delta = 2\Delta$ |
| High spin $(t_{2g})^3(e_g)^2$ | $2/5\Delta \times 3 - 3/5\Delta \times 2 = 0$ |

The transfer of electrons leading from a high to a low spin configurations results in more stabilization energy, but in increased electron–electron repulsion since the additional electron which initially was localized along the x, y, and z axes now resides in the t_{2g} set and is localized in the space between the axes, which it has to share with the other three electrons. The *additional* electron–electron repulsion energy associated with moving the electron from e_g to t_{2g} is described as the *spin pairing energy*, P.

Therefore, the distinction between the high spin and low spin complexes depends on the following inequalities:

$$\Delta > P\text{—strong field case, low spin preferred}$$
$$\Delta < P\text{—weak field case, high spin preferred}$$

Some typical pairing energies and ligand field splitting energies for aqua-complexes are summarized in Tables 5.36 and 5.37.

It is apparent from the Table 5.36 and 5.37 that for complexes of water, which is low in the spectrochemical series, the pairing energies are greater than the ligand field splitting energies. Therefore, aqua complexes of the first row transition metals are rarely observed as low spin complexes. For ligands which are higher in the spectrochemical series then the ligand field splitting energies are greater than the pairing energies and they invariably form low

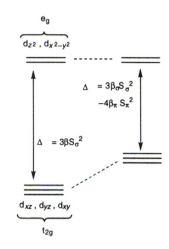

Fig. 5.55 Effect of π-bonding in a π-donor octahedral complex

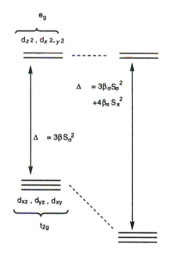

Fig. 5.56 Effect of π-bonding in a π-acceptor octahedral complex

Table 5.36 Pairing energies for some ions /kJ mol^{-1}

Cr^{2+}	281
Mn^{2+}	305
Fe^{2+}	211
Co^{2+}	269
Fe^{3+}	359
Co^{3+}	251

Table 5.37 Ligand field stabilization energies /kJ mol^{-1}

$[V(OH_2)_6]^{2+}$	141
$[Cr(OH_2)_6]^{2+}$	167
$[Mn(OH_2)_6]^{2+}$	93
$[Fe(OH_2)_6]^{2+}$	124
$[Co(OH_2)_6]^{2+}$	111
$[V(OH_2)_6]^{3+}$	215
$[Cr(OH_2)_6]^{3+}$	214
$[Mn(OH_2)_6]^{3+}$	251
$[Fe(OH_2)_6]^{3+}$	164
$[Co(OH_2)_6]^{3+}$	218

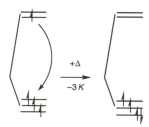

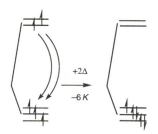

Fig. 5.57 Changes in exchange and ligand field stabilization energies for spin pairing in d^4 and d^5 configurations

spin complexes. The CN^- ligand which is very high in the spectrochemical series always forms low spin complexes.

A major contributor to P is the difference in *exchange energy*. The differences in exchange energy for d^6 and d^7 octahedral complexes are summarized in the Table 5.38 together with the differences in ligand field stabilization energies. The changes in exchange and ligand field stabilization energies for the d^4 and d^5 cases are illustrated in Fig. 5.57.

Table 5.38 Exchange and ligand field stabilization energies

Electron configuration	High spin	Low spin	Difference in exchange energy	Difference in ligand field stabilization energy
d^4	$6K$	$3K$	$3K$	Δ
d^5	$10K$	$4K$	$6K$	2Δ
d^6	$10K$	$6K$	$4K$	2Δ
d^7	$11K$	$7K$	$4K$	Δ

It is more favourable for d^6 complexes to adopt a low spin configuration because they gain a stabilization energy of 2Δ and lose exchange energy of $4K$. As a consequence, the low spin configuration is observed for ligands relatively low in the spectrochemical series for the d^6 ions Co^{3+} and Fe^{2+}. For example, $[Co(OH_2)_6]^{3+}$ provides a rare example of a low spin aqua complex. Complexes with d^5 ions gain the same stabilization energies when they become low spin, but lose $6K$ in exchange energy and therefore ligands higher in the spectrochemical series are required to form low spin d^5 complexes of Fe^{3+} and Mn^{2+}.

Consequences of orbital splittings in octahedral complexes
Magnetic properties
Since high spin and low spin configurations have different total numbers of unpaired spins (for example, the low spin configuration $(t_{2g})^4$ is associated with a total spin $S = 1$; but the high spin configuration $(t_{2g})^3(e_g)^1$ gives a total spin $S = 2$) they may be distinguished by their paramagnetic moments, which are proportional to $\sqrt{(4S(S+1))} = [n(n + 2)]^{1/2}$ if orbital angular momentum effects are ignored. Examples of some octahedral high spin and low spin complexes are given below.

	High spin	Low spin
d^4	$t_{2g}^3 e_g^1$	t_{2g}^4
	$[MnF_6]^{3-}$	$[Mn(CN)_6]^{3-}$
d^5	$t_{2g}^3 e_g^2$	t_{2g}^5
	$[Fe(C_2O_4)_3]^{3-}$	$[Fe(CN)_6]^{3-}$
d^6	$t_{2g}^4 e_g^2$	t_{2g}^6
	$[Fe(OH_2)_6]^{2+}$	$[Fe(CN)_6]^{4-}$

Table 5.39 summarizes the magnetic moments of octahedral transition metal complexes with d^1–d^9 electronic configurations. The examples

illustrate clearly the alternative high spin (hs) and low spin (ls) possibilities for d^4, d^5, d^6, and d^7 configurations. It is noteworthy that for these examples the high spin variant is associated with ligands that are low in the spectrochemical series, such as F^-, and the low spin with ligands such as CN^-. Table 5.39 also suggests that the spin-only formula which neglects orbital contributions to the magnetic moment works very well for the majority of the examples. Indeed, the variations observed for d^4 (ls), d^5 (ls), d^6 (hs), d^7 (hs), and d^8 may be related to the orbital contribution which has been neglected.

Table 5.39 Magnetic properties of some octahedral complexes ($n =$ number of unpaired spins). Entries for high spin configurations are tinted

d configuration		$S = \frac{1}{2}n$	$\mu_{eff} = 2(S(S+1))^{1/2}$	Example	μ_{eff} observed
d^1	$(t_{2g})^1$	$\frac{1}{2}$	1.73	K_3TiF_6	1.70
d^2	$(t_{2g})^2$	1	2.83	K_3VF_6	2.79
d^3	$(t_{2g})^3$	$\frac{3}{2}$	3.87	CrF_3	3.85
d^4 hs	$(t_{2g})^3(e_g)^1$	2	4.90	K_3MnF_6	4.95
d^4 ls	$(t_{2g})^4$	1	2.83	$K_3Mn(CN)_6$	3.2
d^5 hs	$(t_{2g})^3(e_g)^2$	$\frac{5}{2}$	5.92	Na_3FeF_6	5.85
d^5 ls	$(t_{2g})^5$	$\frac{1}{2}$	1.73	$K_3Fe(CN)_6$	2.4
d^6 hs	$(t_{2g})^4(e_g)^2$	2	4.90	K_3CoF_6	5.53
d^6 ls	$(t_{2g})^6$	0	0	$K_4Fe(CN)_6$	0
d^7 hs	$(t_{2g})^5(e_g)^2$	$\frac{3}{2}$	3.87	CoI_2	5.03
d^7 ls	$(t_{2g})^6(e_g)^1$	$\frac{1}{2}$	1.73	$[Co(diars)_3](ClO_4)_2$	1.92
d^8	$(t_{2g})^6(e_g)^2$	1	2.83	$(NH_4)_2Ni(SO_4)_2.6H_2O$	3.3
d^9	$(t_{2g})^6(e_g)^3$	$\frac{1}{2}$	1.73	$(NH_4)_2Cu(SO_4)_2.6H_2O$	1.9

Electronic spectra

The d orbital splittings described above are associated with energy differences which occur approximately in the visible region of the electromagnetic spectrum. The electronic transitions involve the promotion of electrons within the d sub-shell and are therefore forbidden by the Laporte selection rule, i.e. for an electronic transition in an atom to be allowed, $\Delta l = +/-1$, i.e. s → p, p → d, d → f, etc. This selection rule carries over to metal complexes as long as they retain a centre of symmetry, i.e. a centre of inversion retains the distinctions between s and p, p and d, orbitals, etc. The second important selection rule is the spin selection rule which states that $\Delta S = 0$. In general this selection rule can be achieved for d–d electronic transitions for all transition metal ions except high spin d^5.

Weak d–d transitions are observed for the great majority of transition metal complexes in the visible region, because although the transitions are electric dipole forbidden the vibrations associated with the metal–ligand bonds can destroy the centre of symmetry in the octahedral molecules at the instant of the transition and consequently the bands are not completely forbidden.

Octahedral d^1 and d^9 complexes have only one possible transition involving the e_g and t_{2g} orbitals. The other metal ions give rise generally to

For a more detailed discussion of the electronic spectral and magnetic properties of transition metal compounds see S. F. A. Kettle, *Physical Inorganic Chemistry*, Spektrum, Oxford, 1996 two to four spin allowed, Laporte forbidden, transitions. These transitions are commonly responsible for the characteristic colours of transition metal compounds. Table 5.40 contains some typical orbital splitting energies obtained from electronic spectral data.

Table 5.40 Some typical Δ values for octahedral complexes /kJ mol^{-1}

	6F$^-$	6H$_2$O	6NH$_3$	6CN$^-$
Ti^{3+}	209	240		280
V^{3+}	193	221		318
Cr^{3+}	181	214	258	419
Fe^{3+}	167			416
Co^{3+}	155	218	274	404
Fe^{2+}		124	132	
Co^{2+}		111	129	
Ni^{2+}		102		

Thermodynamic effects

Table 5.41 summarizes the stabilization energies (SE) of high spin d^0–d^{10} octahedral complexes. The effects rise from d^0 and d^5 to a maximum at d^3 and d^8. These differences in stabilization energies as a function of the d electron configuration are reflected in a range of thermodynamic data.

The hydration enthalpies for M^{2+} and M^{3+} ions and the enthalpies of formation of the halides MX_2 and MX_3 are expected to become more negative across the series because of the contraction in radius resulting from the increases in effective nuclear charge. This is indeed observed as shown in Figs 5.58 and 5.59, but superimposed on the anticipated smooth curves are perturbations which give the curves a 'double humped' appearance. The additional contribution which results in this appearance can be related to the stabilization energies given in Table 5.41. If these relevant stabilization energies are subtracted the resulting points more closely fit a smooth curve which passes through the points for d^0, d^5, and d^{10} configurations for which the stabilization energies are zero (see Fig. 5.58 for example).

Table 5.41 Stabilization energies for high spin complexes

d configuration	Stabilization energy
d^0	0
d^1, d^6	$\frac{2}{5}\Delta$
d^2, d^7	$\frac{4}{5}\Delta$
d^3, d^8	$\frac{6}{5}\Delta$
d^4, d^9	$\frac{3}{5}\Delta$
d^5, d^{10}	0

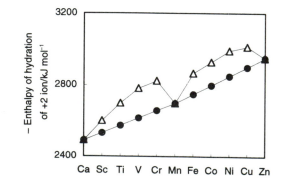

Fig. 5.58 Hydration enthalpies for M^{2+} ions of the first transition series

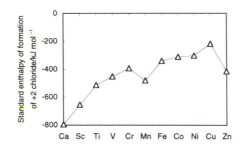

Fig. 5.59 Enthalpies of formation for MCl_2 of the first transition series

Covalent radii

Table 5.42 presents the metallic and covalent radii of the transition metals in their II and III formal oxidation states. The corresponding ionic radii are illustrated in Fig. 5.60. The overall contraction in covalent and ionic radii across the series results from the increase in effective nuclear charge, but superimposed on it are the orbital splitting effects. Specifically, the occupation of the t_{2g} orbitals which point away from the ligands allows the ligands to approach the metal more closely, but occupation of the e_g orbitals makes the approach of the ligands less favourable. In molecular orbital terms, the e_g orbitals are antibonding and consequently their occupation leads to longer metal–ligand bonds. Therefore, the radii rise for d^4 and d^5 high spin complexes and d^9 and d^{10} complexes where the e_g orbitals are populated.

The radius contraction associated with the change in spin state from high spin to low spin is especially important in the functioning of haemoglobin. The transformation of deoxyhaemoglobin (Fe^{II}, high spin d^6) to dioxygen-haemoglobin (Fe^{II}, low spin d^6) results in a movement of the iron atom from above the porphyrin plane to the centre of the plane. In the high spin state the radius of the iron atom is too large for it to be accommodated in the central cavity of the porphyrin. The movement of the iron atom triggers a series of motions within the haemoglobin molecule which increases the ability of the other three iron atoms to pick up dioxygen. See D. E. Fenton, *Biocoordination Chemistry*, OUP, 1996 for a fuller discussion.

Table 5.42 The metallic radii and M^{II} and M^{III} covalent radii for some first row transition elements

Metal	Metallic radius/pm	Covalent radii/pm and d electron configurations	
		M^{II}	M^{III}
Ti	132	166 $(t_{2g})^2$	147 $(t_{2g})^1$
V	122	159 $(t_{2g})^3$	144 $(t_{2g})^2$
Cr	117	160 $(t_{2g})^3(e_g)^1$	142 $(t_{2g})^3$
Mn	117	163 $(t_{2g})^3(e_g)^2$	145 $(t_{2g})^3(e_g)^1$
Fe	116	158 $(t_{2g})^4(e_g)^2$	145 $(t_{2g})^3(e_g)^2$
Co	116	155 $(t_{2g})^5(e_g)^2$	141 $(t_{2g})^4(e_g)^2$
Ni	115	149 $(t_{2g})^6(e_g)^2$	140 $(t_{2g})^5(e_g)^2$
Cu	117	153 $(t_{2g})^6(e_g)^3$	

The different characteristics of the t_{2g} and e_g orbitals also leads to dramatic differences in radii between high and low spin complexes. Some typical values are summarized in Table 5.43. It is noteworthy that the low spin complexes always have the smaller radii, because the depopulation of the e_g antibonding orbitals results in strengthening of the metal–ligand bonds.

More generally when a transition metal complex is associated with ligands which do not cause large d orbital splittings then it is possible to obtain complexes where the antibonding components of the d manifold are occupied. If the orbitals are equally populated with electrons, the symmetry of the complex is retained, but the bond lengths become longer as a result of the population of orbitals which are metal–ligand antibonding. If the orbitals are

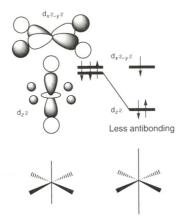

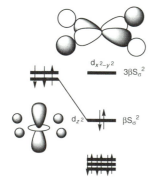

Fig. 5.61 Effects of tetragonal distortion on the energies of the d_{z^2} and $d_{x^2-y^2}$ orbitals

unequally populated then the complex generally undergoes a distortion which removes the degeneracy of the antibonding orbitals.

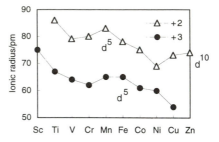

Fig. 5.60 Ionic radii of the +2 and +3 ions of the first transition series

Table 5.43 Radii of high and low spin configurations of d^{4-7} metal ions

d configuration	High spin	Radius/pm	Low spin	Radius/pm
d^4 Cr^{2+}	$(t_{2g})^3(e_g)^1$	160	$(t_{2g})^4$	150
d^5 Mn^{2+}	$(t_{2g})^3(e_g)^2$	163	$(t_{2g})^5$	147
d^6 Fe^{2+}	$(t_{2g})^4(e_g)^2$	158	$(t_{2g})^6$	141
d^7 Co^{2+}	$(t_{2g})^5(e_g)^2$	155	$(t_{2g})^6(e_g)^1$	145

If the antibonding e_g orbitals are unequally occupied, the symmetry of the octahedral complex is reduced. For example, in a d^9, Cu^{2+}, ion the e_g set of orbitals are occupied by three electrons. The occupation of a pair of electrons in d_{z^2} and a single electron in $d_{x^2-y^2}$ leads to more antibonding character along the z axes of the complex and the complexes undergo a tetragonal distortion (see Fig 5.61). Similar tetragonal distortions are observed in high spin octahedral d^4, Cr^{2+} complexes, where there is only a single electron in the e_g set. When there is a pair of electrons in the e_g orbitals the effect of occupying the antibonding orbitals is spread evenly, because both d_{z^2} and $d_{x^2-y^2}$ are occupied. The bond lengths in the octahedron are lengthened, but the symmetry of the complex is not reduced.

An extreme form of a tetragonal distortion is one which leads to the complete loss of the two axial ligands and the formation of a square planar complex. The d orbital splitting diagram for a square planar complex is shown in Fig. 5.62. The d_{xz}, d_{yz}, and d_{xy} orbitals retain the non-bonding character associated with the octahedron, because they do not overlap with the ligand orbitals. The $d_{x^2-y^2}$ orbital overlaps with the ligands in the square plane in an identical manner to the octahedron, but the d_{z^2} orbital is stabilized by $2\beta S_\sigma^2$ because of the loss overlap with the two ligands removed from the z axis to convert the octahedral complex into a square planar complex. Therefore, the square planar complex is associated with three non-bonding d orbitals and a weakly antibonding d_{z^2} orbital. The antibonding character of the latter is reduced further by mixing with the s orbital (see Fig 5.47, p. 322) and consequently the square planar geometry is ideal for accommodating eight electrons in a low spin configuration.

Fig. 5.62 The orbital splitting diagram for a square planar complex

Magnetic and conductivity properties in infinite solids

The magnetic and electronic spectral properties of transition metal complexes described above might be expected to be transferable to solid state compounds having infinite structures. For example, the colour and magnetic properties of an oxide MO should resemble those of the aqua complex $[M(OH_2)_6]^{2+}$, because they both have the metal in an octahedral environment of oxygen atoms. As noted previously, the metal oxides, MO, all have the sodium chloride structure although at times they are not simple stoichiometric compounds. In fact the properties of the oxides, and particularly those of the early transition metals may differ significantly from those of the aqua complexes, e.g. VO is black whereas $[V(OH_2)_6]^{2+}$ is violet, but both NiO and $[Ni(OH_2)_6]^{2+}$ are green.

In the sodium chloride structure the octahedral environment around the metal leads to the e_g–t_{2g} splitting discussed above. The non-bonding t_{2g} set point towards the matching orbitals on the adjacent metal ions and the overlap between the t_{2g} orbitals is small but sufficiently significant for a band structure to develop. The band structure is wider for the earlier metals in the transition series because the effective nuclear charge they experience is smaller and the orbitals are larger and overlap more effectively. These effects are illustrated for TiO and NiO in Figs 5.63 and 5.64. The resultant band is able to accommodate a total of 6 electrons and consequently for an ion such as Ti^{2+} it is partially filled and the resultant oxide is an electrical conductor whose resistivity equals $10^3 \, (\Omega cm)^{-1}$.

The other early transition metal oxides are also conductors because the band width is sufficiently large to give rise to effective delocalization of the electron density through the lattice and the bands are partially filled. For Ni^{2+} however, the band width is much narrower (smaller orbital overlaps) and the t_{2g} band is completely filled. The e_g orbitals which contain two electrons for the d^8 Ni^{2+} ion also give rise to a band structure, but since these orbitals point at the oxide anions rather than adjacent metals they do not create a band which is effectively delocalized over all the metal atoms. Therefore, NiO shows no metallic conduction properties and behaves as an insulator.

The paramagnetic properties which were discussed above are associated with complexes where the interactions between the magnetic fields at individual ions are not influenced by those at other centres. In metal oxides the unpaired electrons on individual metal centres interact resulting in concerted alignments of the magnetic properties throughout the lattice. This ordering has of course to compete with thermal randomizing effects and therefore the magnetic properties in such compounds are generally temperature dependent. If the ordering occurs in such a way that the electron spins on the metals all orientate preferentially in a parallel fashion the result collective magnetic properties are described as *ferromagnetic*. If the electron spins on the metals preferentially align in an anti-parallel fashion, as shown in Fig. 5.65, the resultant collective magnetic property is described as *anti-ferromagnetism*. If the interactions which result in these alignments are strong the transition from ferromagnetic or anti-ferromagnetic behaviour to paramagnetic behaviour occurs at a high temperature, whereas if they are weak very low temperatures are required for this transition.

See O. Kahn, *Adv. Inorg. Chem.*, 1995, **43**, 180 for a discussion of recent attempts at designing molecular magnets.

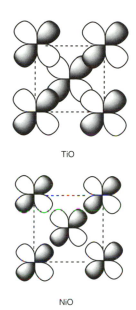

Fig. 5.63 A diagram illustrating the differences in overlap between metal d orbitals in TiO and NiO

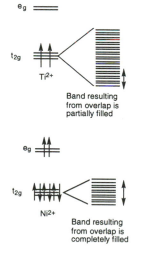

Fig. 5.64 d-band formation and occupation in Ti^{2+} and Ni^{2+}

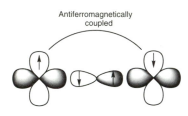

Antiferromagnetically
coupled

Fig. 5.65 A diagram illustrating the anti-ferromagnetic coupling of orbitals

For a more detailed discussion see P. A. Cox, *Transition Metal Oxides*, Clarendon Press, Oxford, 1992

For a specific compound the transition temperature is described as the Curie temperature, T_C, if it is ferromagnetic and the Néel temperature, T_N, if it is anti-ferromagnetic. For ferromagnetic materials the interaction between the compound and the magnetic field is enhanced below the Curie temperature because the spins are aligned parallel in a concerted fashion and their collective interaction with the magnetic field is greater than that of the individual spins. In contrast for anti-ferromagnetic materials the collective interaction is reduced because the spins on adjacent metal ions align in an anti-parallel fashion.

An important mechanism by which the spins of metal ions align in solids is the *super exchange mechanism*. In the metal oxides MO which have the sodium chloride structures the electrons in the e_g orbitals can interact with the orbital pair in the p orbitals of the oxide to polarize the electron spin on the adjacent metal ion in an anti-parallel fashion, i.e. leading to an anti-ferromagnetic interaction. Consequently the metal oxides MnO, Fe_{1-x}, CoO, and NiO which have electrons occupying the e_g orbitals exhibit anti-ferromagnetic interactions. The corresponding Néel temperatures are :

MnO	$Fe_{1-x}O$	CoO	NiO
122 K	198 K	293 K	523 K

The increasing Néel temperature across the series reflects the stronger metal–ligand interactions across the series as the metal ions contract in size.

Structures of isostoichiometric compounds

The MO oxides described above all share the sodium chloride structure and consequently the only important variation in parameters involves the metal–oxygen distances, which of course are reflected in the dimensions of the unit cell. The metal chalcogenides of the transition metals generally adopt the alternative (MX, 6:6) (nickel arsenide) structure which has the metal ions in distorted octahedral environments and the anions in trigonal prismatic environments. The NiAs structure can be derived from a hexagonal close-packed arrangement of anions with the metal ions occupying the octahedral vacancies. For an idealized close packed structure the resultant hexagonal unit cell is $a = b \neq c$ and $\alpha = \gamma = 90°$ and $\beta = 60°$ and the ratio of a/c should equal 1.633. The data given in Table 5.44 for transition metal selenides indicate that this ideal ratio is not usually observed.

Table 5.44 The numbers of valence electrons and the a/c ratios for some transition metal selenides

	TiSe	VSe	CrSe	$Fe_{1-x}Se$	CoSe	NiSe
No. of valence electrons	6	7	8	10 ($x = 0$)	11	12
c/a ratio	1.69	1.67	1.64	1.64	1.46	1.46

Therefore a significant compression of the unit cell occurs as the total number of valence electron is increased. In the NiAs structure the adjacent metal ions along the c axis direction are sufficiently close for the metal orbitals to overlap and the data above indicate that the metal–metal bonding interaction in the series is becoming progressively stronger.

Metal–metal interactions are also important in influencing the observed structures of metal carbides and nitrides. In the older literature the structures of such compounds are frequently described as interstitial compounds, because the metal lattice in the carbides of Groups 4–6 is identical to that of the lattice of the parent metal. The structures of some common transition metal nitrides and carbides are summarized in Table 5.45. It is significant that as the total number of electrons is increased the compounds show an increased preference for the WC structure.

For a more detailed discussion see R. Hoffmann, *Solids and Surfaces*, VCH Publications Inc., New York, 1988

Table 5.45 Structure of metal carbides and nitrides

Structural type	Number of valence electrons				
	8	9	10	11	12
NaCl	TiC	VC	VN		
	ZrC	NbC	MoC_{1-x}	CrN	
	HfC	TaC	WC_{1-x}		
WC			MoC	MoN	RuC
			WC	WN	OsC
			TaN		

Metal–metal bonded compounds

The transition metals form a very wide range of compounds with metal–metal bonds and their occurrence and classification illustrates the importance of defining clearly the d-electron count and total valence electron count. The compounds are also referred to as cluster compounds if three or more metal atoms are joined together by metal–metal bonds and polyhedral molecules if the metal atoms define a three dimensional structure. The early transition metals form metal–metal bonded compounds when in intermediate oxidation states of 2–4 and in the presence of π-donor ligands such as Cl, Br, I, S, and OR. In contrast the later transition elements form the majority of their metal–metal bonded compounds in lower oxidation states and in combination with π-acid ligands.

For a more extensive introduction to cluster chemistry see D. M. P. Mingos and D. J. Wales, *Introduction to Cluster Chemistry*, Prentice-Hall, New Jersey, 1990

Earlier it was noted that the maximum heats of atomization for the transition elements are observed in the centre of the transition series, where the d and s bands are approximately half filled. The molecular transition metal cluster compounds, which may be regarded as very small fragments of the bulk metallic structure, attempt to emulate this favourable bonding situation by using the electron donating and accepting properties of the ligands. Metal ions which have fewer than 6 valence electrons form cluster compounds with π-donor ligands which can effectively increase the total electron count around each metal atom, and metal ions which have more than 6 valence electrons form cluster compounds with π-acid ligands which can remove some of the excess of electron density. This distinction also influences whether the compounds are classified on the basis of the metal d electron count or the EAN rule.

The great majority of cluster compounds can be accommodated within the framework of the EAN rule. If the metals are bonded together by conventional two-centre two-electron bonds and adhere to the 18 electron rule then the following formula defines the number of metal–metal bonds, x, as:

$$x = \frac{18n - N}{2}$$

where n is the number of metal atoms and N is the total number of valence electrons in the molecule. This formula is analogous to that described for main group compounds in Chapter 4, i.e.

$$x = \frac{8n - N}{2}$$

Two-centre
two-electron bond

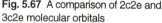

Three-centre
two-electron bond

Fig. 5.67 A comparison of 2c2e and 3c2e molecular orbitals

$(18n - N)$ defines the number of antibonding metal–metal orbitals and therefore since the three-centre two-electron bond has twice as many antibonding molecular orbitals as the two-centre two-electron metal–metal bond then it is necessary to divide $(18n - N)$ by 4 rather than 2.

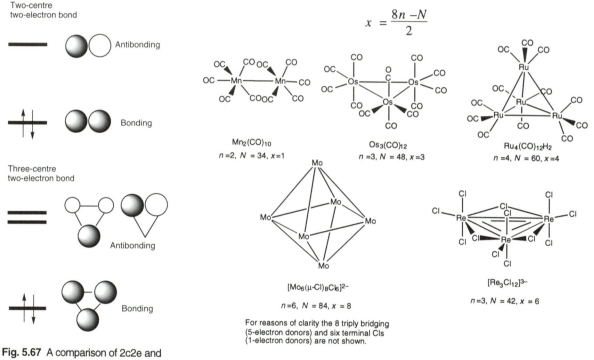

Fig. 5.66 Examples of cluster compounds which conform to the EAN rule.

Some examples of the application of this formula are given in Fig. 5.66. It is evident that the majority of the examples are later transition metals with π-acid ligands although there are some examples of clusters of the early transition metals with π-donor ligands.

In some cluster compounds three-centre two-electron bonds localized on the triangular faces of the cluster are preferred and the formula has to be modified in order to take into account that these metal–metal bonding interactions are associated with one bonding and two antibonding orbitals as shown in Fig. 5.67. In these situations the relevant formula is:

$$x = \frac{18n - N}{4}$$

and some examples of its use are given in Fig 5.68.

In all of these examples the applicability of the EAN rule may also be demonstrated by completing an electron count around each metal atom.

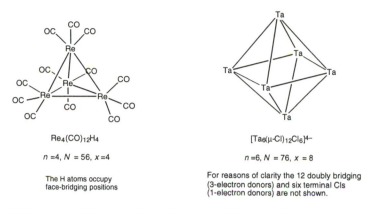

Re$_4$(CO)$_{12}$H$_4$

$n = 4$, $N = 56$, $x = 4$

The H atoms occupy
face-bridging positions

[Ta$_6$(μ-Cl)$_{12}$Cl$_6$]$^{4-}$

$n = 6$, $N = 76$, $x = 8$

For reasons of clarity the 12 doubly bridging
(3-electron donors) and six terminal Cls
(1-electron donors) are not shown.

Fig. 5.68 Examples of clusters which conform to the EAN rule and form three-centre two-electron bonds on each of the triangular faces.

Although the EAN rule works remarkably well for clusters with π-acid ligands, it is generally less applicable to cluster compounds of the earlier transition metals with π-donor ligands. For these clusters the important parameter is the d electron count, since this determines the number of metal–metal bonds which the metal is capable of forming. The examples shown in Fig. 5.69, provide a series of related dimeric complexes with d^1–d^3 electronic configurations which progressively form metal–metal bonds with formal bond orders of 1–3. Of course with the increase in bond order the metal atoms are drawn closer together.

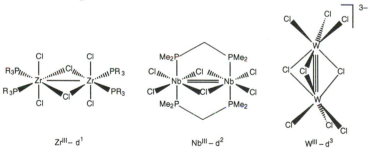

Zr^{III}– d^1 Nb^{III} – d^2 W^{III} – d^3

Fig. 5.69 Examples of metal–metal bonded compounds with π-donor ligands

The formal metal–metal bond orders in this series of molecules could not be determined accurately from the EAN rule. The importance of the d electron count in influencing the number of metal–metal bonds is illustrated by considering d^4 metal ions in more detail.

In all these examples the total number of two-centre two-electron metal–metal bonds is equal to the total number of d electrons for the metals in their formal oxidation states divided by two.

A d^4 metal ion, e.g. Mo^{2+}, Tc^{3+}, or Re^{3+} is capable of forming a total of four metal–metal bonds. This may be achieved as follows:

Four single metal–metal bonds. In the octahedral cluster anion $[Mo_6Cl_8Cl_6]^{2-}$ each Mo^{II} ion forms four metal–metal bonds along the edges of the octahedron. In the related cluster compounds $[Ta_6Cl_{12}Cl_6]^{4-}$ each metal atom forms four three-centre two-electron bonds on the faces of the octahedron. This leads to a total of 8 three-centre two-electron bonds on the 8 faces of the octahedron. The structures of these cluster ions are illustrated in the margin.

Two pairs of double bonds. In the triangular cluster anion (structure shown in the margin) each Re^{IV} (d^4) ion forms a pair of double bonds to the adjacent metal atoms.

A triple and a single bond. The rectangular cluster compound $[Mo_4Cl_8(PEt_3)_4]$ provides an example of this bonding arrangement (see Fig. 5.70).

A double and two single bonds. The trigonal prismatic cluster $[Tc_6Cl_{12}]^{2-}$ illustrated in Fig. 5.70 shows this bonding permutation.

A quadruple bond. In $[Re_2Cl_8]^{2-}$ the d^4 Re^{IV} ions form a multiple bond of order four, by utilizing the σ, 2π, and δ bonding interactions shown in Fig. 5.71.

In these compounds metal–metal bonding arrangements which have no analogues in main group chemistry are generated, because the metal ions are able to utilize the d_δ metal orbitals in addition to d_σ and d_π. The early transition metal ions with d^3 electron configurations form triply bonded compounds, e.g. $Mo_2(OR)_6$ and $[W_2Cl_9]^{3-}$, which utilize the d_σ and two d_π metal orbitals.

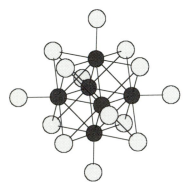

$[Mo_6Cl_8Cl_6]^{2-}$

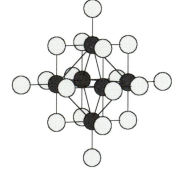

$[Ta_6Cl_{12}Cl_6]^{4-}$

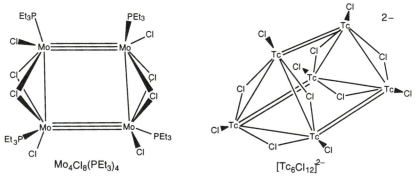

$Mo_4Cl_8(PEt_3)_4$ $[Tc_6Cl_{12}]^{2-}$

Fig. 5.70 Structures of clusters $Mo_4Cl_8(PEt_3)_4$ and $[Tc_6Cl_{12}]^{2-}$

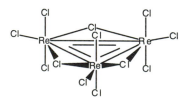

$[Re_3Cl_{12}]^{3-}$

Isostoichiometric relationships

The discussion above has centred on compounds where the number of d electrons allow the maximum formal metal–metal bond order. There are also situations where the same basic structural pattern is maintained, but the population of the metal–metal bonding and antibonding orbitals is varied. For example, in dinuclear metal–metal bonded compounds based on ML_4 and ML_5 fragments the d_σ, d_π, and d_δ bonding orbitals are matched by a set of antibonding d_{δ}^*, d_{π}^*, and d_{σ}^*, molecular orbitals, which are illustrated in Fig. 5.71. These molecular orbitals may be populated by additional electrons.

Examples of compounds with 8–16 d electrons are summarized in Table 5.46. Population of the antibonding components of the metal–metal orbitals leads to a reduction in formal bond order and generally to a lengthening of the metal–metal bond. Examples of such compounds are given in Table 5.46. The situation is therefore analogous to that described in Chapter 4 for diatomic molecules of the main group elements.

Table 5.46 Examples of (dimeric) isostoichiometric compounds

Compound and the number of d electrons involved in metal–metal bonding		Formal bond order	M–M distance /pm	Electronic configuration
$[Re_2Br_8]^{2-}$	8	4	222.6	$\sigma^2\pi^4\delta^2$
$[Tc_2Cl_8]^{2-}$	8	4	215.1	$\sigma^2\pi^4\delta^2$
$Re_2Cl_4(PMe_3)_4$	10	3	224.7	$\sigma^2\pi^4\delta^2\delta^{*2}$
$Tc_2Cl_4(PMe_2Ph)_4$	10	3	212.8	$\sigma^2\pi^4\delta^2\delta^{*2}$
$[Os_2Cl_8]^{2-}$	10	3	218.2	$\sigma^2\pi^4\delta^2\delta^{*2}$
$[Ru_2(\mu_2\text{-acetato})_4(OH_2)_2]^+$	11	2.5	224.8	$\sigma^2\pi^4\delta^2\delta^{*2}\pi^{*1}$
$[Ru_2(\mu_2\text{-acetato})_4(thf)_2]$	12	2	226.1	$\sigma^2\pi^4\delta^2\delta^{*2}\pi^{*2}$
$[Rh_2(\mu_2\text{-acetato})_4(\text{S-thf})_2]$	14	1	241.3	$\sigma^2\pi^4\delta^2\delta^{*2}\pi^{*4}$
$[Pt_2(\mu_2\text{-acetato})_4(OH_2)_2]^{2+}$	16	0	257.8	$\sigma^2\pi^4\delta^2\delta^{*2}\pi^{*4}\sigma^{*2}$

For complexes of the later transition elements with π-acid ligands the bond order of the metal–metal bonds is determined by the total valence electron count and the formula introduced above is widely applicable. For a dimeric metal–metal bonded compound this means that 34 valence electrons are associated with a formal bond order of 1, 32 valence electrons for a formal bond order of 2, and 30 for a formal bond order of 3. Specific examples of such molecules are illustrated in Fig. 5.72. It is noteworthy that in these compounds there is no formal link between the d electron count and the number of metal–metal bonds formed by the metal atoms.

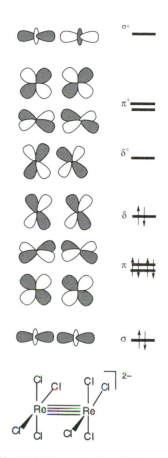

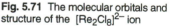

Fig. 5.71 The molecular orbitals and structure of the $[Re_2Cl_8]^{2-}$ ion

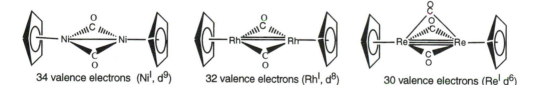

34 valence electrons (Ni^I, d^9)   32 valence electrons (Rh^I, d^8)   30 valence electrons (Re^I d^6)

Fig. 5.72 Multiple bonding in metal dimers with π-acid ligands

The similarity in the formula which defines the number of element–element bonds for transition metal clusters with π-acid ligands and the main group elements means that they both form series of isostructural cluster compounds with either polyhedral or ring geometries. This structural analogy is summarized in Table 5.47 and specific examples are illustrated in Fig. 5.73.

For a more thorough discussion of multiple bonding see F. A. Cotton and R. A. Walton, *Multiple Bonds Between Metal Atoms*, Clarendon Press, Oxford, 1993.

Table 5.47 Total valence electron counts in main group and transition metal cluster compounds. (n is the number of atoms)

Compound class	Total valence electron count for main group example	Total valence electron count for transition metal example
Chain	$6n + 2$	$16n + 2$
Ring	$6n$	$16n$
Three connected polyhedron	$5n$	$15n$
Four connected polyhedron	$4n + 2$	$14n + 2$
closo-deltahedra	$4n + 2$	$14n + 2$
nido-deltahedra	$4n + 4$	$14n + 4$
arachno-deltahedra	$4n + 6$	$14n + 6$

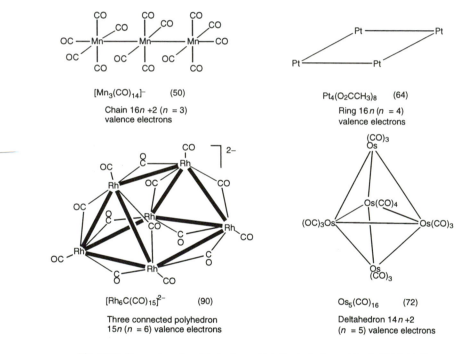

$[Mn_3(CO)_{14}]^-$ (50)

Chain $16n + 2$ ($n = 3$)
valence electrons

$Pt_4(O_2CCH_3)_8$ (64)

Ring $16n$ ($n = 4$)
valence electrons

$[Rh_6C(CO)_{15}]^{2-}$ (90)

Three connected polyhedron
$15n$ ($n = 6$) valence electrons

$Os_5(CO)_{16}$ (72)

Deltahedron $14n + 2$
($n = 5$) valence electrons

Fig. 5.73 Examples of metal clusters which illustrate the entries in Table 5.47

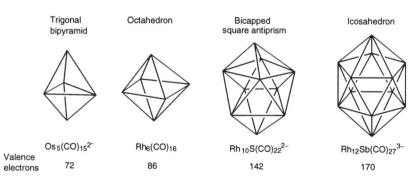

	Trigonal bipyramid	Octahedron	Bicapped square antiprism	Icosahedron
	$Os_5(CO)_{15}{}^{2-}$	$Rh_6(CO)_{16}$	$Rh_{10}S(CO)_{22}{}^{2-}$	$Rh_{12}Sb(CO)_{27}{}^{3-}$
Valence electrons	72	86	142	170

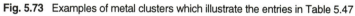

Fig. 5.74 Examples of deltahedral metal carbonyl cluster compounds

A wide range of deltahedral boranes and carboranes have been made and synthesized and they are characterized by $4n + 2$ valence electrons. There are analogous examples of deltahedral metal carbonyl clusters with $14n + 2$ valence electrons and these are illustrated in Fig. 5.74. Related examples of *nido*- and *arachno*-metal carbonyl clusters derived from the *closo*-deltahedra by the loss of vertices are illustrated in Fig. 5.75.

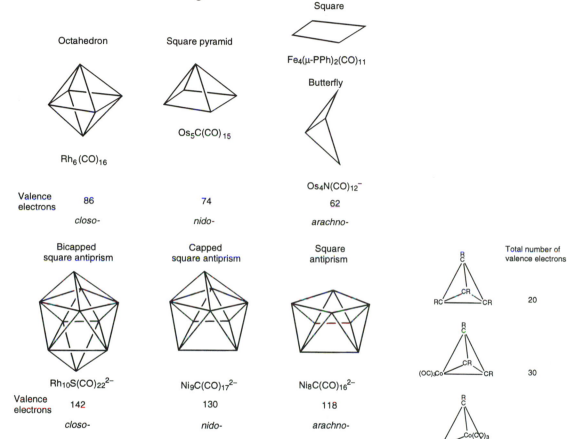

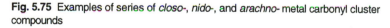

Fig. 5.75 Examples of series of *closo*-, *nido*-, and *arachno*- metal carbonyl cluster compounds

Since there are isostructural series of main group and transition metal polyhedral molecules which differ by $10n$ valence electrons, i.e. corresponding to a filled d shell, there must be intermediate molecules with transition metal and main group component which are isostructural. Such a series of compounds is illustrated in Fig. 5.76. In this series C–H fragments are progressively being replaced by $Co(CO)_3$ fragments and therefore the two fragments must have similar bonding capabilities. Specifically both fragments have three out-pointing hybrid orbitals, for C–H based on sp^3 hybrids, and for $Co(CO)_3$ based on d^2sp^3 hybrids each of which is occupied by a single electron. Fragments of this type which exhibit similar bonding capabilities are described as *isolobal* and the symbol ⟵o⟶ is used to indicate this similarity. Further examples of such isolobal relationships are

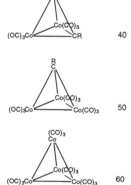

Fig. 5.76 An isostructural series of molecules with main group and transition metal fragments at the vertices of a tetrahedron.

illustrated in Fig. 5.77. The *isolobal* relationships defined in this figure are extremely useful for rationalizing the structures of cluster compounds containing both main group and transition metal components.

It is significant that with these metal carbonyl fragments of the later transition metals the d_δ orbitals are not used and therefore they form a maximum of three out-pointing hybrids, thus making them analogous to the main group fragments.

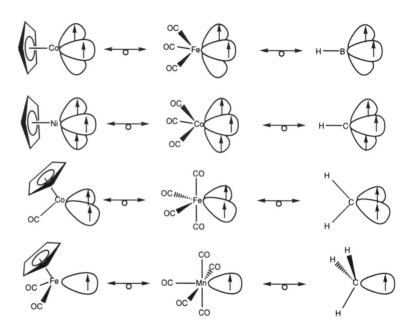

Fig. 5.77 Examples of the isolobal analogy.

5.6 Chemical properties of compounds

Variations of properties across the series where the oxidation state remains constant

The transition metals, particularly the first series, are the chameleons of the Periodic Table. If they are combined with ligands which are large and electronegative and not capable of forming strong multiple bonds then isostoichiometric series of compounds are observed. In these compounds the ligands overlap only weakly with the 3d orbitals of the transition metals and consequently the orbitals remain essentially core-like and do not contribute greatly to the structural and chemical properties of the metals. The lone pairs of electrons remain localized mainly on the ligand and the bonding has significant ionic character. Examples are given in Tables 5.48, 5.49, and 5.50 of such compounds of the transition elements.

Table 5.48 The dihalides of the first transition series (stable indicated by √, × indicates that the compound is unstable or unknown)

	Sc	Ti	V	Cr	Mn	Fe	Co	Ni	Cu	Zn
MF_2	×	×	√	√	√	√	√	√	√	√
MCl_2	×	√	√	√	√	√	√	√	√	√
MBr_2	×	√	√	√	√	√	√	√	√	√
MI_2	×	√	√	√	√	√	√	√	√	√

Table 5.49 The trihalides of the first transition series

	Sc	Ti	V	Cr	Mn	Fe	Co	Ni	Cu	Zn
MF_3	√	√	√	√	√	√	√	√	×	×
MCl_3	√	√	√	√	*	√	×	×	×	×
MBr_3	√	√	√	√	×	√	×	×	×	×
MI_3	×	√	√	√	×	√	×	×	×	×

* decomposes at $-40°C$

Table 5.50 Chlorides of the second and third transition series

	Chlorides of the second series								
	Zr	Nb	Mo	Tc	Ru	Rh	Pd	Ag	Cd
MCl	×	×	×	×	×	×	×	√	×
MCl_2	√	×	√	×	×	×	√	×	√
MCl_3	√	√	√	×	√	√	×	×	×
MCl_4	√	√	√	√	√	×	√	×	×
MCl_5	×	√	√	×	×	×	×	×	×
MCl_6	×	×	√	×	×	×	×	×	×

	Chlorides of the third series								
	Hf	Ta	W	Re	Os	Ir	Pt	Au	Hg
MCl	×	×	×	×	×	×	×	√	√
MCl_2	×	√	√	×	√	×	√	×	√
MCl_3	√	√	√	√	√	√	×	√	×
MCl_4	×	√	√	×	√	√	√	×	×
MCl_5	×	√	√	√	√	×	×	×	×
MCl_6	×	×	√	√	√	×	×	×	×

The decreased stability of the higher oxidation compounds across the series is noteworthy and is discussed on pages 285 and 296

The occurrence of isostochiometric compounds for the transition metals enables comparisons to be made across the series. For example, the metal dihalides are known from Ti to Zn and the properties of the dichlorides are summarized in Table 5.51.

The similarities are striking across the series, but the following differences are also noteworthy.

1.   The halides become less reducing across the series.
2.   They become more soluble and more readily form $[M(H_2O)_6]^{2+}$ in solution across the series.
3.   The geometries of complexes of metal ions in solution are octahedral, except for Cu^{2+} and Cr^{2+} which are distorted towards square planar and Zn^{2+} which is tetrahedral. In the solid state the octahedral coordination

also predominates, but at the end of the series tetrahedral coordination is observed for $ZnCl_2$.

4. Titanium dichloride is very reducing and reacts violently with water. In this relatively low oxidation state it forms complexes with ligands such as dmpe, $(CH_3)_2PCH_2CH_2P(CH_3)_2$. Vanadium dichloride forms a wide range of octahedral complexes with thf, py, EtOH; chromium dichloride forms a wide range of octahedral complexes $CrCl_2L_4$. Manganese shows a preference for tetrahedral complexes in addition to octahedral, e.g. $MnCl_2(py)_2$ and $MnCl_2(en)_2$. This pattern persists for Fe, Co, and Ni which form both octahedral and tetrahedral complexes, e.g. $FeCl_2(py)_2$, $CoCl_2(py)_2$, $NiCl_2(py)_2$, $CoCl_2(py)_4$, $NiCl_2(en)_2$. Copper(II) forms tetragonally distorted complexes and zinc(II) forms octahedral and tetrahedral complexes.

pyridine = py

tetrahydrofuran = thf

Table 5.51 Some properties of the dichlorides of the first row transition elements

Compound	Properties
$TiCl_2$	$CdCl_2$ structure with Ti octahedrally coordinated, red-brown solid, m.p. 1025°C, reducing agent, reacts violently with $H_2O \rightarrow H_2$ and burns in air. Paramagnetic($\mu_{eff} = 1.08\ \mu_B$)
VCl_2	CdI_2 structure with V octahedrally coordinated, light green, sol, EtOH, m.p. 1350°C Paramagnetic ($\mu_{eff} = 2.41\mu_B$)
$CrCl_2$	Distorted rutile structure, with tetragonally distorted octahedral chromium sites. Hygroscopic, colourless, lustrous crystals, m.p. 815°C. Blue aqueous solutions, reducing agent. Antiferromagnetic at low temps. ($\mu_{eff} = 5.13\ \mu_B$)
$MnCl_2$	Pink solid, $CdCl_2$ structure octahedral Mn^{II}, m.p. 654°C. v. sol in H_2O, readily forms adducts with nitrogen and oxygen donors. Paramagnetic ($\mu_{eff} = 5.73\ \mu_B$)
$FeCl_2$	$CdCl_2/CdI_2$ structures, colourless hygroscopic crystals, sol. in H_2O, insol. Et_2O. m.p. 676°C. Paramagnetic ($\mu_{ef} = 5.67\mu_B$)
$CoCl_2$	Pale blue $CdCl_2$ structure, sol. in H_2O (turns pink), soluble in polar organic solvents. m.p. 735°C. Paramagnetic ($\mu_{eff} = 5.59\ \mu_B$)
$NiCl_2$	Golden yellow, $CdCl_2$ structure, sol. in H_2O and polar organic solvents, m.p. 1001°C ($\mu_{eff} = 3.32\ \mu_B$)
$CuCl_2$	Chains of chloride bridged square planar $CuCl_4$ units, yellow brown, deliquescent. Soluble in H_2O, EtOH, acetone. m.p. 630°C. Paramagnetic ($\mu_{eff} = 1.97\ \mu_B$)
$ZnCl_2$	Tetrahedral Zn in close packed Cl lattice. Colourless, hygroscopic, sol. in EtOH, Et_2O, acetone, v. sol. in H_2O. m.p. 318°C. Diamagnetic ($\mu_{eff} = 0$)

The second and third row transition metals also form isostoichiometric halides, but the structural variations are greater especially in lower oxidation states. The majority of MCl_4 (M = 3rd row transition metal) compounds have chain structures based on edge-sharing octahedra, but $ReCl_4$ has zig-zag chains based on face-sharing Re_2Cl_9 bioctahedra. The MCl_3 compounds show greater variations because of metal–metal bond formation. $TaCl_3$ and WCl_3 for example have M_6 octahedral clusters and $ReCl_3$ has a triangular cluster with chlorine bridges, whereas $OsCl_3$ has $OsCl_6$ octahedral coordination centres.

Variation in properties with oxidation state

Given the wide range of oxidation states exhibited by transition metals it is of interest to evaluate how the properties of a related series of compounds varies as the oxidation state is decreased. In general, the halides and oxides in the highest (Group) oxidation state are molecular, but as the oxidation state is reduced they become progressively more coordinatively unsaturated and adopt polymeric structures. Table 5.52 summarizes the properties of some tungsten chlorides in oxidation states +6 to +2.

Table 5.52 Comparison of the properties of tungsten chlorides

Compound	Properties
WCl_6	Blue black, m.p. 275°C, b.p. 337°C; has octahedral WCl_6 molecules in the solid state. It is readily hydrolysed by H_2O, but soluble in organic solvents because of its molecular nature.
WCl_5	Dark green-blue, disproportionates to WCl_4 and WCl_6 on heating, m.p. 248°C, b.p. 286°C. Dimeric pairs of octahedra with bridges in solid state, trigonal bipyramidal monomers in gas phase. Readily hydrolysed and forms a complex anion $[WCl_6]^-$ with Cl^-.
WCl_4	Black needles, decomp. 300°C; edge sharing octahedra giving polymeric structure in solid state. Readily hydrolysed and oxidized, reacts with Cl^- to give $[WCl_6]^{2-}$.
WCl_3	Black brown, m.p. 550°C; decomposes on heating to WCl_2 and WCl_5. In solid state exists as an octahedral cluster with W–W bonds, *viz*, $[W_6Cl_{12}]Cl_6$.
WCl_2	Grey solid - decomp. >500°C, reducing agent. In solid state exists as an octahedral cluster with W–W bonds, $[W_6Cl_8]Cl_4$ The clusters are joined by Cl bridges to form sheets.

The following general points are noteworthy.

1. In the highest oxidation state octahedral coordination can be achieved by using the six available ligands. The resulting complex is molecular, making it relatively volatile and soluble in organic solvents. However, the relatively high residual positive charge on the central metal and the ability of the metal to expand its coordination number makes the compound susceptible to hydrolysis.

2. As the oxidation state is reduced the residual positive charge on the metal is reduced but so is the number of ligands. The latter effect is more significant and the metal manages to retain octahedral coordination by forming halide bridges. This leads to a dimer for WCl_5 and a polymer for WCl_4. As the degree of polymerization increases the m.p. increases, and the solubility particularly in non-donor organic solvents decreases. However, very good donor solvents are able to cleave the halogen bridges to give complexes such as $WCl_4(CH_3CN)_2$, $WCl_4(py)_2$, and $WCl_4(thf)_2$. Also, Cl^- may result in bridge cleavage and the formation of anions such as $[WCl_6]^-$ and $[WCl_6]^{2-}$. Less sterically demanding and better donating ligands can result in higher coordination numbers, e.g. $WCl_4(PMe_3)_4$, $WCl_4(PMe_2Ph)_3$, and $WCl_4(PPh_3)_3$.

3. As the oxidation state is reduced further an alternative mode of relieving the coordinative unsaturation is used, namely the formation of metal–metal bonds. These are sufficiently strong for the second and third transition series metals to compete with metal–halogen bonds. Both WCl_3 and WCl_2 have octahedral clusters of metal atoms—in each the tungsten atoms form five tungsten-chlorine bonds and metal–metal bonds to four adjacent metals. In $[W_6Cl_8]^{4+}$ the chloro-ligands bridge triangular faces of the metal octahedron and in $[W_6Cl_{12}]^{6+}$ the chloro-ligands bridge the edges of the octahedron. These clusters are sufficiently robust that the addition of the ligands does not break down the cluster—it remains intact and one of the ligands, L, coordinates to each of the metals, giving $[W_6Cl_8L_6]^{4+}$ and $[W_6Cl_{12}L_6]^{6+}$.

If tungsten is compared with chromium the following differences are apparent.

1. Chromium does not form stable $CrCl_6$ and $CrCl_5$ species.
2. $CrCl_4$ is a brown gas (m.p. -38°C) which is readily hydrolysed and only stable in the presence of an excess of chlorine. It forms tetrahedral molecules, whereas WCl_4 has a polymeric structure.
3. $CrCl_3$ is a red violet solid (m.p. 1152°C) which has hexagonal or cubic close packed layer structures based on $CrCl_6$ octahedra, whereas the corresponding tungsten compound is based on a metal–metal bonded octahedral cluster.
4. $CrCl_2$ is a hygroscopic colourless solid which forms aqueous reducing solutions in H_2O. It has a distorted rutile structure based on distorted $CrCl_6$ chlorine bridged octahedra. WCl_2 has a structure based on octahedral clusters.

Note that the Cr^{II} and Cr^{III} halides metal–metal bonding does not feature in the same way that has been noted for W^{II} and W^{III}. This aspect is discussed in more detail in Section 5.7.

Stabilities of complexes

For a given ligand the divalent metal ions of the transition series have stability constants which follow the order:

$$Mn^{2+} < Fe^{2+} < Co^{2+} < Ni^{2+} < Cu^{2+} > Zn^{2+}$$

This order, known as the Irving–Williams order, reflects the decreasing sizes of the ions across the series. Superimposed on this are ligand field stabilization effects which reach a maximum at d^8 for high spin complexes. Some typical data for EDTA complexes are presented in Table 5.53.

Table 5.53 Stability constants of EDTA complexes

Metal ion	log K(MY)	Metal ion	log K(MY)
Ca^{2+}	10.96		
V^{2+}	12.70	V^{3+}	25.9
Mn^{2+}	14.04		
Fe^{2+}	14.33	Fe^{3+}	25.1
Co^{2+}	16.31		
Ni^{2+}	18.62		
Cu^{2+}	18.80		
Zn^{2+}	16.50		

The higher values of the stability constants for M^{3+} ions relative to those of M^{2+} ions are also noteworthy.

For a more detailed discussion see R. D. Hancock and A. E. Martell, *J. Chem. Educ.*, 1996, **73**, 654

Hard and soft character

Acids and bases may be classified as hard or soft. **Hard acids** are generally small and highly charged cations of electropositive metals and **hard bases** have electronegative donor atoms which are not highly polarizable. **Soft acids** are larger cations of less electropositive metals, often in a lower oxidation state and **soft bases** have less electronegative and more polarizable donor atoms. Some typical hard and soft acids and bases are summarized in Table 5.54.

Table 5.54 Some hard and soft acids and bases (underlining indicates donor atom)

Hard		Soft		Borderline	
Acids	Bases	Acids	Bases	Acids	Bases
Be^{2+}	R_2O	Ga^{3+}	R_2S	Fe^{2+}	Br^-
Al^{3+}	ROH	Cu^+	R_3P	Co^{2+}	py
H^+	NR_3	Ag^+	R_3As	Ni^{2+}	
Mn^{2+}	OH^-	Au^+	RNC	Cu^{2+}	
Sc^{3+}	F^-	Hg^+/Hg^{2+}	C_2H_4		
Cr^{3+}	Cl^-	Pd^{2+}	CO		
Fe^{3+}	SO_4^{2-}	Pt^{2+}/Pt^{4+}	CN^-		
Ti^{4+}	$CH_3CO_2^-$	Rh^+	I^-		
	$SC\underline{N}^-$	Ir^+	$\underline{S}CN^-$		

In general soft acids occur towards the right hand side of the transition series. Metals also become softer as the oxidation state is reduced. The presence of filled d orbitals which can back donate to the vacant π^* orbitals of ligands such as CO and C_2H_4 enhances their soft character.

Oxygen and sulfur are typical hard and soft atoms in crown ether ligands, e.g. the 18-crown-6 (O_6) and (O_4S_2) shown in the margin. The following equilibrium constants (values of log K_{298} at 298K) for K^+ and Ag^+ illustrate the effect of introducing the 'soft' S donor atoms.

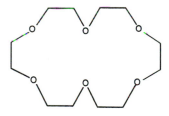

18-crown-6 (O_6)

	18-crown-6 (O_6)	18-crown-6 (O_4S_2)
K^+	6.0	1.1
Ag^+	1.6	4.4

Ag^+ forms a stronger complex with the sulfur containing ligands whereas K^+ forms a stronger complex with $[O_6]$.

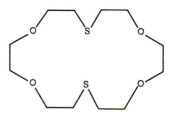

18-crown-6 (O_4S_2)

Examples of the applications of the hard-soft principle

Titanium(IV) exhibits many typical hard characteristics. $TiCl_4$ is readily hydrolyzed and forms alkoxides $Ti(OR)_4$. It forms a wide range of complexes with Me_3N, MeCN, Ph_3PO, and THF which are somewhat more stable than comparable complexes with PR_3 and SR_2. Vanadium(III) chloride is also converted into an alkoxide, $V(OR)_3$, with ROH and forms stable complexes with NH_3, MeCN, and oxygen ligands.

In contrast, platinum(II) does not readily form alkoxides, ethers, and hydroxo-complexes. It does form complexes with NH_3 and MeCN of the type $PtCl_2(NH_3)_2$ and $PtCl_2(MeCN)_2$, but it also forms very stable complexes with phosphines and thioethers, e.g. $PtCl_2(PR_3)_2$ and $PtCl_2(SR_2)_2$. It also forms stable complexes with CO and ethene, e.g. $PtCl_2(CO)_2$ and $[PtCl_3(C_2H_4)]^-$. Therefore, a knowledge of the hard or soft character of the metal ion allows educated guesses to be made about the ligands which are likely to form more stable complexes with a specific metal ion.

The hard and soft classification of metals is also useful for rationalizing the types of minerals found in nature for the transition elements. Fig. 5.78 illustrates that the elements to the left of the transition series are found as oxide or halide minerals whereas those on the right are found either as sulfide ores or in the native state.

For biological implications see J. J. R. Frausto da Silva and R. J. P. Williams, *The Biological Chemistry of the Elements*, OUP, Oxford, 1991

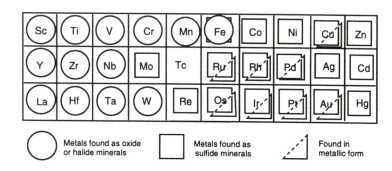

Fig. 5.78 Mineral types found in nature; occurrence of transition metals as minerals or in metallic unreacted forms

The hard and soft acid base principle is also very useful for understanding the environments found for metal ions in biologically active molecules.

Rates of substitution of metal complexes

Table 5.55 summarizes the rates of exchange of aqua-complexes across the transition series.

Table 5.55 Rates of water ligand exchange for some transition metal cations

Metal ion	Number of d electrons	k/s^{-1}	Metal ion	Number of d electrons	k/s^{-1}
Cr^{2+}	4	7×10^9	Cr^{3+}	3	3×10^{-6}
Mn^{2+}	5	3×10^7	Fe^{3+}	5	3×10^3
Fe^{2+}	6	3×10^6			
Co^{2+}	7	1×10^6			
Ni^{2+}	8	3×10^4			
Cu^{2+}	9	8×10^9			

Clearly the $M^{2+}(aq)$ ions are much more labile than the $M^{3+}(aq)$ ions. Within the $M^{2+}(aq)$ series the rates depend very much on the d configuration of the ions. The most inert ion is Ni^{2+}(aq), d^8, which has the maximum ligand field stabilization energies for a high spin complex. Either ligand addition or loss would result in some loss of this stabilization energy. The d^5, d^6, and d^7 ions have exchange rates of approximately 10^6–10^7 s^{-1} which is in the range anticipated for M^{2+} aqueous ions on the basis of size effects.

For these ions the ligand field stabilization energies for the octahedron are zero (Mn^{2+}, d^5) or small (Fe^{2+}, d^6, Co^{2+}, d^7). In contrast for the d^4 and d^9 complexes of Cr^{2+} and Cu^{2+} are particularly labile (rates $\approx 10^9$ s^{-1}) because complexes with these electronic configurations are known to have non-regular octahedral geometries. Generally, these complexes have tetragonally distorted geometries. This distortion arises from a stabilization of the d_{z^2} orbital in the e_g set of orbitals of an octahedron (see page 336).

For a fuller discussion of ligand field stabilization energies see page 339

Volatilities of compounds

In common with the compounds of the post-transition elements volatile compounds of the transition metals are obtained when the metal is neutral, and coordinatively saturated, so that it exists in a monomeric form in the gas phase and the solid. It is important that ligands surrounding the metal do not generate strong van der Waals interactions or form hydrogen bonds.

Coordinative saturation can only be achieved for simple monoanionic ligands in high oxidation states and therefore the hexafluorides and the hexa(methyl compounds) tend to be particularly volatile. Table 5.56 contains some relevant melting and boiling points. Mono-anionic ring compounds, e.g. C_5H_5 are able to achieve coordinative saturation by forming sandwich compounds (see margin for an illustration). Alternatively, chelating monoanionic ligands may be used, e.g. derivatives of acetylacetonates. The van der Waals interactions are reduced by replacing the CH_3 groups by CF_3 and the corresponding melting points are reduced in the $[M^{III}(acac)_3]$ derivatives summarized in the Table 5.56.

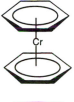

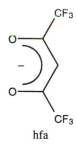

hfa

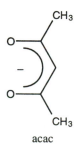

acac

Table 5.56 Melting and boiling points of some volatile transition metal compounds

	m.p. (°C)	b.p. (°C)
WF_6	2	17
OsO_4	41	subl. 130
$Fe(\eta\text{-}C_5H_5)_2$	173	subl. above 100
$Fe(\eta\text{-}C_5Me_5)_2$	291	subl.$_{0.1}$ 100–200
$Fe(CO)_5$	−20	103
$Ni(CO)_4$	−25	43
$Cr(CO)_6$	152-5	subl. in vacuo
$W(CO)_6$	150(dec.)	subl. in vacuo
$Cr(\eta\text{-}C_6H_6)_2$	284-5	subl. 160 in vacuo
$Cr((\eta\text{-}C_7H_7)(\eta\text{-}C_5H_5)$	230(dec)	subl.$_{0.1}$ 80–100
$Cr(CO)_3(\eta\text{-}C_6H_6)$	162-3	subl.$_{0.2}$ 130
$Cr(acac)_3$	216	subl. 240
$Cr(hfa)_3$	84-5	subl. in vacuo

subl.$_{0.1}$ = sublimes at 0.1 mm Hg pressure; subl.$_{0.2}$ = sublimes at 0.2 mm Hg pressure

Dianionic ligands such as O^{2-} can result in monomeric volatile high oxidation state compounds, e.g. OsO_4 and RuO_4. More generally OR and NR_2 ligands lead to polymeric derivatives unless the R groups are sufficiently bulky to prevent oligomerization reactions. For example,

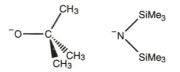

Low oxidation state molecular and volatile compounds may be obtained by using neutral π-acid ligands such as CO, C_6H_6, and PMe_3. Some relevant examples and their m.ps are given in Table 5.56.

These volatile compounds have been extensively studied in recent years as molecular precursors for forming thin films of semi-conducting, magnetic, or ferroelectric compounds on silicon chips.

Substituent effects

Electronic effects

Transition metals have a large range of available oxidation states and are capable of complexing to a wide range of ligands. The differing electron donating and accepting properties of these ligands lead to modifications of the chemical properties which may exceed those brought about by changes in the oxidation state. The steric requirements of ligands may also be modified by changing the sizes of the organic substituents on the ligand.

Relative stabilization of oxidation states

Changing the ligands can also influence the redox properties of transition metal complexes and thereby affect the relative stabilities of oxidation states of transition metal complexes. Table 5.57 summarizes the standard redox potentials for a range of cobalt(III)/cobalt(II) complexes in aqueous solution.

Table 5.57 Redox potentials for some Co^{III}/Co^{II} complexes

Oxidized form			Reduced form	$E^\ominus/V$
$Co^{3+}(aq)$	$+e$	$\rightarrow$	$Co^{2+}(aq)$	1.94
$[Co(edta)]^-$	$+e$	$\rightarrow$	$[Co(edta)]^{2-}$	0.38
$[Co(bipy)_3]^{3+}$	$+e$	$\rightarrow$	$[Co(bipy)_3]^{2+}$	0.34
$[Co(NH_3)_6]^{3+}$	$+e$	$\rightarrow$	$[Co(NH_3)_6]^{2+}$	0.10
$[Co(en)_3]^{3+}$	$+e$	$\rightarrow$	$[Co(en)_3]^{2+}$	-0.80
$[Co(CN)_6]^{3-}$	$+e$	$\rightarrow$	$[Co(CN)_5]^{4-}$	-1.00

All of the complexes are conventional coordination complexes with commonly available ligands yet $[Co(H_2O)_6]^{3+}$ is a stronger oxidizing agent than chlorine ($E^\ominus(Cl_2/Cl^-) = 1.36$ V) and $[Co(CN)_5]^{4-}$ is almost as strongly reducing as manganese metal and more strongly reducing than zinc metal. Therefore, merely by changing the ligands the properties of these metals in these two oxidation states have been radically altered. In this series of compounds the variations in $E^{\ominus}$ can be directly related to the ability of the ligands to interact with the d orbitals. Ligands which are high in the spectrochemical series such as CN^- stabilize the Co^{3+} oxidation state preferentially, because this oxidation state has a low spin d^6 configuration and achieves the maximum ligand field stabilization energy.

Table 5.58 summarizes similar data for the complexes of iron(II) and iron(III). Dioxygen ($E^{\ominus} = 1.23$ V) is a sufficiently strong oxidizing agent to oxidize $Fe^{II}(aq)$ to $Fe^{III}(aq)$, although the process is a little slow.

Table 5.58 Redox potentials for some Fe^{III}/Fe^{II} complexes

Oxidized form			Reduced form	$E^{\ominus}/V$
$[Fe(CN)_6]^{3-}$	+ e	$\rightarrow$	$[Fe(CN)_6]^{4-}$	0.36
$[Fe(H_2O)_6]^{3+}$	+ e	$\rightarrow$	$[Fe(H_2O)_6]^{2+}$	0.77
$[Fe(tmp)_3]^{3+}$	+ e	$\rightarrow$	$[Fe(tmp)_3]^{2+}$	0.83
$[Fe(phen)_3]^{3+}$	+ e	$\rightarrow$	$[Fe(phen)_3]^{2+}$	1.13
$[Fe(5-NO_2phen)_3]^{3+}$	+ e	$\rightarrow$	$[Fe(5-NO_2phen)_3]^{2+}$	1.25

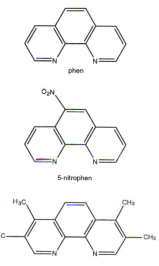

phen

5-nitrophen

tmp = tetramethylphen

The oxidation process becomes more thermodynamically favourable if the H_2O ligands are replaced by CN^-, presumably because the CN^- ligand forms a more stable complex with the more highly charged Fe^{3+} ion. However, the phenanthroline ligands which possess delocalized empty π^* molecular orbitals which are able to accept electron density from the filled metal d orbitals stabilize the Fe^{2+} oxidation state preferentially. Electron withdrawing nitro-groups on the phenanthroline ligands encourage the electron donation from metal to ligand π^* and stabilize the Fe^{2+} state more effectively than the unsubstituted phenanthroline or the electron donating methyl substituted ligands. The discriminating complexing powers of ligands is a general aspect of transition metal chemistry and carries with it the following implications.

1. Specific oxidation states may be stabilized by suitably chosen ligands. Broadly speaking high oxidation states may be stabilized by π-donor ligands which have high electronegativities such as F^-, O^{2-}, N^{3-}, NR^{2-}, and O_2^{2-} and low oxidation states are stabilized by π-acceptor ligands such as CO, C_2H_4, CNR, and NO^+, phenanthroline, and bipyridyl.

2. Since the geometric preferences of metals in different oxidation states may differ, a ligand with a particular rigid backbone which creates a well defined coordination environment can lead to a preferential stabilization of that oxidation state.

3. Similarly, the size of an ion may change significantly on oxidation or reduction particularly if the oxidation state change is associated with a low spin to high spin transformation. Therefore, a ligand which has a rigidly defined cavity size may preferentially distinguish between the oxidation states.

4. The differential solubilities of compounds in the two oxidation states may also influence the redox couple. For example, the solubilities of metal hydroxides of the transition elements are very sensitive to changes in oxidation states. Both $M(OH)_2$ and $M(OH)_3$ are highly

insoluble for the first row transition metals, but the $M(OH)_3$ are much less soluble (sometimes by twenty orders of magnitude) than the corresponding $M(OH)_2$. Consequently, addition of bases to aqueous M^{2+} ions usually encourages aerial oxidation of the $M(OH)_2$ precipitate which is initially formed. Similarly, the insolubility of CuI results in the stabilization of Cu^I relative to Cu^{II} when I^- is added to Cu^{2+}(aq):

M(OH)₂ and M(OH)₃ are better described as hydrated metal oxides

$$Cu^{2+}(aq) \ + \ 2I^- \ \rightarrow \ CuI\downarrow \ + \ {}^1\!/_2 I_2$$

These important discriminatory effects may be used for the selective extraction of transition elements and are also widely utilized in biological systems to fine tune the properties of transition metal ions in proteins and enzymes.

Lewis acid and base behaviour

In general terms the replacement of π-acid ligands in a low oxidation compound by ligands which are good σ-donors but poor π-acceptors makes the complex a better Lewis base. For example,

$$Ni(CO)_4 + H^+ \ \rightarrow \ \text{No reaction}$$
$$Ni(PPh_3)_4 + H^+ \ \rightarrow \ [NiH(PPh_3)_4]^+$$

In this example replacement of CO by the poorer π-acceptor PPh_3 increases the basicity of the metal centre. The donor and acceptor properties of the phosphine are greatly influenced by the electronegativities of the groups attached to phosphorus atoms, e.g.

$$\sigma\text{-donor ability} \quad PMe_3 > PPh_3 > P(OMe)_3 > PF_3$$
$$\pi\text{-acceptor ability} \quad PMe_3 < PPh_3 < P(OMe)_3 < PF_3$$

So a compound such as $Ni(PF_3)_4$ behaves like $Ni(CO)_4$ rather than $[Ni(PPh_3)_4]$ and is a poor Lewis base.

The ligand effect is also reflected in the K_a acid dissociation constants given in Table 5.59.

$$HM(CO)_4 \ \underset{}{\overset{K_a}{\rightleftharpoons}} \ H^+ \ + \ M(CO)_4^-$$

Table 5.59 Acid dissociation constants for some metal carbonyl hydrides

	K_a
$HCo(CO)_4$	~1
$HCo(CO)_3(PPh_3)$	1×10^{-7}
$HFe(CO)_4^-$	$\sim 4 \times 10^{-14}$

The electron releasing PPh_3 group discourages dissociation of the proton from $HCo(CO)_3(PPh_3)$ very significantly and the observed K_a is halfway between that of $HCo(CO)_4$ and the negatively charged $HFe(CO)_4^-$ anion.

The electronic properties of the ligand also influence the ability of the metal to undergo oxidative addition reactions.

Relative rates: Y = I > Br > Cl
X = F > Cl > Br > I
L = PEt₃ > PEt₂Ph > PEtPh₂ > PPh₃ > (PhO)₃P

In general, the rate of the oxidative–addition reaction increases with the donating properties of the ligands because the reaction involves transfer of an electron pair from the metal to the substrate (CH_3X in this case).

Ionization energies of molecular compounds

For compounds whose redox properties and acid–base properties cannot be studied in solution the effect of the ligands on the metal's properties can be estimated by using gas phase photoelectron spectroscopy. The molecules have to be sufficiently volatile to produce a reasonable vapour pressure, then the electrons from the highest occupied molecular orbitals may be ionized using ultraviolet light. In this technique the ionization energy of a molecule as a whole is measured, i.e. the enthalpy change for the reaction:

$$ML_n \rightarrow ML_n^+ + e^-$$

where both ML_n and ML_n^+ are in their ground electronic states. The ionization energies of some molybdenum compounds are summarized in Table 5.60. In these compounds the first ionization energy results from the removal of an electron which has a high proportion of d orbital character.

The manner in which the ligand donating properties transcend differences in oxidation state is illustrated by the similarity in ionization energies for $Mo(PF_3)_6$ and $MoCl_5$ whose formal oxidation states differ by five. Conversely $Mo(PF_3)_6$ and $Mo(\eta\text{-}C_6H_6)_2$ have the same formal oxidation state, but first ionization energies which differ by nearly 300 kJ mol^{-1}. Chemically, this means that $Mo(\eta\text{-}C_6H_6)_2$ is a much stronger Lewis base and more likely to undergo oxidative addition reactions and form stable oxidized products than $Mo(PF_3)_6$ or $Mo(CO)_6$.

Table 5.60 Ionization energies of some molybdenum compounds /kJ mol^{-1}

$Mo(PF_3)_6$	888
$Mo(CO)_6$	820
$Mo(\eta\text{-}C_4H_6)_3$	695
$Mo(\eta\text{-}C_6H_6)_2$	533
$MoCl_5$	894

Acid-base character of oxides

The oxidation state of the metal and the ligand environment influences the acid–base properties of the metal ion. In a well defined series of compounds such as the oxides the behaviour is reminiscent of that described earlier for the post-transition metals. The oxides become more acidic as the metal oxidation state is increased. Some illustrative data are given in Table 5.61.

Table 5.61 Properties of some oxides of Ti, V , and Cr

Oxidation state	+2	+3	+4	+5	+6
	TiO	Ti_2O_3	TiO_2		
	VO	V_2O_3		V_2O_5	
	CrO	Cr_2O_3			CrO_3
	Basic	Amphoteric		Acidic	

The oxides in +2 oxidation states dissolve in acids to form $[M(OH_2)_6]^{2+}$ whereas the high-oxidation state oxides form oxyanions with alkalis, e.g. $[CrO_4]^{2-}$, $[VO_4]^{3-}$, and $[TiO_4]^{4-}$.

Tendency to polymerize

For a series of compounds with the same ligands the compounds tend to become more molecular as the oxidation state of the metal increases. Some examples are given in Table 5.62. The lower oxidation state compounds have a deficiency of ligands and therefore polymerize, e.g. halides form infinite structures with halide bridges.

Table 5.62 Properties of Ti and V chlorides which vary with oxidation state

High melting point solids, insoluble in organic solvents		Liquids, soluble in organic solvents	
$TiCl_2$	$TiCl_3$	$TiCl_4$	
VCl_2	VCl_3	VCl_4	$VOCl_3$

Similarly, if the halide is changed the tendency to polymerize changes. For example, TiF_4 (b.p. 284°C) is an infinite solid, $TiCl_4$ (b.p. 136°C), $TiBr_4$ (b.p. 230°C), and TiI_4 (b.p. 377°C) are molecular solids with the tetrahedral molecules held only by van der Waals forces. The oxo-anions $[MO_4]^{x-}$ show an increased tendency to polymerize as the charge increases in an isoelectronic series:

$$[MnO_4]^-$$

$$[CrO_4]^{2-} \xrightarrow{\text{H}^+} [Cr_2O_7]^{2-}$$

$$[VO_4]^{3-} \xrightarrow{\text{H}^+} [V_3O_9]^{3-} + [V_{10}O_{28}]^{6-}$$

A similar trend was noted for oxo-polyanions of the post-transition elements in Chapter 3. As the negative charge increases on $[MO_4]^{x-}$ the oxo-ligands become more nucleophilic and more able to attack a metal site of another anion. Also the resultant polynuclear oxo-polyanion is able to delocalize the greater negative charge over the increased number of centres.

Steric effects

The organic substituents on phosphine ligands not only change the electronic properties of the ligands but may also influence their steric requirements. Attempts have been made to quantify these effects for phosphines by measuring the pK_a values of the corresponding conjugate acids $[HPR_3]^+$, which provide an estimate of the donor properties of the ligands, and the Tolman cone angle which estimates the steric requirements of the ligands. The latter is estimated from the cone which radiates from the metal and just touches the van der Waals surfaces of the hydrogen atoms on the surface of the ligand as shown in Fig. 5.79. Table 5.63 summarizes the pK_a values and Tolman cone angles for a wide range of phosphine ligands. It is significant that the cone angle may be doubled by varying the organic group from (OMe) 107° to (2,4,6-$Me_3C_6H_2$) 212° and the pK_a values from 0.03 (PPh_2H) to 11.40 (PBu^t_3). These variations have enabled chemists to 'fine-tune' the reactivities of metal complexes such as $RhCl(PPh_3)_3$, $RhH(CO)(PPh_3)_3$, and $Ni(P(OPh)_3)_4$ which are homogeneous catalysts.

The dissociation of phosphine ligands in $Pd(PR_3)_4$ appear to be dominated by steric effects and the dissociation constant K defined below:

$$Pd(PR_3)_4 \underset{}{\overset{K}{\rightleftharpoons}} Pd(PR_3)_3 + PR_3$$

increases in the order:

$$PPh(Bu^t)_2 > PCy_3 > P(Pr^i_3) > PPh_3 \sim PEt_3 > PMePh_2 \sim PMe_2Ph \sim PMe_3$$

Table 5.63 Tolman cone angles for phosphine ligands

Phosphine	Cone angle	pK_a
$P(OMe)_3$	107	2.6
PMe_3	118	8.65
$PhP(OMe)_2$	120	2.64
$PhPMe_2$	122	6.5
Ph_2PH	126	0.03
$P(OPh)_3$	128	−2.00
$P(OPr^i)_3$	130	4.08
Ph_2POMe	132	2.69
PEt_3	132	8.69
PBu^n_3	132	8.43
Ph_2PMe	136	4.57
PPh_3	145	2.73
$P(p\text{-}MeC_6H_4)_3$	145	3.84
$P[(p\text{-}MeO)C_6H_4)_3]$	145	4.59
PPr^i_3	160	–
PBz_3	165	6.0
$P(m\text{-}MeC_6H_4)_3$	165	3.30
PCy_3	170	9.70
PBu^t_3	182	11.40
$P[(2,6\text{-}MeO)_2C_6H_3)_3]$	184	9.33
$P[(2,4,6\text{-}MeO)_3C_6H_2)_3]$	184	11.02
$P(o\text{-}MeC_6H_4)_3$	194	3.08
$P(2,4,6\text{-}Me_3C_6H_2)_3$	212	7.3

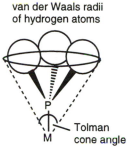

van der Waals radii
of hydrogen atoms

Fig. 5.79 An illustration of the Tolman cone angle and its dependence upon the van der Waals radii of the ligands

D. White and N. J. Coville, *Adv. Organomet. Chem.*, 1994, **25**, 36 and T. E. Müller and D. M. P. Mingos, *Trans. Met. Chem.*, 1995, **26**, 1

Changing the donor atom in these ligands influences the electronic characteristics of the ligand less than the steric effects.

The relative tendency of Group 15 ligands to replace amines has been estimated by measuring K for the following reaction:

$$W(CO)_5(amine) + L \xrightarrow{K} W(CO)_5L + amine$$

K increases in the order:

$$Ph_3Bi < (PhO)_3P \sim Ph_3Sb < Ph_3As < Ph_3P < (BuO)_3P < PCy_3 < PBu_3$$

This suggests the stabilities of the phosphine complexes are favoured by small ligands with good donor substituents.

The cone angles for other commonly encountered ligands are given in Table 5.64. The coordination of sterically demanding ligands to metals has the following consequences.

Table 5.64 Tolman cone angles (°) for some common ligands

H	75
F	92
Cl	102
Br	105
I	107
Me	90
Et	102
Pr^i	114
Bu^t	126
C_5H_5	136
CO	95

1. Low coordination numbers

The ligands N(SiMe₃)₂ and PCy₃ are particularly effective at promoting low coordination numbers and some relevant examples are shown in Fig. 5.80.

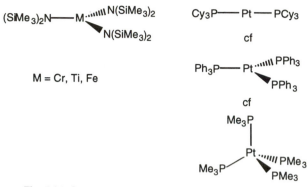

M = Cr, Ti, Fe

Fig. 5.80 Some complexes with low coordination numbers

See F. G. N. Cloke, *Chem. Soc. Rev.*, 1993, **22**, 17 for an illustration of the application of steric effects for isolating low oxidation state sandwich compounds.

2. Multiple bonds

Large substituents protect the multiple bond from nucleophiles and thereby make the compound more inert than those with smaller ligands. The following reaction occurs readily, but this is not the case with the corresponding reaction for the related unsubstituted cyclopentadienyl compound.

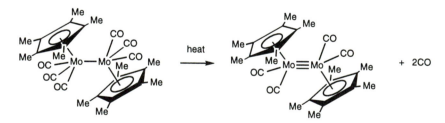

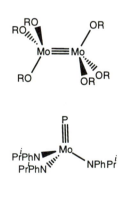

The sterically demanding alkoxide ligands, OCH₂SiMe₃, and OBuᵗ, have similarly been used to stabilize the following multiply bonded compounds.

3. Protecting reactive ligands on the metal centre

For example, the phosphido-ligand P³⁻ has recently been stabilized in the protected environments provided by NPhPrⁱ ligands.

Similarly, the dihydrogen complex was first isolated in a molybdenum compound where it is protected by the PCy₃ ligands. Examples of the complexes are illustrated in the margin

The variation of the organic substituent may also be used to kinetically trap specific metal-ligand combinations. For example, the transition metal alkyls decompose readily by the following β-elimination pathway which results in an unstable hydrido-olefin complex:

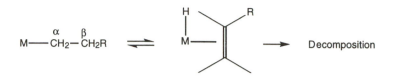

This decomposition route may be blocked either by choosing organic ligands which do not have β-hydrogen atoms, e.g. CH_2SiMe_3, CH_3, CH_2Ph, or by introducing additional ligands at the metal centre which are so strongly bound and inert that they block the hydrogen migration, e.g.

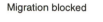

5.7 Vertical trends associated with the transition elements

Atomic properties

Atomic sizes

The radii of the atoms in a particular column of transition metals increase significantly between the first and second rows, but then there is only a slight increase between second and third rows. This may be attributed to the interpositioning of the 4f shell between the second and third row transition elements and the phenomenon is described as the *lanthanide contraction*. As a consequence the coordination chemistry of the second and third row elements show remarkable similarities, but their redox and substitution chemistry, which is less influenced by size effects, remains distinctive.

The increase in radii down a column of a transition series and particularly between the first two rows has a direct influence on the preferred coordination numbers. Table 5.65 provides some specific examples of these general trends.

Table 5.65 Coordination geometries of some Group 5 fluoro-complexes

	V	Nb	Ta
Octahedral (6)	$[VF_6]^-$	$[NbF_6]^-$	$[TaF_6]^-$
Capped trigonal prismatic (7)		$[NbF_7]^{2-}$	$[TaF_7]^{2-}$
Square antiprismatic (8)		$[NbF_8]^{3-}$	$[TaF_8]^{3-}$

In general complexes with coordination numbers 7, 8, and 9 are much more common for the second and third row elements and particularly those at the left of the transition series. The relative ease with which the metals expand their coordination numbers is reflected in the behaviour of the oxo-anions of chromium, molybdenum, and tungsten as a function of pH. All

Summary of important differences down a column

1. The second and third row transition metals provide examples of compounds in higher oxidation states.

2. Down a column the metals show a greater variety of metal–metal bonded compounds.

3. The metal–metal and metal–ligand bond enthalpies increase down a column.

4. The compounds become more substitutionally inert.

5. The metals become more noble.

6. The d orbital splitting energies increase, therefore high spin complexes are not observed for the second and third row transition metal complexes.

7. The coordination numbers of complexes of the heavier elements are greater.

three tetrahedral metal oxo-anions oligomerize as the solution is made more acid. The $[CrO_4]^{2-}$ ion oligomerizes at low pH values predominantly to $[Cr_2O_7]^{2-}$, although under more strongly acid conditions $[Cr_3O_{10}]^{2-}$ and $[Cr_4O_{13}]^{2-}$ are also formed. All these anions have structures based on tetrahedral CrO_4 moieties sharing corners. The tetrahedral $[MoO_4]^{2-}$ on acidification forms $[Mo_7O_{24}]^{6-}$ as the predominant species, and $[Mo_8O_{26}]^{4-}$ and $[Mo_{36}O_{112}]^{8-}$ at lower pH values. In addition, the anions $[Mo_7O_{24}]^{6-}$, $[Mo_6O_{19}]^{2-}$, and $[Mo_{10}O_{34}]^{x-}$ have also been isolated from these solutions. The majority of these anions have MoO_6 octahedra which are linked by shared corners and edges (some examples of the structures are shown in the margin). Some MoO_4 tetrahedra are retained, but they are very much in the minority. Moreover, $[Mo_6O_{12}(OH_2)_{16}]^{8-}$ contains some pentagonal bipyramidal (7-coordinate centres). The $[WO_4]^{2-}$ ion under similar conditions forms $[W_6O_{21}]^{5-}$ and $[W_{12}O_{42}H_2]^{10-}$ which are also based on edge and corner sharing octahedra. The higher coordination numbers accessible to Mo and W therefore have a large effect on the structures of the isopolyanions.

The size differences also influence the changes in geometry which accompany the crystallization process. For example, $TiBr_4$ is tetrahedral in the gas phase and in the solid, and $ZrBr_4$ and $HfBr_4$ are tetrahedral in the gas phase, but adopt the SnI_4 structure with edge sharing octahedra in the solid state.

Ionization energies and electronegativity effects

The first ionization energies of the transition elements belonging to a common Group generally increase. This trend is opposite to that noted earlier for the pre-transition elements and the majority of the post transition elements. The data presented in Table 5.66 for nickel, palladium, and platinum are reasonably representative.

$[Mo_7O_{24}]^{6-}$

$[Mo_6O_{19}]^{2-}$

Table 5.66 Ionization energy data for nickel, palladium, and platinum / kJ mol^{-1}

	I_1	I_2	I_3	I_4	ΣI_{1-4}
Ni	737	1752	3393	5404	11285
Pd	804	1874	3176	4729	10583
Pt	865	1791	2799	3957	9412

It is significant that, although the I_1 energies follow the order Pt > Pd > Ni, the sum of the first four ionization energies, given in the last column, follow the order Ni > Pd > Pt.

Nickel is more electropositive than palladium and platinum and dissolves in dilute acids to form Ni^{II}, whereas *aqua regia* is required for the heavier members of the Group. Conversely, there are many stable compounds of platinum(IV), a few for palladium(IV), and very few for nickel(IV). It is simplistic to relate the occurrence of compounds directly to the ionization energies because other terms such as enthalpies of atomization, hydration enthalpies, and lattice enthalpies contribute to the Hess's Law thermodynamic cycle, but nonetheless the fact that the ionization energies of the heavier elements increase less as the charge on the atom increases is a major contributing factor to the general observation that these elements form

β-$[Mo_8O_{26}]^{4-}$

more compounds in higher oxidation states. Similar trends in ionization energy data are apparent for the earlier groups of transition elements.

The evidence for the relative stabilization of the higher oxidation states for the heavier transition elements is both phenomenological, e.g. the existence of RuO_4 and OsO_4 and the non-isolation to date of FeO_4, and observable in thermodynamic data, for example in standard reduction potential data summarized in Table 5.67.

Table 5.67 Standard reduction potentials for compounds in higher oxidation states.

Oxidized form		Reduced form	E°/ V	Oxidized form		Reduced form	E°/ V
$[MnO_4]^-$	$\rightarrow$	Mn^{3+}	1.51	$[CrO_4^{2-}]^*$	$\rightarrow$	Cr	−0.8
$[TcO_4]^-$	$\rightarrow$	Tc^{2+}	0.50	$[MoO_4]^{2-*}$	$\rightarrow$	Mo	−0.91
$[ReO_4]^-$	$\rightarrow$	Re^{3+}	0.42	$[WO_4^{2-}]^*$	$\rightarrow$	W	−1.07
FeO_4	$\rightarrow$	FeO_4^-	reduced form not known	$[FeO_4]^{2-*}$	$\rightarrow$	Fe^{3+}	2.20
RuO_4	$\rightarrow$	RuO_4^-	1.04	$[RuO_4]^{2-*}$	$\rightarrow$	Ru^{2+}	1.56
OsO_4	$\rightarrow$	OsO_4^-	0.83	$[OsO_4]^{2-*}$	$\rightarrow$	Os^{2+}	0.99

*Under standard basic conditions, i.e. $a(OH^-) = 1$

The data of Table 5.67 confirm the following relative oxidizing abilities:

$$[MnO_4]^- > [TcO_4]^- > [ReO_4]^-$$
$$[FeO_4]^{2-} > [RuO_4]^{2-} > [OsO_4]^{2-}$$
$$RuO_4 > OsO_4$$

It is noteworthy that the relative stabilization of higher oxidation states down the column is in marked contrast to that noted for the post-transition metals where the higher oxidation state compounds become less stable down the column, i.e. the inert pair effect. These differences have their origins in the different trends in atomic properties and particularly the sizes of the orbitals and their ability to overlap strongly with the orbitals of other atoms. For the post transition elements the homonuclear and heteronuclear bond enthalpy contributions generally decrease down the column, whereas for the transition metals they increase.

It also follows that the +2 and +3 oxidation states, which are ubiquitous in the aqueous chemistry of the first row transition metals, are less important for the second and third row elements and particularly for the metals on the left hand side. These oxidation states are important for the later transition metals, e.g. Ru, Ir, Rh, Pd, Pt, and Au, but with chloro-, bromo-, thiocyanato-ligands rather than as aqua-complexes.

As noted previously (page 279) the electronegativities of the transition elements follow the order: first row > second row < third row. However, the differences in electronegativity for a particular column of elements are not very large. Also, the quoted electronegativity, based on ground state properties of the atom, is likely to be more representative of the properties of the atom in its lower oxidation state compounds than its higher oxidation state. Since the properties of the metal atoms are so strongly influenced by the ligands coordinated to it then even within one oxidation state the effective electronegativity of the metal may vary considerably as a function of the donor and acceptor properties of the ligands.

The diagrams shown in Fig. 5.81 give schematic illustrations of the commonly observed oxidation states of the three rows of the transition elements and an indication of those that are the most common in aqueous solutions. These diagrams reinforce the generalizations made in the preceding paragraphs.

Charge transfer transitions and colours

Charge transfer transitions are also discussed on page 101, where it was noted that the observed band position depends primarily on the electronegativity difference between the ligand and the metal. Charge transfer bands are commonly observed in transition metal complexes and in higher oxidation state compounds they generally originate from ligand–metal charge transfer processes. In lower oxidation state compounds then it is also possible to observe metal to ligand charge transfer bands. The colours of some compounds where charge transfer transitions dominate the spectral characteristics of the compounds are summarized in Table 5.68.

Table 5.68 Colours of some transition metal compounds

Ligand to metal charge transfer		
MnO_4^-	TcO_4^-	ReO_4^-
Purple	Colourless	Colourless
CrO_4^{2-}	MoO_4^{2-}	WO_4^{2-}
Yellow	Colourless	Colourless
MnO_3F	TcO_3F	ReO_3F
Dark green	Yellow	Yellow
Metal to ligand charge transfer		
$Fe(CO)_5$	$Ru(CO)_5$	$Os(CO)_5$
Yellow	Colourless	Colourless
Metal-metal transitions		
$Co_4(CO)_{12}$	$Rh_4(CO)_{12}$	$Ir_4(CO)_{12}$
Black	Orange	Pale yellow
$Fe_3(CO)_{12}$	$Ru_3(CO)_{12}$	$Os_3(CO)_{12}$
Dark green	Orange	Pale yellow

It is significant that for both types of transition the energy of the band moves towards the ultraviolet for the second and third row transition elements. It is also noticeable that the second and third row elements have very similar colours reflecting their similar electronegativities. The relatively small increase in electronegativity across the series leads to only a minor change in colour as in the following series: WF_6, ReF_6, OsF_6, (colourless) IrF_6 (yellow), PtF_6 (red).

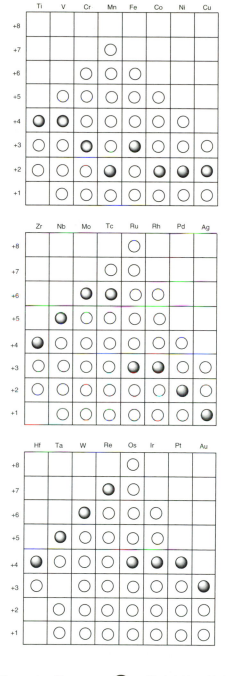

Fig. 5.81 Common oxidation states for transition metal compounds

For the sake of completeness the colours of some cluster compounds where the metal–metal transitions are primarily responsible for the colours are also given in this table. The compounds become less highly coloured down the column because the energy separation between the bonding and antibonding molecular orbitals becomes larger because the 4d and 5d orbitals overlap with each other more effectively than the 3d.

Finally, Table 5.69 gives a series of isostructural octahedral tungsten compounds where the oxidation state of the metal is varied from VI to III.

Table 5.69 Colours of isostructural tungsten compounds where the oxidation state varies

WCl_6	WCl_6^-	WCl_6^{2-}	WCl_6^{3-}
W^{VI}	W^V	W^{IV}	W^{III}
Blue/black	Dark green	Orange/brown	Yellow

As the oxidation state on the central tungsten ion increases then the charge transfer transition moves into the visible, i.e. the electronegativity of the metal is increasing, thereby lowering the energy required for the formal transfer of an electron from the ligand to the metal.

Orbital size effects

In the previous section it was noted that for the first row transition elements the 3d orbitals are very contracted and only overlap effectively with ligand orbitals if the ligand approaches the metal at distances significantly shorter than the sum of the covalent radii. When the principal quantum number is increased to 4 and 5 the additional radial nodes in the wave functions move the maximum in the radial distribution functions further away from the nucleus as shown in Fig. 5.82.

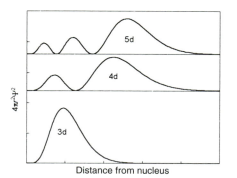

Fig. 5.82 Radial distribution functions for 3d, 4d, and 5d orbitals. The 4d and 5d functions are plotted further up the rds axis for reasons of clarity

The 4d and 5d orbitals overlap more effectively with ligand orbitals when the metal–ligand distances are equal to approximately the sum of the covalent radii. Also, the metals in these rows generally overlap more effectively with metal atoms located at distances which approximate to those found in the bulk metal. Besides forming stronger σ-covalent bonds the improved overlaps of the second and third row metals also result in stronger multiple bonds either to main-group atoms or to other metals.

The bonding abilities of the valence orbitals of the third row transition metals are also influenced by relativistic effects. These have the effect of contracting the 6s orbitals and thereby making them better acceptor orbitals and expanding the 5d orbitals. The latter effect also improves the overlapping capabilities of these orbitals. It may seem implausible that both the expansion and the contraction of orbitals can lead to better overlaps. However, it has to be remembered that the best overlap is achieved when the contributed orbitals are well matched in size. If one of the orbitals is more diffuse than this optimum then contraction leads to improved overlaps. Conversely, if an orbital is too small then its expansion also leads to better overlaps.

The following analogy may be helpful. A particular size of garment may not fit two individuals who have very different weights. The smaller individual could attempt to achieve a better fit by putting on weight, whereas the larger would have to lose weight.

Orbital splittings and pairing energies

The superior overlaps between the 4d and 5d orbitals and ligand orbitals leads to larger ligand field splitting energies, Δ. Some of the relevant comparative data are summarized in Table 5.70.

Table 5.70 Orbital splittings(Δ) in octahedral complexes /kJ mol^{-1}

Complex	Δ(kJ mol^{-1})	Complex	Δ(kJ mol^{-1})	Complex	Δ(kJ mol^{-1})
$[MnF_6]^{2-}$	261	$[Co(en)_3]^{3+}$	287	$[Ni(PF_3)_4]^+$	102
$[TcF_6]^{2-}$	340	$[Rh(en)_3]^{3+}$	419	$[Pd(PF_3)_4]^+$	222
$[ReF_6]^{2-}$	392	$[Ir(en)_3]^{3+}$	490	$[Pt(PF_3)_4]^+$	252

As anticipated the largest increase occurs between the first and second transition series and a smaller increase is observed between the second and third row. The more diffuse 4d and 5d orbitals are able to accommodate the electrons in a larger volume of space and consequently they experience less electron–electron repulsion. Therefore, electron–electron repulsion is reduced as is the pairing energy. The increased Δ values, the reduced electron–electron repulsion, and pairing energies all come together to favour the formation of low spin rather than high spin complexes. Therefore, there are very few examples of high spin complexes of the second and third row elements. The relative magnitudes of these important energy terms are summarized in Table 5.71.

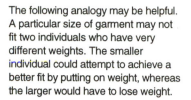

en

Table 5.72 gives some relevant magnetic moments for *tris*-diketonates of iron, ruthenium, and osmium. The iron complex is d^5 high spin ($S = \frac{5}{2}$) whereas the ruthenium and osmium complexes are d^5 low spin ($S = \frac{1}{2}$).

Table 5.72 Magnetic moments of Group 8 *tris*-diketonates

	μ_{eff}/μ_B
$[Fe(acac)_3]$	5.90 (high spin)
$[Ru(acac)_3]$	1.95 (low spin)
$[Os(acac)_3]$	1.81 (low spin)

Table 5.71 Important energy terms for transition element complexes /kJ mol^{-1}

	Ligand field splitting energy, Δ	Electron–electron repulsion energy	Spin-pairing energy	Spin-orbit coupling energy
3d^n	179	144	72	5
4d^n	259	96	48	12
5d^n	299	60	36	30

diars =

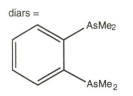

See also C. D. Holt, *Prog. Inorg. Chem.*, 1993, **40**, 503 for a more detailed discussion of metal–ligand bond enthalpies.

Table 5.74 Variation of E(M–Cl) with oxidation state

	E /kJ mol^{-1}
TiCl$_2$	508
TiCl$_3$	458
TiCl$_4$	429
WCl$_4$	414
WCl$_5$	373
WCl$_6$	345

The assumption which arises from using the solutions of the Schrödinger equation for the hydrogen atom in polyelectronic atoms and the associated separation of spin and orbital angular momentum becomes increasingly less reliable as the atomic number increases. The extent of coupling between the spin and orbital angular momentum—the spin-orbit coupling constant—increases down the transition series. This has the effect of making the magnetic properties of the transition metals more complicated, because the spin only formula is no longer reliable. Some representative data illustrating this point are given in Table 5.73.

Table 5.73 Magnetic moments for transition metal compounds belonging to the same periodic group

	μ_{eff}/μ_B	$2\sqrt{(S(S+1))}$
[Fe(diars)$_2$Cl$_2$]$^+$	2.34	1.73
[Ru(diars)$_2$Cl$_2$]$^+$	1.91	1.73
[Os(diars)$_2$Cl$_2$]$^+$	1.42	1.73
[Fe(diars)$_2$Cl$_2$]$^{2+}$	2.98	2.83
[Os(diars)$_2$Cl$_2$]$^{2+}$	1.25	2.83

Bond enthalpy contributions

Table 5.74 summarizes the bond enthalpy contributions for some metal chlorides in the gas phase. The decrease in E(M–Cl) with increasing oxidation state and coordination number is a trend which was noted previously for the representative elements.

Table 5.75 gives comparable data for metals in the same column of the Periodic Table. The bond enthalpy contributions generally increase down the columns for the majority of bond types and reflect the increased orbital overlaps between the metal d orbitals and the ligand orbitals.

Table 5.75 Bond enthalpy contributions /kJ mol^{-1}

	E(M–CO)		E(M–C)		E(M–CO)	E(M–M)
Cr(CO)$_6$	107	Ti(CH$_2$Ph)$_4$	260	Fe$_3$(CO)$_{12}$	117	82
Mo(CO)$_6$	151	Zr(CH$_2$Ph)$_4$	310	Ru$_3$(CO)$_{12}$	172	117
W(CO)$_6$	178			Os$_3$(CO)$_{12}$	190	130
		(OC)$_5$MnCH$_3$	150	Co$_4$(CO)$_{12}$	136	83
Fe(CO)$_5$	117	(OC)$_5$ReCH$_3$	220	Rh$_3$(CO)$_{12}$	166	114
	180			Ir$_4$(CO)$_{12}$	110	130

	E(M–F)	E(M–Cl)	E(M–Br)	E(M–I)	E(M–CH$_3$)	E(M–NMe$_2$)	E(M–OPr^i)
TiX$_4$	586	429	367	244	260	307	444
ZrX$_4$	648	491	425	350	310	350	518
HfX$_4$	605	497			330	370	534

Effective atomic number (EAN) rule

The formation of stronger bonds by the heavier transition elements also leads to a more pronounced tendency of such elements to form compounds which conform to the *effective atomic number rule*. Table 5.76 compares the range of compounds which conform to the effective atomic number rule for chromium and molybdenum. For molybdenum such compounds exist for the oxidation states −4 to +6, whereas for chromium only from oxidation states −4 to +2. Compounds of Cr^{IV} and Cr^{VI} tend to be associated with oxo- and fluoro- ligands and generally do not obey the EAN rule.

Table 5.76 Examples of compounds of chromium and molybdenum which conform to the EAN Rule.

Formal oxidation state	Cr	Mo	Coordination number and geometry
−4	$[Cr(CO)_4]^{4-}$	$Mo(NO)_4$	4–tetrahedral
−2	$[Cr(CO)_5]^{2-}$	$[Mo(CO)_5]^{2-}$	5–trigonal bipyramidal
0	$Cr(CO)_6$	$Mo(CO)_6$	6–octahedral
2	$[Cr(CO)_2(diars)_2X]^+$	$MoCl_2(CO)_3(PR_3)_2$	7–capped octahedron, pentagonal bipyramidal
4		$[Mo(CN)_8]^{4-}$	8–dodecahedral
6		$MoH_6(PR_3)_3$	9–tricapped trigonal prismatic

This phenomenon can have surprising consequences for organometallic compounds. For example, the first row transition metals form a wide range of sandwich compounds which have electron counts which span from 14 to 20. Although many of these compounds do not obey the EAN rule the conventional sandwich structure is observed and the major structural difference is the distance between the metal and the centre of the sandwich ring. For second and third row metals the driving force for the conforming to the EAN rule is sufficiently large that sometimes it is achieved with the loss of significant π-electron delocalization energy as aromatic rings are distorted. For example, $[Ru(C_6H_6)_2]$ has the structure shown in the margin where the aromaticity of one of the rings is lost in order to enable the metal to conform to the EAN Rule.

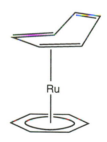

Physical and chemical properties of the metals

Melting points, boiling points, and densities
The melting points, boiling points, and densities of the transition metals all increase significantly on descending a column. The first two properties reflect the larger enthalpies of atomization, which in turn result from the improved metal–metal overlaps for the 4d and 5d elements discussed in the previous section. The stronger metal–metal bonding exhibited by the second and third row elements also has the following implications.
1. The second and third row elements form a wider range and more stable compounds containing metal–metal bonds.
2. Such elements form compounds with stronger multiple metal–metal bonds.

3. The compounds of such elements are less commonly paramagnetic because the energy gain from forming a metal–metal bond is sufficiently large to overcome the pairing energy of the electrons on adjacent metal atoms.

Standard reduction potentials

The higher enthalpies of atomization of the metals, their larger ionization energies, and the less favourable hydration enthalpies for the larger ions make the second and third row transition metals poorer reducing agents. Indeed, the metals at the right-hand bottom section of the transition series are described as the noble metals for this reason. The standard electrode potential data summarized in Table 5.77 illustrates the thermodynamic basis of this general phenomenon.

Table 5.77 Standard electrode potentials for the transition metals

Oxidized form			Reduced form	E°/ V $(M^{2+/3+}/M)$/V
$Mn^{3+}(aq)$	$+ 3e$	$\rightarrow$	$Mn(s)$	-0.28
$Re^{3+}(aq)$	$+ 3e$	$\rightarrow$	$Re(s)$	~ 0.3
$Mn^{2+}(aq)$	$+ 2e$	$\rightarrow$	$Mn(s)$	-1.19
$Tc^{2+}(aq)$	$+ 2e$	$\rightarrow$	$Tc(s)$	0.40
$Cr^{3+}(aq)$	$+ 3e$	$\rightarrow$	$Cr(s)$	-0.74
$Mo^{3+}(aq)$	$+ 3e$	$\rightarrow$	$Mo(s)$	-0.2
$Co^{2+}(aq)$	$+ 2e$	$\rightarrow$	$Co(s)$	-0.29
$Rh^{2+}(aq)$	$+ 2e$	$\rightarrow$	$Rh(s)$	0.6
$Ni^{2+}(aq)$	$+ 2e$	$\rightarrow$	$Ni(s)$	-0.25
$Pd^{2+}(aq)$	$+ 2e$	$\rightarrow$	$Pd(s)$	0.92
$Pt^{2+}(aq)$	$+ 2e$	$\rightarrow$	$Pt(s)$	~ 1.2

The second and third row elements are therefore generally unreactive towards non-oxidizing acids and only dissolve with difficulty in oxidizing acids. The simultaneous presence of an oxidizing acid and anions capable of coordinating and stabilizing the metal ions which are formed is commonly required to dissolve these metals. For example, *aqua regia*, a 1:3 combination by volume of concentrated HNO_3 and concentrated HCl, is used to dissolve the platinum metals.

The standard reduction potentials given in Table 5.78 provide the thermodynamic basis for the way in which complex formation increases the ease of dissolving the noble metals.

The final entry in Table 5.78 underlines the widespread use of CN^- in the recovery and purification of gold in the mining and electroplating industries. Gold dissolves in cyanide solution if air or hydrogen peroxide are present as the oxidant.

As a consequence of the standard electrode potentials discussed above and the rather unfavourable hydration enthalpies for such large metal ions the second and third row transition metals have a poorly developed aqueous chemistry particularly in the +2 and +3 oxidation states, i.e. there are very few $[M(OH_2)_6]^{x+}$ ions. Much of the aqueous chemistry is based on anionic complexes with halide and oxo ligands.

Table 5.78 Effect of complex formation on the standard electrode potentials for the later transition metals

Oxidized form		Reduced form	$E^{\oplus}/V$
$Pd^{2+}(aq)$	$\rightarrow$	$Pd(s)$	0.92
$PdCl_4^{2-}$	$\rightarrow$	$Pd(s)$	0.60
$PdBr_4^{2-}$	$\rightarrow$	$Pd(s)$	0.49
$Pt^{2+}(aq)$	$\rightarrow$	$Pt(s)$	~1.2
$PtCl_4^{2-}$	$\rightarrow$	$Pt(s)$	0.76
$PtBr_4^{2-}$	$\rightarrow$	$Pt(s)$	0.70
PtI_4^{2-}	$\rightarrow$	$Pt(s)$	0.40
$Au^+(aq)$	$\rightarrow$	$Au(s)$	1.69
$AuCl_2^-$	$\rightarrow$	$Au(s)$	1.15
$AuBr_2^-$	$\rightarrow$	$Au(s)$	0.96
$Au(SCN)_2^-$	$\rightarrow$	$Au(s)$	0.66
$Au(CN)_2^-$	$\rightarrow$	$Au(s)$	0.60

The reactions of the metals with halogens commonly lead to compounds which are not simple binary halides MX_n, but are actually cluster compounds, i.e. compounds which retain metal–metal bonds. For example, the reaction of rhenium metal with chlorine gives a compound which analyses as $ReCl_3$, but which has a structure based on rhenium triangles linked by halide bridges to give an infinite structure. For these metals the metal–metal bond strengths are comparable to metal-halogen bonds and therefore the metallic structure is not broken down completely but retained in part.

Catalytic properties

In the previous section the lack of electropositive character and the 'nobility' of the heavier transition metals is stressed. Therefore, at first acquaintance, it is surprising that many of the platinum metals are used as catalysts for the heavy chemical industry and in automobile catalytic converters. This apparent contradiction may be resolved by recognizing that many of the heavier transition metals are able to chemisorb unsaturated organic and inorganic molecules. The atoms on a metal surface behave similarly to low oxidation state complexes and utilize synergic bonding interactions to coordinate such molecules. Table 5.79 summarizes the relative tendencies of metals to chemisorb such molecules. It is particularly significant that the post transition metals Ag........Bi are only capable of chemisorbing O_2, whereas the majority of transition metals are able to chemisorb O_2, C_2H_2, C_2H_4, CO, and H_2. An effective catalyst is not only able to bring molecules together on a metal surface, but also has to provide a low energy pathway for the molecules on the surfaces in order for them to interchange bonds and finally to release the molecules of product which are formed as a result.

The most effective catalysts have intermediate chemisorption abilities and thereby enable the molecules to dock on the surface and undergo reaction without holding on to them so strongly that further reactions are energetically unfavourable. Therefore, the most effective hydrogenation, reforming, and carbonylation catalysts tend to be based on Rh, Pd, and Pt. For the hydrogenation of dinitrogen to ammonia and carbon monoxide to hydrocarbons iron tends to be the favoured metal catalyst, since the earlier transition metals chemisorb N_2 and CO too strongly. Specific examples of catalysts used by the chemical industry are given in Table 5.80.

Table 5.79 The relative abilities of metals to chemisorb gases ($\sqrt{}$ indicates a strong absorption, $\times$ indicates unobservable absorption)

	N_2	CO	O_2	H_2	C_2H_2	C_2H_4	CO_2
Ag, Zn, Cd, In, Ge, Sn, Pb, As, Sb, Bi	$\times$	$\times$	$\sqrt{}$	$\times$	$\times$	$\times$	$\times$
Ti, Zr, Hf; V, Nb, Ta; Cr, Mo, W; Fe, Ru, Os; Ni, Pd, Pt; Co, Rh, Ir	$\sqrt{}$	$\sqrt{}$	$\sqrt{}$	$\sqrt{}$	$\sqrt{}$	$\sqrt{}$	$\sqrt{}$
Al, Au	$\times$	$\sqrt{}$	$\sqrt{}$	$\times$	$\sqrt{}$	$\sqrt{}$	$\times$

Table 5.80 Specific examples of catalysts used by the chemical industry

Reaction	Catalyst and support
Ammonia synthesis from N_2	Fe / Al_2O_3
SO_2/SO_3 oxidation for sulfuric acid synthesis	V/K_2SO_4/SiO_2
Ammonia oxidation to NO for nitric acid synthesis	Pt alloy gauze
Methanation CO/CO_2 + H_2 $\rightarrow$ CH_4 + H_2O	Ni / Al_2O_3
Methanol synthesis from CO + H_2	Cu/ ZnO/Al_2O_3
Desulfurization of hydrocarbon feed stocks	Co / Mo sulfides/Al_2O_3
Oxidation of methanol to formaldehyde	Ag granules
Water gas shift reaction CO + H_2O $\rightarrow$ CO_2 + H_2	Cu/ ZnO/Al_2O_3 low temp. Fe_3O_4/Cr_2O_3 high temp.
Ethylene oxidation	Ag / Al_2O_3
Ethyne hydrogenation	Pd / carbon

Table 5.81 First order rate constants (at 25°C) for substitution of Cl by H_2O

Complex ion	k / 10^7 s^{-1}
$[Co(NH_3)_5Cl]^{2+}$	17
$[Rh(NH_3)_5Cl]^{2+}$	0.6
$[Ir(NH_3)_5Cl]^{2+}$	0.001

Relative rates of reaction

Since many reactions of transition metal coordination compounds proceed by a dissociative pathway, the fact that the mean bond enthalpies generally increase down a column also means that the complexes become more inert with respect to substitution reactions.

The rates of nucleophilic reactions similarly decrease down the column and some representative data are given in Table 5.81.

Similar considerations apply to inner sphere electron transfer reactions where the rate determining step involves ligand replacement to form a bridged intermediate. However, outer sphere electron transfer reactions where the coordination spheres remain intact and electron tunnelling occurs have similar rates.

However, for various reasons which have not been explained, the rates for carbonyl substitution reactions generally follow the orders:

Cr < Mo > W, Co < Rh > Ir and Ni < Pd > Pt

The stronger metal-ligand bonding for the second and especially the 3rd row transition metals also makes their coordination complexes more stereochemically rigid, i.e. the energies between alternative coordination geometries become larger and, therefore, rearrangements in which these alternative geometries participate as intermediates or transition states require more energy. For example, the activation energy for the intramolecular stereochemical rearrangement of octahedral $[RuH_2(PPh(OEt)_2)_4]$ which occurs through a trigonal prismatic intermediate is larger than that for $[FeH_2(PPh(OEt)_2)_4]$.

More generally, the fluxional rearrangements of organometallic compounds have activation barriers which increase down the column. The bonds between the ligand become stronger down a column and therefore more energy is required to distort the structure to form the intermediate geometry which occurs in the transition state.

For example, the haptotropic rearrangements for $Fe(CO)_3(\eta^4\text{-}C_8H_8)$, which involve the metal migrating around the cyclo-octatetraene ring have a lower free energy of activation than those for $Ru(CO)_3(\eta^4\text{-}C_8H_8)$.

The relative inertness of second and third row transition metal compounds towards substitution and radical reactions leads to the isolation of many more organometallic and hydrido-compounds for these elements. For example, $Cr(CH_3)_6$ and $Mn(CH_3)_7$ are unknown, but $W(CH_3)_6$ and $Re(CH_3)_6$ have been isolated. Similarly, ethyl compounds of the first row transition metals are generally very unstable because of their tendency to decompose via β-elimination pathways. However, $[Rh(C_2H_5)(NH_3)_5]^{2+}$ may be readily isolated because the d^6 Rh^{III} ion is substitutionally inert and does not readily lose an ammonia ligand which is the necessary first step for providing an empty coordination site for the β–migration process to occur. Hydrido-complexes of the second and third row elements are similarly more readily isolated and more numerous (see page 286 for some typical examples).

Summary of important differences down a column

1. The second and third row transition metals provide examples of compounds in higher oxidation states.

2. Down a column the metals show a greater variety of metal–metal bonded compounds.

3. The metal–metal and metal–ligand bond enthalpies increase down a column.

4. The compounds become more substitutionally inert.

5. The metals become more noble.

6. The d orbital splitting energies increase, therefore high spin complexes are not observed for the second and third row transition metal complexes.

7. The coordination numbers of complexes of the heavier elements are greater.

8. The second and third row transition metals have more similar chemical and physical properties, e.g. covalent radii, coordination numbers, and magnetic properties.

5.8 Lanthanides

The fifteen lanthanide elements and some of their atomic properties are summarized in Table 5.82. Lanthanum has the electronic configuration $[Xe]6s^25d^1$ and the lanthanide group of elements are the fourteen which succeed it and have electrons which populate the 4f shell. The 4f shell becomes progressively more stable across the series relative to the 5d orbitals and consequently the ground state electronic configurations are $[Xe]\,6s^24f^x$ (x = 2 to x = 14 for elements Ce to Lu) with the exception of gadolinium which has the configuration $6s^25d^14f^7$ in preference to $6s^24f^8$ because of exchange energy effects associated with the half-filled 4f shell. Lutetium is the final element in the series with the electronic configuration $6s^25d^14f^{14}$. In common with the transition elements the 4f orbitals are stabilized more than the 5d and 6s when positive ions are formed and, therefore, the commonly observed M^{3+} ions and the less commonly observed M^{2+} and M^{4+} ions, have the ground state electronic configurations: $4f^n$ with vacant 6s and 5d orbitals.

Representational drawings of the angular parts of the wavefunctions for the f orbitals, appropriate for lanthanide and actinide complexes with cubic symmetries, are illustrated in Fig. 5.83. Particularly noteworthy is the occurrence of the three nodal surfaces for each orbital.

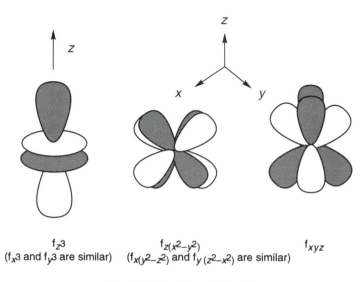

f_{z^3}
(f_{x^3} and f_{y^3} are similar)

$f_{z(x^2-y^2)}$
($f_{x(y^2-z^2)}$ and $f_{y(z^2-x^2)}$ are similar)

f_{xyz}

Fig. 5.83 The cubic set of f orbitals

The following r_{max} values (in pm) indicate the relative maxima in the radial distribution functions of the 4f and 5f valence orbitals and compares them with those of a typical third row transition metal.

Ce	U	W
[Xe]6s^2 4f^2	[Rn]7s^2 5d^{1}4f^3	[Xe]6s^2 5d^{4}4f^{14}
4f: 38	5f: 56	
5d: 116	6d: 129	5d: 79
6s: 207	7s: 194	6s: 147

The participation of the 4f orbitals (for the lanthanides) and the 5f orbitals for the actinides in covalent bond formation is clearly an important issue and it remains a topic which is actively investigated. The previous discussion of transition metals suggested that the 3d orbitals are far more contracted than the 4s and 4p orbitals and consequently they behave in a semi-core-like manner and participate only strongly in covalent bonding when the metal–ligand distances are short. The 4f and 5f orbitals, and particularly the former, are even more contracted and are even less available for bonding. The core-like behaviour of the 4f orbitals in lanthanide complexes is supported by the following circumstantial evidence.

1. The lanthanides form extensive series of isostoichiometric compounds and complexes with closely related structural and chemical preoperties, particularly in the +3 oxidation state.

2. The splittings of the 4f orbitals induced by the ligand field environment are an order of magnitude smaller than those observed for the 3d orbitals in transition metal complexes. The f–f transitions observed in complexes are related to the emission spectra of the corresponding free ions in the gas phase, but are red-shifted indicating a somewhat smaller positive charge in the complex. The f–f transitions in complexes are very weak and sharp indicating only weak d–f orbital mixings occurring during the vibrations of the complex.

3. The lanthanides do not form stable complexes with π-acceptor and π-donor ligands which indicate that f_π–p_π multiple bonding is weak.

4. The rates of nucleophilic substitution of the aqua-ions in solution do not show any significant discontinuities as a function of the number of f electrons, i.e. there is no lanthanide equivalent of d^3 and d^6 (low spin) and d^8 inert complexes.

5. The complexes are generally stereochemically non-rigid and there is no equivalent to the EAN (18 electron) rule governing their stoichiometry. This suggests the absence of strongly directional covalent bonds.

The increased principal quantum number makes the 5f orbitals less contracted in actinide complexes and their participation in covalent bonding is greater. Specifically, the actinides do form stable complexes with π-donor ligands and particularly with O^{2-}, and provide some examples of π-acceptor complexes. Although the role of the 5f orbitals remains energetically ill-defined it is convenient to invoke it to account for some anomalous observations, e.g. the occurrence of cubic coordination in $[UF_8]^{3-}$ and the sandwich structures based on cyclooctatetraene ligands.

Atomic and ionic sizes

The metallic and ionic radii for the lanthanides are summarized in Table 5.82. The latter show a steady contraction in size across the series similar to that noted for the transition elements. The increase in effective nuclear charge across the series is responsible for these trends. This 'lanthanide contraction' has a direct influence on the properties of the M^{3+} ions and an indirect effect on the properties of the transition elements. The very similar radii for M^{x+} ions of the metals belonging to the same groups in the second and third transition series, e.g. Mo^{3+} and W^{3+}, has been attributed to the larger effective nuclear charge due to the filled 4f shell for the third-row elements.

Ionization energies

The first three ionization energies, I_1, I_2, and I_3, are given in Table 5.82. The first point to note is that the variation in I_1, is much smaller than that for the transition series of elements, e.g. the difference in I_1, between Ti and Cu is approximately 200 kJ mol^{-1}. This arises because the 4f orbitals are even more core-like than the 3d orbitals and therefore the increase in effective nuclear charge across the series is much smaller. The chemical consequences of the smaller variation in ionization energies is that the properties of the lanthanide elements are much more uniform than those of the transition elements.

The data for I_1, I_2, and I_3, show discontinuities which arise primarily from electron–electron repulsion and exchange energy effects. The discontinuities at the half-filled shells may be attributed to exchange energy effects. There are also smaller discontinuities at the quarter-filled and three-quarter-filled 4f shells which are particularly noticeable for I_3. These breaks are caused by the dependence of the interelectronic repulsion energy on the direction of the orbital rotation of the electrons.

Table 5.82 Electronic configurations, metallic radii, M^{3+} ionic radii, and the first three ionization energies (kJ mol^{-1}) of the lanthanide elements

Element	6s	5d	4f	r_{met}/ pm	$r(M^{3+})$/pm	I_1	I_2	I_3
La	2	1	0	188	122	538	1067	1850
Ce	2	0	2	183	107	527	1047	1949
Pr	2	0	3	183	106	523	1018	2086
Nd	2	0	4	182	104	530	1035	2132
Pm	2	0	5	181	106	536	1052	2152
Sm	2	0	6	180	100	543	1068	2258
Eu	2	0	7	204	98	547	1085	2404
Gd	2	1	7	180	97	593	1166	1990
Tb	2	0	9	178	93	565	1112	2114
Dy	2	0	10	177	91	572	1126	2213
Ho	2	0	11	177	89	581	1139	2204
Er	2	0	12	176	89	589	1151	2194
Tm	2	0	13	175	87	597	1163	2285
Yb	2	0	14	194	86	603	1176	2415
Lu	2	1	14	173	85	524	1341	2022

Physical properties of the metals

The metals are generally manufactured either by electrolytic reduction or the reduction of LnF_3 with calcium above 1000°C. They are silvery white and their properties vary fairly smoothly across the series. The enthalpies of atomization, $\Delta_{at}H$ of the metals, shown in Fig. 5.83 parallel I_3 quite closely for these elements. The 4f orbitals are so contracted that they do not contribute greatly to metal–metal bonding in the solid state and the conduction band is dominated by the overlap of the 5d and 6s orbitals of the metals.

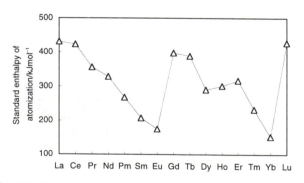

Fig. 5.84 Standard enthalpies of atomization for the lanthanide elements

All the metals except Eu and Yb contribute three electrons to the conduction band derived from 5d and 6s and therefore, compared with the atom in the gas phase, the promotion $4f^{n+1}6s^2 \rightarrow 4f^n5d^16s^2$ is relevant.

The promotion energy required to transfer an electron from 4f to 5d is related to the value of I_3 for the atom. Europium and ytterbium contribute only two electrons to the conduction band because their electronic configurations $4f^7 6s^2$ and $4f^{14} 6s^2$ make the donation of an additional electron and the associated loss of exchange energies unfavourable (i.e. they have high values of I_3).

The melting and boiling points of the metals, given in Table 5.83, follow in general the trends in $\Delta_{at} H^{\oplus}$ illustrated in Fig. 5.84. The discontinuities at Eu and Yb are particularly noteworthy.

Densities and resistivities

The lanthanides are much less dense than the third row transition elements which immediately succeed them. The densities increase steadily across the series with discontinuities at Eu and Yb which have lower densities because they contribute only two electrons to their conduction bands and therefore adopt less compact structures. Cerium has a density higher than might be expected, presumably because it contributes an additional electron to the conduction band (Ce, $[Xe]4f^1 5d^1 6s^2$).

The resistivities of the metals fall within a reasonably narrow range, with only Gd and Tb having larger than expected values and Yb having a lower than expected value.

Structures of the metals

The lanthanides are polymorphic and some are found with as many as four different crystal structures dependent upon conditions of temperature and pressure. All the lanthanide metals adopt close-packed or body-centred cubic structures in their polymorphs. The majority of the later metals in the series crystallize at room temperature with the hexagonal close-packed structures. Some of the earlier lanthanides adopt more complex close-packed structures based on close-packed sequences which alternate between cubic close-packed (*c*) and hexagonal close-packed (*h*). A 9-layer repeating sequence based on *chh* is the room temperature structure for Sm and some of the earlier lanthanides have the packing sequence *chchch*. Europium is body centred cubic and Yb is cubic close-packed.

The metallic radii of the 4f metals fall steadily from 182 pm (Pr) to 172 pm (Lu), but for reasons discussed above Eu (206 pm) and Yb (194 pm) have anomalously large radii.

Chemical properties

The low enthalpies of atomization and ionization energies of the metals make them electropositive. They react slowly with cold water, but react rapidly on heating to liberate dihydrogen. They tarnish in air and burn easily to give the oxides M_2O_3, except for Ce which gives CeO_2. Ytterbium forms a protective oxide coating and therefore does not react further with air below 1000°C. The metals react with most non-metals, e.g. H_2, N_2, C, Si, P, S, F_2, Cl_2, Br_2, and I_2, although for the less reactive non-metals high temperatures are required.

Table 5.83 Melting and boiling points (°C) of the lanthanide elements

Element	Melting point	Boiling point
La	921	3457
Ce	799	3426
Pr	931	3512
Nd	1021	3068
Pm	1168	2727
Sm	1077	1791
Eu	822	1597
Gd	1313	3266
Tb	1356	3123
Dy	1412	2562
Ho	1474	2695
Er	1529	2863
Tm	1545	1947
Yb	824	1193
Lu	1663	3395

Within the free electron model most lanthanides may be considered to be represented as $Ln^{3+}(e)_3$ except for Yb and Eu where $Ln^{2+}(e)_2$ are more appropriate and Ce where $Ln^{4+}(e)_4$ is more appropriate, the $(e)_n$ nomenclature indicating the number of free electrons, *n*, which contribute to the metallic bonding.

The standard reduction potentials given in Table 5.84 are remarkably alike and underline the close relationship in the properties of these metals. Europium is again anomalous and is less reducing (a less negative $E^{\ominus}$ value) than its neighbours reflecting the higher value of I_3 for this metal noted previously and originating from its $4f^7 6s^2$ ground state electronic configuration.

In general the reactivity order for the lanthanides is: **Eu**, La, Ce, Pr, Nd, Sm, **Yb**, Gd, Tb, **Y**, Ho, Er, Tm, Lu, with those which diverge from the periodic order are given in bold. Eu resembles Ca and is rapidly oxidized in air, whereas the later metals are stable in dry air. They all react quickly with dilute acids and slowly with the common halogens at room temperature.

Oxidation states

The earlier metals oxidize at room temperature especially when moist air is present, but the later metals are fairly stable in dry air. All the metals dissolve in dilute acids. They react with the halogens and a wide range of non-metals.

The chemistry of the lanthanide metals is dominated by the formation of compounds in the +3 oxidation state, but there are a sufficient number of compounds in the +2 and some compounds in the +4 oxidation states to add interest to the chemistry of these elements.

The only well-characterized M^{2+}(aq) ions are those of Eu^{2+}(aq) $(4f^7)$, Yb^{2+}(aq) $(4f^{14})$, and $(4f^6)$ and, in general, the order of stability of M^{2+} is:

$$Nd^{2+}, Pm^{2+}, < Sm^{2+} < Eu^{2+} \gg Gd^{2+} < Tm^{2+} < Yb^{2+}$$

Eu^{2+}(aq) can be formed by zinc reduction of Eu^{3+}(aq). Sm^{2+}(aq) and Yb^{2+}(aq) can be formed by electrolytic reduction, but are very oxygen-sensitive. The M^{2+} ions can be stabilized in non-aqueous solvents.

Maximum stability for M^{2+}(aq) is therefore achieved at the half-filled (Eu^{2+}) and completely filled 4f shells. In addition, stability increases progressively as the number of 4f electrons increases reflecting the increase in I_3 across the series. Thus, the stability order should be:

$$La < Ce < Pr < Nd < Pm < Sm < Eu \gg Gd < Tb < Dy > Ho > Er < Tm < Yb$$

For metal ions in the +4 oxidation state the stability order is:

$$Ce \gg Pr > Nd > Pm > Sm > Eu > Gd \ll Tb > Dy > Ho$$

Cerium is the only lanthanide with an extensive chemistry in the +4 oxidation state. The aqueous ion can be obtained by dissolving CeO_2 in acid. Complexes of Ce^{4+} with neutral ligands containing oxygen donor atoms are readily formed. The Ce^{4+} ion is widely used in synthetic organic and inorganic chemistry and analytical procedures as a one-electron oxidizing agent.

Table 5.84 Standard reduction potentials for the lanthanides

Lanthanide	$E^{\ominus}$ $(M^{3+}/M)/V$
La	−2.37
Ce	−2.34
Pr	−2.35
Nd	−2.32
Pm	−2.29
Sm	−2.30
Eu	−1.99
Gd	−2.29
Tb	−2.30
Dy	−2.29
Ho	−2.33
Er	−2.31
Tm	−2.31
Yb	−2.22
Lu	−2.30

Note the much smaller variations compared to those across a d-block transition metal series

Coordination numbers and geometries

The relatively large size of the lanthanide ions leads to a common coordination number of 9, although lower coordination numbers are observed if the steric demands of the ligands are great, e.g. $La(N(SiMe_3)_2)_3$ is pyramidal and $[Lu(2,6\text{-dimethylphenyl})_4]^-$ is tetrahedral. Higher coordination numbers up to 12 have been structurally characterized if the ligand has a small bite angle, e.g. as in $[Ce(NO_3)_6]^{2-}$. Nine-coordination is also observed in $[M(OH_2)_9]^{3+}$ ions which, in the solid state, have tricapped trigonal prismatic coordination geometries.

The $[Ln(OH_2)_9]^{3+}$ ions are observed in the solid state as $CF_3CO_2^-$, $EtOSO_3^-$, and BrO_3^- salts for the earlier lanthanides and $[Ln(OH_2)_8]^{3+}$ later in the series, i.e. Sm, Eu, and onwards. The ClO_4^- salts are, however, octahedral $[Ln(OH_2)_6]^{3+}$. There are a significant number of seven-coordinate complexes, e.g. $[LnX_3(thf)_4]$, many of which have pentagonal bipyramidal structures. The β-diketonate complexes $[Ln(acac)_3]$ are coordinatively unsaturated and are observed as hydrates, e.g. $[Ln(acac)_3(OH_2)_2]$, (La–Ho, Y), and $[Ln(acac)_3(OH_2)]$, (Yb). Complexes with EDTA and DTPA are common, e.g. $[Ln(edta)(OH_2)_3]^-$ and the DTPA complex of Gd, the latter is an important contrast enhancing agent in magnetic resonance imaging. The gadolinium(III) ion has paramagnetic properties which shorten the proton relaxation times in the water molecules of body tissue.

The contraction in ionic radii across the lanthanide series is associated with a decrease in coordination numbers for isostoichiometric compounds. For example, in M_2S_3 the La–Dy compounds are 8 or 7-coordinate, the Dy to Tm compounds are 7 or 6-coordinate, and the Yb and Lu compounds are 6-coordinate octahedral. In MCl_3, the La–Gd compounds are 9-coordinate, Tb and Dy are 8-coordinate, and Tb and Lu are 6-coordinate. For $LnBr_3$ Ln = La–Pr are nine-coordinate, Ln = Nd–Eu are eight-coordinate, and Ln = Gd–Lu are 6-coordinate.

For these high coordination numbers the energy differences between alternative geometries are small and therefore a range of structures is observed. For example, for nine coordination tricapped trigonal prismatic and capped square antiprismatic geometries have been determined.

Organometallic compounds

The organometallic chemistry of the lanthanides is generally associated with Ln^{3+}, although for Sm, Eu, and Yb organometallic compounds in the +2 oxidation state are also prevalent. Many of the compounds have either cyclopentadienyl or the more sterically crowded pentamethyl-cyclopentadienyl groups as spectator ligands. $LnCp_3$ are known for the whole series and they are crystalline compounds which show significant volatilities, even though they have polymeric structures involving the sharing of rings between the metals. The lanthanide contraction influences the mode of polymerization across the series. $LaCp_3$ has three η^5 Cp rings and one η^2 Cp ring which bridges between metal atoms, whereas $YbCp_3$ has three η^5 Cp rings and only weak interactions with the Cp rings of adjacent molecules and $LuCp_3$ has two η^5 Cp rings and two η^1 Cp rings. The polymeric structures in these compounds are broken up when neutral ligands such as PR_3, RNC, and THF

For a more detailed discussion of lanthanide complexes and their applications in magnetic resonance imaging and as luminescent probes see D. Parker and J. A. G. Williams, *J. Chem. Soc., Dalton Trans.*, 1996, 3613.

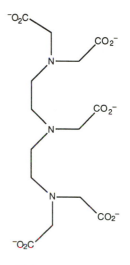

DTPA, diethylenetriaminepentaacetate

The first organometallic compounds of the lanthanides were reported by Wilkinson and Birmingham in 1954, viz. $Ln(\eta\text{-}C_5H_5)_3$

For a more detailed discussion of organolanthanide chemistry see *Topics in Current Chemistry*, 1996, 179, which has contributions from R. Anwarder, F. T. Edelmann, and W. H. Herrmann

Summary

1. They are electropositive metals reacting with water on heating.

2. In their compounds the +3 oxidation state dominates.

3. They do not form compounds with multiple bonds in high oxidation states, e.g. Ln=O or Ln≡N, or low oxidation states, e.g. LnCO.

4. The compounds are labile in solution undergoing rapid nucleophilic substitution reactions.

5. The complexes of the metal ions show a wide range of coordination numbers (3–12) and in general the coordination numbers are higher than those for comparable d-block transition metal ions. 9-coordination is particularly common for simple ligands.

6. The 4f orbitals participate little in bonding and are effectively core-like. This leads to extensive series of isostoichiometric compounds, small ligand field splittings and sharp f–f transitions in their electronic spectra.

7. The coordination complexes are sterochemically non-rigid and their geometries are defined primarily by ligand–ligand repulsion effects and steric repulsions within the ligands.

8. The metal ions are hard Lewis acids and exhibit a preference to coordinate to ligands with N, O, and F donor atoms.

are added and the resulting complexes have pseudo-tetrahedral structures, Cp_3LnL.

The compounds Cp_2LnCl are very useful reagents for the synthesis of σ-bonded organometallic compounds and hydrido complexes, e.g. Cp_2LnR and Cp_2LnH. The corresponding C_5Me_5 compounds are very important because they undergo C–H activation reactions with CH_4:

$$(C_5Me_5)_2LuCH_3 + {}^*CH_4 \longrightarrow (C_5Me_5)_2Lu^*CH_3 + CH_4$$

The cyclo-octatetraene sandwich compounds, $[(\eta\text{-}C_8H_8)_2Ln]^-$, have been made and are related to the related actinide sandwich compounds discussed below. More surprisingly, the neutral sandwich compounds, $[Ln(\eta\text{-}C_6H_3Bu'_3)_2]$, have been made using metal vapour synthesis techniques. These compounds have the metal in a formal oxidation state of zero and are only formed if bulky substituents are present on the benzene rings. The bond dissociation enthalpies of the sandwich compounds (200–300 kJ mol^{-1}) suggest that the metal–ring bonding is even stronger than that in $Cr(\eta\text{-}C_6H_6)_2$ (165 kJ mol^{-1}). The homoleptic alkyls, $Ln\{CH(SiMe_3)_2\}_3$, are known for the majority of lanthanides and they have trigonal three-coordinate geometries. In these low-coordination number compounds the metal ions are highly electrophilic and this induces interactions between the metal and C–H bonds of the ligands.

5.9 Actinides

The fifteen actinide elements have properties intermediate between those of the lanthanides and the heavier transition metals. Few of them are found in nature and many have been synthesized by transmutation reactions. They are all radioactive and the half lives of their most stable isotopes are summarized in Table 5.85. Uranium and thorium were discovered in 1789 and 1828 respectively, and took on a special significance during World War II when it was realized that nuclear fission properties of uranium could be utilized in the formation of atomic bombs. The fission characteristics of ^{235}U were discovered when the isotope was exposed to slow neutrons and formed the basis of the development of nuclear energy. This isotope occurs only in 1% abundance in natural uranium and therefore the great power associated with this process could not be unlocked until chemists found successful methods for enriching uranium. Diffusion processes based on volatile UF_6 were successfully used to achieve the required enrichment. The more abundant isotope of uranium ^{238}U is non-fissionable.

The peaceful and military exploitation of nuclear energy led to a detailed study of these elements and the synthesis of the remaining members of the actinide series using nuclear bombardment reactions. These reactions are discussed in Chapter 1. The electronic configurations of the actinide elements are summarized in Table 5.86. The seventh row of elements in the Periodic Table begins with the radioactive elements francium (Fr) and radium (Ra) with $7s^1$ and $7s^2$ electronic configurations. The elements actinium (Ac, $[Xe]7s^25f^06d^1$) and thorium (Th $[Xe]7s^25f^06d^2$) begin the process of filling the 6d shell, but as with the lanthanides the 5f shell becomes more stable and its occupation begins with the element protactinium (Pa). The 6d and 5f orbitals have very similar energies and consequently Th, Pa, U, and Np have electrons occupying both 6d and 5f orbitals and only after these elements does the *aufbau* preference dominate and the remaining elements have $7s^2 5f^n$ configurations. In the positively charged ions of these elements, the 5f orbital is stabilized more than 6d and the elements all have $5f^n 6d^0$ configurations. The radial distribution functions of the 5f orbitals show maxima which lie further away from the nucleus than the 4f orbitals of the lanthanides and consequently they are less core like and participate more in covalent bonding (see the r_{max} values on p. 374).

Table 5.86 Electronic configurations and common oxidation states for the actinides.

Actinide element	Symbol	Z	Electronic configuration	Common oxidation states
Actinium	Ac	89	$7s^2 5f^0 6d^1$	3
Thorium	Th	90	$7s^2 5f^0 6d^2$	3, 4
Protactinium	Pa	91	$7s^2 5f^2 6d^1$ (or $5f^1 6d^2$)	3, 4, 5, 6
Uranium	U	92	$7s^2 5f^3 6d^1$	3, 4, 5, 6
Neptunium	Np	93	$7s^2 5f^4 6d^1$ or $7s^2 5f^5$	3, 4, 5, 6, 7
Plutonium	Pu	94	$7s^2 5f^6$	3, 4, 5, 6, 7
Americium	Am	95	$7s^2 5f^7$	2, 3, 4, 5, 6
Curium	Cm	96	$7s^2 5f^7 6d^1$	3, 4
Berkelium	Bk	97	$7s^2 5f^9$	3, 4
Californium	Cf	98	$7s^2 5f^{10}$	2, 3
Einsteinium	Es	99	$7s^2 5f^{11}$	2, 3
Fermium	Fm	100	$7s^2 5f^{12}$	2, 3
Mendelevium	Md	101	$7s^2 5f^{13}$	2, 3
Nobelium	No	102	$7s^2 5f^{14}$	2, 3
Lawrencium	Lr	103	$7s^2 5f^{14} 6d^1$	3

Table 5.85 Half-lives of the most stable isotopes of the actinides

Element	Isotope	Half-life
Act	^{227}Ac	22 years
Th	^{232}Th	10^{10} years
Pa	^{231}Pa	3×10^4 years
U	^{238}U	4.5×10^9 years
Np	^{237}Np	2×10^6 years
Pu	^{244}Pu	8×10^7 years
Am	^{243}Am	7×10^3 years
Cm	^{247}Cm	1.56×10^7 years
Bk	^{247}Bk	1.4×10^3 years
Cf	^{251}Cf	890 years
Es	^{252}Es	1.29 years
Fm	^{257}Fm	101 days
Md	^{258}Md	56 days
No	^{259}No	58 minutes

Atomic and ionic sizes

The metallic radii of the actinides, which are summarized in Table 5.87, decrease to U, then an increase is observed to Cm, followed by another decrease at Bk. Clearly, the metallic band structure for these elements is complex since it involves the 5f, 6d, and 7s orbitals, which have very different maxima in their radial distribution functions.

Therefore, a simple pattern which results in a maximum in bonding at the half-filled shell does not emerge. Similar discontinuities are observed in the melting points of the metals. The ionic radii for the M^{3+} ions which have a regular pattern of $5f^n$ configurations show a more regular decrease in radii which is expected from the increase in effective nuclear charges across the series.

Table 5.87 Metallic and ionic radii of the actinides, together with the melting point of the parent metal

Actinide	Metallic radius/ pm	Ionic radius (M^{3+}) / pm	Melting point of metal (°C)
Ac	190	126	110
Th	180	118	1760
Pa	164	117	1572
U	154	114	1132
Np	155	115	639
Pu	159	112	1173
Am	173	111	1350
Cm	174	110	986
Bk	170	109	900
Cf	186		
Es	(186)		
Fm	(194)		
Md	(194)		
No	(194)		
Lr	(171)		

The metals

The metals are silvery in appearance and their densities and melting points vary significantly. U, Np, and Pu are very dense, whereas Am and Cm are much less dense. When finely divided the metals are pyrophoric and in the bulk form they tarnish in air. The standard electrode potentials of the metals M^{3+}(aq)/M vary from 2.13 to 1.50 V with the exception of Th ($E^{\oplus} = 1.17$ V) and consequently they are reasonably electropositive metals which dissolve common mineral acids, e.g. HCl and HNO_3. The M^{3+}(aq)/M $E^{\oplus}$ values become more negative in the order:

$$U < Np < Pu < Am$$

i.e. they become more electropositive and this has no analogy in lanthanide chemistry, where the metals become less electropositive across the series. The M^{3+}(aq) ion becomes more stable in the order U < Np < Pu < Am so that U^{3+} is a strong reducing agent.

Oxidation states

The actinides show a much wider range of oxidation states than the lanthanides as is shown by Table 5.88.

Table 5.88 Oxidation states of the actinide elements in aqueous solution. The most important oxidation state(s) are shown in bold

Ac	Th	Pa	U	Np	Pu	Am	Cm	Bk	Cf	Es	Fm	Md	No
3	**3**	**3**	**3**	**3**	**3**	**3**	**3**	**3**	**3**	**3**	**3**	**3**	**3**
	4		4	4	4	4	4	**4**					
		5	5	5	5	5							
			6	**6**	6	6							

More specifically, the elements U, Np, Pu, and Am have compounds with oxidation states III–VI and for Np and Pu also the VII oxidation state. For these elements the 6d and 5f orbitals are very close in energy, especially for the early actinides, and the participation of the 6d orbitals makes these elements behave more like the transition metals. On the right-hand side of the actinide series where the 6d orbitals are less available and the ionization energies of the atoms are larger, the 2 and 3 oxidation states predominate. The +2 oxidation state is only really important for Am $(7s^25f^7)$, where the third ionization of the metal atom is relatively high because of the loss of exchange energy associated with going from f^7 to f^6. It is therefore the lanthanide analogue of Eu^{2+}. The ions Md^{2+} and No^{2+} are known in aqueous solutions and for the other actinides the M^{2+} ions are known only in the solid state as halides, e.g. AmX_2, CfX_2, and EsX_2. The M^{3+} actinides closely resemble the M^{3+} lanthanides, but whereas the latter are stable across the series the M^{3+} actinides, when M = Th, Pa, U, Np, Pu, and No are unstable towards disproportionation processes. The +3 oxidation state is characteristic of all the actinides except Th and Ra. It is the only state for Ac and is the predominant one for the subsequent metals except No. The ionic radii of the M^{3+} actinides are only 5 pm larger than those for the corresponding lanthanides and consequently the same high coordination numbers are observed and many of the halides are isomorphous. Nine- and eight-coordination is commonly observed in the MF_3 and MCl_3 compounds. Eight- coordination is also observed for the earlier actinides, but the latter show a preference for 6 coordination.

The +4 oxidation state is the most stable for Th, in aqueous solutions, and for U, Np, Pu, and Bk it is commonly observed in aqueous solutions. For the other metals up to Es this oxidation state is relatively unimportant. The enhanced stability of Bk^{4+} may be correlated with the occurrence of the half-filled shell. Pu^{4+} and Am^{4+} disproportionate to An^{3+} and An^{5+} in aqueous solutions unless stabilized by F^-. The MO_2 compounds have the fluorite structure and consequently the metal ions are 8-coordinate. The MF_4 compounds are isostructural with the lanthanides and also have 8-coordination, usually square antiprismatic. The corresponding chlorides are dodecahedral.

The +5 oxidation state is most commonly observed for Pa which has a $7s^25f^26d^1$ electronic configuration and the actinides from U to Cf and is particularly stable for Pa and Np. The ions exist as AnO_2^+ in aqueous solutions. These compounds may often be reduced to related Pa^{4+} compounds in aqueous solutions by zinc reduction. Pa_3O_5 and PaX_5 (X = F, Cl, Br, and I) resemble the corresponding compounds of Ta, but in the solid state higher coordination numbers are observed, e.g. $PaCl_5$ has pentagonal bipyramids linked by halide bridges. The halides are readily hydrolysed and therefore behave similarly to the higher oxidation halides of the transition metals.

The +6 oxidation state is observed for U, Np, Pu, Am, and Cm and most commonly in the form of the fluorides, MF_6, and their dioxo-dications MO_2^{2+}. The +6 oxidation state is quite oxidizing for Am and Cm. The MF_6 compounds are volatile compounds which are readily sublimed and the volatility of the uranium compound is important in the enrichment of ^{235}U

by gaseous diffusion. They are readily hydrolysed and form MO_2F_2 and MOF_4. The MO_2^{2+} ions have the relative stabilities:

$$U > Np > Pu > Am$$

They all have a *trans*-O–M–O arrangement and there is considerable evidence that the strong bonding in this unit is a result of multiple bond formation involving the 6d and 5f orbitals. The MO_2^{2+} unit is generally accompanied by a group of ligands in a perpendicular plane leading overall to octahedral, pentagonal bipyramidal, or hexagonal bipyramidal geometries.

Compounds in the +7 oxidation state are only known for Pu ($7s^25f^6$) and Np($7s^25f^46d^1$) in strongly basic oxidizing solutions, e.g. $[PuO_6]^{5-}$ and $[NpO_4(OH)_2]^{3-}$. It is a rather unstable oxidation state for Pu and Am and even for Np the ions are sufficiently oxidizing to oxidize water under dilute conditions.

Complex formation

The actinide elements form a wide range of high coordination number complexes, particularly with oxygen donor ligands. The complexes are generally more stable than corresponding lanthanide complexes. Coordination numbers can rise up to 12, for example $[Th(NO_3)_6]^{2-}$. For example, UCl_4 and UBr_4 form neutral complexes with Me_2SO, R_3PO, py, and anionic complexes of the general type, $[UX_6]^{2-}$, (X = F–I); $[UF_7]^{3-}$, $[UF_8]^{4-}$. With chelating ligands the complexes with acetylacetonates are particularly stable and valuable for MOCVD studies, e.g. $[U(acac)_4]$ and $[Th(acac)_4]$.

Other related ligands which form stable complexes include RCO_2^-, NO_3^-, SO_4^{2-}, CO_3^{2-}, and $C_2O_4^{2-}$ which have square anti-prismatic geometries. With edta uranium forms $[U(edta)(H_2O)_2]$. The actinide ions are typically hard Lewis acids. The stabilities of the complexes decrease in the order $An^{4+} > An^{3+} \sim AnO_2^{2+} > AnO_2^+ > An^{2+}$ and for a given species the ligand dependency is $CO_3^{2-} > MeCO_2^- > SO_4^{2-} > NO_3^-$. In general ligands with exclusively N-donor atoms do not form very stable complexes.

Almost all the actinides are very toxic because they accumulate in the bones, liver, etc. where their radioactive emissions can cause long term biological damage. The low level of radioactivity associated with U and Th means that their chemistries have been more extensively studied. Plutonium is extremely toxic.

MOCVD—metal organic chemical vapour deposition—is a technique for depositing thin films of oxides, etc. on substrates by the thermal decomposition of volatile compounds.

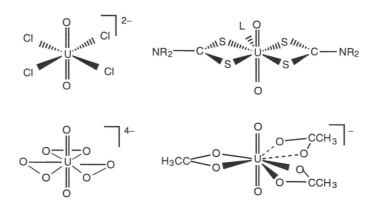

With soft ligands such as phosphines and thioethers the complexes are rarer, but anionic sulfur ligands e.g. dithiocarbamate, lead to more stable complexes, e.g. $U(S_2CNEt_2)_4$. Neutral nitrogen chelating ligands result in stable complexes, e.g. $[UCl_4(phen)_2]$. The An^{3+} and An^{5+} ions generally exhibit coordination numbers 8 and 9 and the compounds are commonly isomorphous with the corresponding lanthanide complexes. The coordination polyhedra are stereochemically non-rigid in solution and alternative geometries are at times observed for closely related complexes. The linear AnO_2 fragment is ubiquitous and the An–O distances lie between 170 and 200 pm, with the shorter bonds being associated with the higher oxidation state An^{6+} ions. Ancilliary ligands coordinate around the equator of the coordination sphere leading to AnO_2L_n where n is four or greater.

Organometallic compounds

The actinides form a range of sandwich compounds with cyclo-octatetraene.

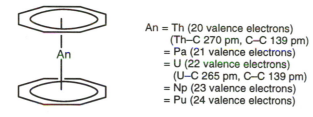

An = Th (20 valence electrons)
(Th–C 270 pm, C–C 139 pm)
= Pa (21 valence electrons)
= U (22 valence electrons)
(U–C 265 pm, C–C 139 pm)
= Np (23 valence electrons)
= Pu (24 valence electrons)

For a more detailed discussion of the bonding in AnO_2L_n complexes and the role of f orbitals see R. G. Denning, *Struct. Bond.*, 1991, **79**, 215.

The first organoactinide compound, $U(\eta\text{-}C_5H_5)_3$ was reported by Wilkinson and Reynolds in 1956 and the first alkyl in 1972 by Marks and Segan. However, the most formative development was the discovery of the sandwich compound $U(\eta\text{-}C_8H_8)_2$ by Streitweiser in 1968.

These complexes exceed the 18 electron valence electron count associated with transition metal sandwich compounds. The 5f orbitals of the actinides overlap sufficiently with the π-orbitals of the rings to involve them significantly in bonding. The sandwich bonding involves the 7s, 7p, 6d, and 5f orbitals and the strength of the bonding interaction is sufficient to remove the degeneracies of the 5f orbitals. A doubly degenerate e_u set is the most important component. Hence, $U(\eta\text{-}C_8H_8)_2$ (f^2) has two unpaired electrons occupying these orbitals. All the compounds are very air and moisture sensitive. Half-sandwich compounds, e.g. $[U(\eta\text{-}C_8H_8)X_2]$ and mixed sandwich compounds are also known, e.g. $An(\eta\text{-}C_8H_8)(\eta\text{-}C_5H_5)$.

A range of cyclopentadienyl and alkyl compounds has also been isolated, e.g. $U(\eta\text{-}C_5H_5)_4$, $U(\eta\text{-}C_5H_5)_3$, $U(\eta\text{-}C_5H_5)_3X$, $U(\eta\text{-}C_5H_5)_2X_2$, and $U(\eta\text{-}C_5H_5)X_3$, where X is commonly Cl, Br, or I, but can also be NR_2, SiR_3, etc. In addition σ-bonded hydrido and alkyl organometallic complexes have been extensively studied in recent years, e.g.

$AnR_4(dmpe)$ ($R = CH_3, CH_2Ph$); An = actinide metal

$An(\eta\text{-}C_5H_5)_3R$ ($R = CH_2Ph, CH_3$)

$[An(CH_3)_7]^{3-}$ (An = Th) — monocapped trigonal prismatic

$[An(\eta\text{-allyl})_4]^{3-}$ (An = Ce, Th)

In addition the ability to coordinate π-acid ligands is greater than for the lanthanides but less than for the transition metals. For example, $U(CO)_6$ is only stable below 20 K, but CO has been shown to coordinate to organometallic actinide complexes, e.g.

$$(\eta\text{-}C_5H_4SiMe_3)_3U + CO \rightarrow (\eta\text{-}C_5H_4SiMe_3)_3U(CO)$$

The CO frequency in this compound is 1976 cm^{-1} which is 170 cm^{-1} lower than that in free CO, indicating some back donation from filled 5f orbitals on uranium to the π-antibonding orbitals of CO. Examples of organoactinides with metal–hydride bonds are also well documented, e.g. $[Th(\eta\text{-}C_5Me_5)_2H_2]_2$.

In common with much of organo-transition metal chemistry the introduction of bulky groups and organic ligands which do not undergo β-elimination processes lead to the more facile isolation of complexes which may be stored under ambient conditions.

Summary
1. All the actinide metals are radioactive; the higher the atomic number, the lower is the nuclear stability. Only four (Ac, Th, Pa, and U) occur naturally; the others are man-made.
2. The early actinides show a wide variety of oxidation states, especially U, Np, Pu, and Am; the later ones (from Cm to Lr) behave like lanthanides, with +3 as the principal oxidation state.
3. Like lanthanides, the actinides exhibit a wide range of coordination numbers (from 3 to 14).
4. Whereas for the lanthanides the 4f orbitals participate little in bonding, the 5f and 6d orbitals are of comparable energy for the early actinides (Ac to Am); for the later elements (Cm to Lr) however, the 5f orbitals fall in energy relative to the 6d with the consequence that the later actinides are more lanthanide-like in their spectroscopic and chemical properties.
5. Unlike the lanthanides, a number of the actinides (Pa, U, Np, Pu, and Am) form some metal–ligand multiple bonds, in particular with the oxo (O^{2-}) ligand.
6. Ligand donor atoms for the actinides are, in general and like the lanthanides, hard (e.g. -N, -O, and -F). There is, however, evidence of more covalent interactions in organo-actinides than in organo-lanthanides.

Fourth Transition Series—Transactinides

The synthesis of new elements using the nuclear bombardment methods described in Chapter 1 have not only led to a completion of the actinide series but the discovery of new elements which belong to the fourth transition series. Elements 104, 105, and 106 were synthesised in Berkeley in the 1970s and were also claimed to have been synthesised in the Dubna Laboratory in the Soviet Union. Elements 107, 108, 109, 110, and 111 were made and characterised in the 1980s and 1990s in Germany using a cyclotron source. Element 110 was simultaneously studied in Berkeley and Dubna. The yields of these elements produced in nuclear fusion reactions involving target nuclei with heavy ions are extremely small and at times the quantity is measured in terms of number of atoms rather than milligrams. Their half-lives also decrease and therefore characterization by conventional chemical techniques becomes very problematical.

Given these difficulties and the limited number of laboratories in the world which are capable of carrying out experiments in this area it is not surprising that there have been conflicting claims concerning the discoverers of these elements and much discussion about the appropriate names for these elements. The recently accepted IUPAC names for these elements are given in Table 5.89.

Some study of the chemistry of elements 104, 105, and 106 have been made using sophisticated tracer techniques based on one atom at a time chemistry and the limited properties which have been studied are consistent with those expected from an extrapolation of the corresponding members of the third transition series, i.e. Hf, Ta, and W. The electronic structures of these atoms have been probed using modern high speed computer programs which take into account the relativistic effects. The calculations have confirmed that the elements 104 through to 112 have their outer electrons occupying the 6d shell making them homologous to the elements hafnium to mercury.

A number of chemists have attempted to predict the detailed properties of these elements by combining Mendeleev's extrapolation principles, using the known properties of the lighter elements in the same group as a starting point, with the results of modern electronic structure calculations, which give reliable results even for atoms with atomic numbers greater than 104. They represent, therefore, a practical example of the usefulness of the methodology developed in this book. A detailed discussion of the predictions arising from these analyses is beyond the scope of this book, but recent review articles by Seaborg and Cotton quoted in the references below provide a good starting point for the reader wishing to confirm for himself the likely properties of these elements. In the process the reader will also discover whether I have been successful in communicating the basic principles required for predicting "essential trends"!

Table 5.89 The IUPAC accepted names for the transactinide elements

Element	Atomic number	Symbol
Lawrencium	103	Lr
Rutherfordium	104	Rf
Dubnium	105	Db
Seaborgium	106	Sg
Bohrium	107	Bh
Hassium	108	Hs
Meitnerium	109	Mt

Further reading

F. A. Cotton and G. Wilkinson, *Advanced Inorganic Chemistry*, 5th Edn., John Wiley and Sons, New York, 1988.

D. F. Shriver, P. W. Atkins and C. H. Langford, *Inorganic Chemistry*, 2nd Edn., Oxford University Press, Oxford, 1994.

N. N. Greenwood and A. Earnshaw, *Chemistry of the Elements*, 2nd Edn, Butterworth-Heinemann, 1997.

R. G. Wilkins, *Kinetics and Mechanisms of Transition Metal Complexes*, VCH, New York, 1991.

P. Cooma, *Coord. Chem. Rev.*, 1993, **123**, 1.

S. J. Ashcroft and C.T. Mortimer, *Thermochemistry of Transition Metal Compounds*, Academic Press, New York, 1979.

S. A. Cotton, *Lanthanides and Actinides*, Macmillan, London, 1991.

S. A. Cotton, *The Chemistry of Precious Metals*, Chapman and Hall, London, 1997.

A. P. B. Lever, *Inorganic Electronic Spectroscopy*, Elsevier, Amsterdam, 1984.

E. C. Constable, *Metals and Ligand Reactivity*, Ellis Horwood, Chichester, 1990.

S. F. A. Kettle, *Physical Inorganic Chemistry*, Spektrum, Oxford, 1996.

G. T. Seaborg, *Evolution of the Periodic Table, J. Chem. Soc., Dalton Trans.*, 1996, 3899.

S. A. Cotton, *After the Actinides, Chem. Soc. Rev.*, 1996, **25**, 219.

F. Arnaud-Neu, *Solution Chemistry of Lanthanide Macrocyclic Complexes, Chem. Soc. Rev.*, 1996, **23**, 2435.

INDEX